Small Modular Reactors as Renewable Energy Sources

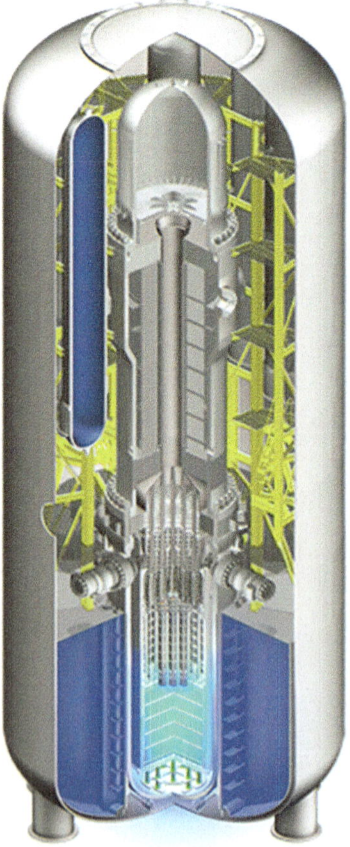

Bahman Zohuri

Small Modular Reactors as Renewable Energy Sources

Springer

Bahman Zohuri
University of New Mexico
Galaxy Advanced Engineering, Inc.
Albuquerque, NM, USA

ISBN 978-3-030-06469-3 ISBN 978-3-319-92594-3 (eBook)
https://doi.org/10.1007/978-3-319-92594-3

Softcover re-print of the Hardcover 1st edition 2019

Printed on acid-free paper

This Springer imprint is published by the registered company Springer International Publishing AG part of Springer Nature.
The registered company address is: Gewerbestrasse 11, 6330 Cham, Switzerland

This book is dedicated to my grandson Darius Nikpour, my son Sasha Zohuri, and my two daughters Dr. Natasha Zohuri and Natalie Zohuri

Preface

The major growth in the electricity production industry in the last 30 years has centered on the expansion of natural gas power plants based on gas turbine cycles. The most popular extension of the simple Brayton gas turbine has been the combined cycle power plant with the Air-Brayton cycle serving as the topping cycle and the Steam-Rankine cycle serving as the bottoming cycle for new generation of nuclear power plants that are known as GEN-IV. The Air-Brayton cycle is an open-air cycle and the Steam-Rankine cycle is a closed cycle. The Air-Brayton cycle for a natural gas driven power plant must be an open cycle, where the air is drawn in from the environment and exhausted with the products of combustion to the environment. This technique is suggested as an innovative approach to Generation-IV International Forum (GIF) nuclear power plants in the form and type of small modular reactors (SMRs). The hot exhaust from the Air-Brayton cycle passes through a heat recovery steam generator (HSRG) prior to exhausting to the environment in a combined cycle. The HRSG serves the same purpose as a boiler for the conventional Steam-Rankine cycle.

In 2007, gas turbine combined cycle plants had a total capacity of 800 GW and represented 20% of the installed capacity worldwide. They have far exceeded the installed capacity of nuclear plants, though in the late 1990s they had less than 5% of the installed capacity worldwide. There are a number of reasons for this. First, natural gas is abundant and cheap. Second, combined cycle plants achieve the greatest efficiency of any thermal plant. And third, they require the least amount of waste heat cooling water of any thermal plant.

A typical gas turbine plant consists of a compressor, combustion chamber, turbine, and an electrical generator. A combined cycle plant takes the exhaust from the turbine and runs it through a heat recovery steam generator (HRSG) before exhausting to the local environment. The HRSG serves the function of the boiler for a typical closed cycle steam plant. The steam plant consists of a steam turbine, a condenser, a water pump, an evaporator(boiler), and an electrical generator. In a combined cycle plant, the gas turbine and steam turbine can be on the same shaft to eliminate the need for one of the electrical generators. However, the two shafts, two generator systems provide a great deal more flexibility at a slightly higher cost. In addition to the closed loop for the steam, an open loop circulating water system is required to extract the waste heat from the condenser. The waste heat extracted by this "circulating" water system is significantly less per megawatt for a combined cycle system as the open Brayton cycle exhausts its waste heat directly to the air.

Uniqueness of the subject and benefits of GEN-IV power plants and argument to defend why we need the nuclear power plant for source of renewable energy, efficiency, cost effectiveness as well as safety-related issues are discussed in this book with every detail oriented both from an experimental and technical point of view. This book also goes over different types of GEN-IV nuclear power plants by giving general configurations and specifications of each modular reactor that are considered as part of the new generation of these nuclear plants that fall into the small modular reactors category.

As part of this book, we envision a world where almost every country has one affordable type of GEN-IV power plant, such as molten salt reactors burning thorium, uranium, and of spent fuel actinides, producing electricity, hydrogen, and desalinized water without any serious accidents.

Small modular reactors (SMRs) are highlighted as a viable alternative to the nuclear power plants (NPPs) that have been used as desalination plant energy sources for the following reasons:

- SMRs have lower investment costs;
- Almost all SMR concepts appear to show increased availability ($90%);
- Because of inherent safety features, most SMRs have good potential for location near population centers, hence lowering the transport costs;

This book also discusses some angles of nuclear energy generated by small modular reactors (SMRs) specifically and why we need nuclear power plants from SMRs point of view that are the new generation and design for nuclear plants. As we know, the major growth in the electricity production industry in the last 30 years has centered on the expansion of natural gas power plants based on gas turbine cycles. The most popular extension of the simple Brayton gas turbine has been the combined cycle power plant with the Air-Brayton cycle serving as the topping cycle and the Steam-Rankine cycle serving as the bottoming cycle for new generation of nuclear power plants that are known as GEN-IV. The Air-Brayton cycle is an open-air cycle and the Steam-Rankine cycle is a closed cycle. The air-Brayton cycle for a natural gas driven power plant must be an open cycle, where the air is drawn in from the environment and exhausted with the products of combustion to the environment. This technique is suggested as an innovative approach to GEN-IV nuclear power plants in form and type of small modular reactors (SMRs). The hot exhaust from the Air-Brayton cycle passes through a heat recovery steam generator (HSRG) prior to exhausting to the environment in a combined cycle. The HRSG serves the same purpose as a boiler for the conventional Steam-Rankine cycle.

In addition, this book shows the uniqueness of these small modular reactors (SMRs) and presents innovative approaches to the implementation of these reactors as part of GEN-IV technologies.

Albuquerque, NM, USA B. Zohuri
2018

Acknowledgment

I am indebted to the many people who aided me, encouraged me, and supported me beyond my expectations. Some are not around to see the results of their encouragement in the production of this book, yet I hope they know of my deepest appreciations. I especially want to thank my true friends that I cannot name all, to whom I am deeply indebted, who have continuously given their support without hesitation. He has always kept me going in the right direction. My many thanks to Distinguished Professor, Dr. Edl Schamiloglu at Electrical and Computer Engineering Department of University of New Mexico, for without his moral and financial support I could not have achieved the goal of writing this book. He is really a person of integrity and true words.

Above all, I offer very special thanks to my late mother and father, and to my children, in particular, my son Sasha and daughters Natalie and Natasha and the newcomer in our family my grandson Darius. They have provided constant interest and encouragement, without which this book would not have been written. Their patience with my many absences from home and long hours in front of the computer to prepare the manuscript are especially appreciated.

Contents

1 Introduction to the Nuclear Power Industry 1
1.1 Fission Energy Generation................................ 1
1.2 The First Chain Reaction................................ 2
1.3 Concepts in Nuclear Criticality 5
1.4 Fundamental of Fission Nuclear Reactors.................. 5
1.5 Reactor Fundamentals 8
1.6 Thermal Reactors....................................... 9
1.7 Nuclear Power Plants and Their Classifications 9
1.8 Nuclear Power Plants and Their Classifications 10
1.8.1 Light-Water Reactors (LWR)....................... 10
1.8.2 Graphite-Moderated Reactors (GMR)................ 10
1.8.3 Heavy-Water Reactors (HWR) 11
1.9 Classified by Coolant Material............................ 14
1.9.1 Pressurized Water Reactors (PWR).................. 14
1.9.2 Boiling Water Reactor (BWR)...................... 16
1.9.3 Gas-Cooled Reactors (GCR) 17
1.10 Classified by Reaction Type............................. 19
1.10.1 Fast-Neutron Reactor (FNR) 20
1.10.2 Thermal-Neutron Reactor 22
1.10.3 Liquid-Metal Fast-Breeder Reactors (LMFBR)................................. 23
1.11 Nuclear Fission Power Generation 27
1.12 Generation IV Nuclear Energy Systems 27
1.13 Technological State of the Art and Anticipated Developments.. 30
1.14 Next-Generation Nuclear Plant (NGNP).................... 32
1.15 Generation IV Systems 34
1.15.1 Very-High-Temperature Reactor (VHTR)............. 35
1.15.2 Molten Salt Reactor (MSR)........................ 37
1.15.3 Sodium-Cooled Fast Reactor (SFR) 39
1.15.4 Supercritical Water-Cooled Reactor (SCWR).......... 40

1.15.5 Gas-Cooled Fast Reactor (GFR) 44
1.15.6 Lead-Cooled Fast Reactor (LFR).................... 45
1.16 Next-Generation of Nuclear Power Reactors for Power Production 47
1.17 Goals for Generation IV Nuclear Energy Systems 48
1.18 Why We Need to Consider the Future Role of Nuclear Power Now.................................... 50
1.19 The Generation IV Roadmap Project........................ 53
1.20 Licensing Strategy Components 56
1.21 Market and Industry Status and Potentials.................. 57
1.22 Barriers... 58
1.23 Needs 59
1.24 Synergies with Other Sectors............................. 60
References... 60

2 Nuclear Power Structure from Past to Present 63
2.1 Introduction 63
2.2 The First Self-Sustaining Chain Reaction 66
2.3 Nuclear Energy Expands and Stagnates for Peace Usages 68
2.4 Government and Nuclear Energy............................ 70
2.5 Going Forward with Nuclear Energy......................... 73
2.6 Small Modular Reactors 73
2.7 Small Modular Reactors: Safety, Security, and Cost Concerns 75
2.7.1 Safety Concepts of the MSR 77
2.7.2 Economies of Scale and Catch....................... 79
2.7.3 Are Small Modular Reactors Safer? 79
2.7.4 Shrinking Evacuation Zones 80
2.7.5 Safety Conclusions of Nuclear Power Plants.......... 80
2.8 Why We Need Nuclear Power Plants.......................... 83
2.8.1 Methodology of Combined Cycle 87
2.8.2 Why We Still Need Nuclear Power 88
2.8.3 Is Nuclear Energy Renewable Source of Energy........ 88
2.8.4 Argument for Nuclear as Renewable Energy........... 90
2.8.5 Argument Against Nuclear Energy as Renewable Energy............................. 90
2.8.6 Today's Safety of Nuclear Power Plant............... 91
2.8.7 Summary 92
References... 93

3 Nuclear Energy Research and Development Roadmap 95
3.1 Introduction 95
3.2 US Industry Opportunities for Advanced Nuclear Technology Development 99
3.3 Benefits of Small Modular Reactors 100

3.4 Cooling Water Requirement for Nuclear Power Reactors 103
3.5 Open-Air Brayton Gas Power Cycle 104
3.5.1 Computer Code Development 107
3.6 Modeling the Nuclear Air Brayton Combined Cycle 110
3.7 A Combined Cycle Power Conversion System for Small Modular LMFBR 111
3.7.1 The Air Brayton Cycle Pros and Cons 112
3.7.2 The Feed Water Heater 112
3.7.3 Results of Modeling 113
3.8 Summary 115
References 116

4 A Combined Cycle Power Conversion System for Small Modular Reactor 117
4.1 Introduction 117
4.2 Nuclear Reactors for Power Production 121
4.3 Future of Nuclear Power Plant System 123
4.4 Next Generation of Nuclear Power Reactions for Power Production 123
4.5 Technology Roadmap for Generation IV Nuclear Energy Systems 126
4.6 Power Conversion Study and Technology Options Assessment 127
4.6.1 Heat Exchanger Components 131
4.6.2 Turbomachinery 131
4.6.3 Advanced Computational Materials Science Proposed for GEN-IV Systems 132
4.6.4 Material Classes Proposed for GEN-IV Systems 135
4.6.5 Generation IV Materials: Challenges 136
4.7 Generation IV Materials: Fundamental Issues 137
4.8 A Combined Cycle Power Conversion System Driven Next Generation IV 138
4.8.1 Modeling the Brayton Cycle 140
4.8.2 Modeling the Rankine Cycle 141
4.8.3 The Combined Brayton-Rankine Cycle 141
4.8.4 Study Results 144
4.9 Open Air Brayton Gas Power Cycle 146
4.9.1 Introduction to Open Air Brayton Gas Power Cycle 146
4.9.2 System Description 147
4.9.3 Nature of Problem Solved 148
4.10 Preliminary Conclusions 156
4.11 Combined Cycle Power Conversion System for Small Modular LMFBR 157
4.11.1 The Air Brayton Cycle Pros and Cons 157

4.11.2 The Feed Water Heater ... 158
4.11.3 Results ... 158
4.12 Why Combined Cycle Power Conversion System Approach ... 161
4.12.1 Typical Cycles ... 162
4.12.2 Typical Methodology ... 163
4.13 Motivation for This Innovative Approach ... 165
4.14 End of Cheap Oil and Future of Nuclear Power ... 174
4.15 The Future of Energy ... 175
4.16 Nuclear Power in the World Today and Time for Change ... 177
4.17 Improved Performance from Existing Reactors ... 184
4.18 Other Nuclear Reactors ... 185
References ... 185

5 Small Modular Reactors, the Next Big Renewable Energy Source ... 187
5.1 Introduction ... 187
5.2 The Limit of Renewable Energy and Small Modular Reactor ... 191
5.3 Small Modular Reactor-Driven Renewable and Sustainable Energy ... 196
5.4 Small Modular Reactor-Driven Hydrogen Energy for Renewable Energy Source ... 197
5.4.1 Hydrogen Energy for Renewable Energy Source ... 201
5.5 The Future of Nuclear Power ... 205
5.6 Key Conclusion ... 208
References ... 209

6 Safety and Nonproliferation Aspect of GEN-IV Reactors ... 211
6.1 Introduction ... 211
6.2 Safety and Reliability Goals ... 215
6.3 Potential Benefits Offered by Small Modular Reactors ... 219
6.4 Generation IV International Forum (GIF) Focus ... 219
6.5 Associated Ongoing Programs ... 223
6.6 Summary ... 224
References ... 227

7 Electricity Production and Renewable Source of Energy, Economics ... 229
7.1 Introduction ... 229
7.2 Electricity Production in the United States ... 231
7.3 Energy Supply, Demand, and Market ... 232
7.4 What Is a Capacity Market? ... 237
7.5 Renewable and Nonrenewable Energy Sources ... 239
7.6 Role of Renewable Energy ... 243
7.7 Frequently Asked Questions ... 245
References ... 245

8 Energy Storage Technologies and Their Role in Renewable Integration 247
8.1 Introduction 247
8.2 The Electric Grid 250
8.3 Power Generation 257
8.4 Transmission and Distribution 257
8.5 Load Management 258
8.6 Types of Storage Technology 260
8.6.1 Kinetic Energy Storage or Flywheels Concept 263
8.6.2 Superconducting Magnetic Energy Storage 266
8.6.3 Batteries 271
8.6.4 Other and Future Batteries in Development 277
8.7 A Battery Inspired Strategy for Carbon Fixation 286
8.8 Saliva-Powered Battery 288
8.9 Summary 289
References 289

Index 291

About the Author

Bahman Zohuri currently works at the Galaxy Advanced Engineering, Inc., a consulting company that he started in 1991 when he left both the semiconductor and defense industries after many years, working as a chief scientist. He presently holds the position of Research Associate Professor at the University of New Mexico, Albuquerque, NM, with the Department of Electrical and Computer Engineering. After graduating from University of Illinois in field of Physics and Applied Mathematics, he joined Westinghouse Electric Corporation where he performed thermal hydraulic analysis and natural circulation for Inherent Shutdown Heat Removal System (ISHRS) in the core of a Liquid Metal Fast Breeder Reactor (LMFBR) as a secondary fully inherent shutdown system for secondary loop heat exchange. All these designs were used for nuclear safety and reliability engineering for self-actuated shutdown system. He designed the mercury heat pipe and electromagnetic pumps for large pool concepts of LMFBR for heat rejection purpose for the reactor around 1978, when he received a patent for it. He later on was transferred to defense division of Westinghouse, where he was responsible for the dynamic analysis and method of launch and handling of MX missile out of canister. The results are applied to MX launch seal performance and muzzle blast phenomena analysis (i.e., missile vibration and hydrodynamic shock formation). He also was involved in analytical calculation and computation in the study of nonlinear ion wave in rarefying plasma. The results are applied to the propagation of "Soliton Wave" and the resulting charge collector traces, in the rarefactions characteristic of the corona of a laser-irradiated target pellet. As part of his graduate research work at Argonne National Laboratory, he performed computation and programming of multi-exchange integral in surface physics and solid-state physics. He holds various patents in areas such as diffusion processes and design of diffusion furnace while he was senior process engineer working for different semiconductor industries such as Intel, Varian, and National Semiconductor corporations. Later on he joined Lockheed Missile and Aerospace Corporation as Senior Chief Scientist and was responsible for R&D and the study of vulnerability, survivability, and both radiation and laser hardening of different components of payload (i.e., IR Sensor) for Defense Support Program (DSP), Boost Surveillance and Tracking Satellite (BSTS), and

Space Surveillance and Tracking Satellite (SSTS) against laser or nuclear threat. While in there he also studied and performed the analysis of characteristics of laser beam and nuclear radiation interaction with materials, transient radiation effects in electronics (TREE), electromagnetic pulse (EMP), system generated electromagnetic pulse (SGEMP), single-event upset (SEU), blast, and thermo-mechanical, hardness assurance, maintenance, device technology.

He did few years of consulting under his company Galaxy Advanced Engineering with Sandia National Laboratories (SNL), where he was supporting development of operational hazard assessments for the Air Force Safety Center (AFSC) in concert with other interested parties. Intended use of the results was their eventual inclusion in Air Force Instructions (AFIs) specifically issued for directed energy weapons (DEW) operational safety. He completed the first version of a comprehensive library of detailed laser tools for airborne laser (ABL), advanced tactical laser (ATL), tactical high energy laser (THEL), mobile/tactical high energy laser (M-THEL), etc.

He also oversaw SDI computer programs, in connection with Battle Management C^3I and artificial intelligence, and autonomous systems. He is the author of several publications and holds several patents, such as for a laser-activated radioactive decay and results of a through-bulkhead initiator. He has published the following works: *Heat Pipe Design and Technology: A Practical Approach* (CRC Press); *Dimensional Analysis and Self-Similarity Methods for Engineering and Scientists* (Springer); *High Energy Laser (HEL): Tomorrow's Weapon in Directed Energy Weapons Volume I* (Trafford Publishing Company); and recently the book on the subject directed energy weapons and physics of high energy laser with Springer. He has other books with Springer Publishing Company: *Thermodynamics in Nuclear Power Plant Systems* and *Thermal-Hydraulic Analysis of Nuclear Reactors*.

He spent several years as a consultant at Galaxy Advanced Engineering serving Sandia National Laboratories, where he supported the development of operational hazard assessments for the Air Force Safety Center in collaboration with other researchers and third parties. Ultimately, the results were included in Air Force Instructions issued specifically for directed energy weapons operational safety. He completed the first version of a comprehensive library of detailed laser tools for airborne lasers, advanced tactical lasers, tactical high-energy lasers, and mobile/tactical high-energy lasers, for example.

He has been awarded three patents and has published 30 textbooks and numerous other journal publications.

Chapter 1
Introduction to the Nuclear Power Industry

Abstract Currently, about half of all nuclear power plants are located in the United States. There are many, different kinds of nuclear power plants, and we will discuss a few important designs in this text. A nuclear power plant harnesses the energy inside atoms themselves and converts this to electricity. All of us use this electricity. In Sect. 1.1 of this chapter, we show you the idea of the fission process and how it works. A nuclear power plant uses controlled nuclear fission. In this chapter, we will explore how a nuclear power plant operates and the manner in which nuclear reactions are controlled. There are several different designs for nuclear reactors. Most of them have the same basic function, but one's implementation of this function separates it from another. There are several classification systems used to distinguish between reactor types. Below is a list of common reactor types and classification systems found throughout the world, and they are briefly explained below according to the three types of classification either (1) classified by moderator material, (2) classified by coolant material, or (3) classified by reaction types.

1.1 Fission Energy Generation

There is strategic as well as economic necessity for nuclear power in the United States and indeed most of the world. The strategic importance lies primarily in the fact that one large nuclear power plant saves more than 50,000 barrels of oil per day. At $30 to $40 per barrel (1982), such a power plant would pay for its capital cost in a few short years. For those countries that now rely on but do not have oil or must reduce the importation of foreign oil, these strategic and economic advantages are obvious. For those countries that are oil exporters, nuclear power represents an insurance against the day when oil is depleted. A modest start now will assure that they would not be left behind when the time comes to have to use nuclear technology.

The unit costs per kilowatt-hour for nuclear energy are now comparable to or lower than the unit costs for coal in most parts of the world. Other advantages are the lack of environmental problems that are associated with coal or oil-fired power plants and the near absence of issues of mine safety, labor problems, and transportation

B. Zohuri, *Small Modular Reactors as Renewable Energy Sources*,
https://doi.org/10.1007/978-3-319-92594-3_1

bottlenecks. Natural gas is a good, relatively clean-burning fuel, but it has some availability problems in many countries and should, in any case, be conserved for small-scale industrial and domestic uses. Thus, nuclear power is bound to become the social choice relative to other societal risks and overall health and safety risks.

Nuclear fission is the process of splitting atoms, or fissioning them. This page will explain.

1.2 The First Chain Reaction

Early in World War II, the scientific community in the United States, including those Europeans now calling the United States their safe home, pursued the idea that uranium fission and the production of excess neutrons could be the source of extraordinary new weapons. They knew Lise Meitner's interpretation, in Sweden, of Hahn's experiments would likely be known in Germany. Clearly there might now be a race commencing for the development and production of a new, super weapon based on the fission of ${}^{235}U_{92}$ or ${}^{239}Pu_{94}$.

By early 1942, it was known that the two naturally occurring isotopes of uranium reacted with neutrons as follows:

$$
\begin{aligned}
&{}^{235}U_{92} + {}^{1}n_{0} \rightarrow \text{fission products} + (2.5)^{1}\, n_{0} + 200\,\text{MeV Energy} \\
&{}^{238}U_{92} + {}^{1}n_{0} \rightarrow {}^{239}U_{92} \\
&{}^{239}U_{92} \rightarrow {}^{239}Np_{93} + \quad {}^{-1} \quad t_{1/2} = 23.5\,\text{min}. \\
&{}^{239}Np_{93} \rightarrow {}^{239}Pu_{94} + \quad {}^{-1} \quad t_{1/2} = 2.33\,\text{days}
\end{aligned}
$$

Each U-235 that undergoes fission produces an average of 2.5 neutrons. In contrast, some U-238 nuclei capture neutrons, become U-239, and subsequently emit two beta particles to produce Pu-239. The plutonium was fissile also and would produce energy by the same mechanism as the uranium. A flow sheet for uranium fission is shown in Fig. 1.1 below [1].

The answers to two questions were critical to the production of plutonium for atomic bombs:

1. Is it possible, using natural uranium (99.3% U-238 and 0.7% U-235), to achieve a controlled chain reaction on a large scale? If so, some of the excess neutrons produced by the fission of U-235 would be absorbed by U-238 and produce fissile Pu-239.
2. How can we separate (in a reasonable period of time) the relatively small quantities of Pu-239 from the unreacted uranium and the highly radioactive fission-product elements?

Although fission had been observed on a small scale in many laboratories, no one had carried out a controlled chain reaction that would provide continuous production of plutonium for isolation.

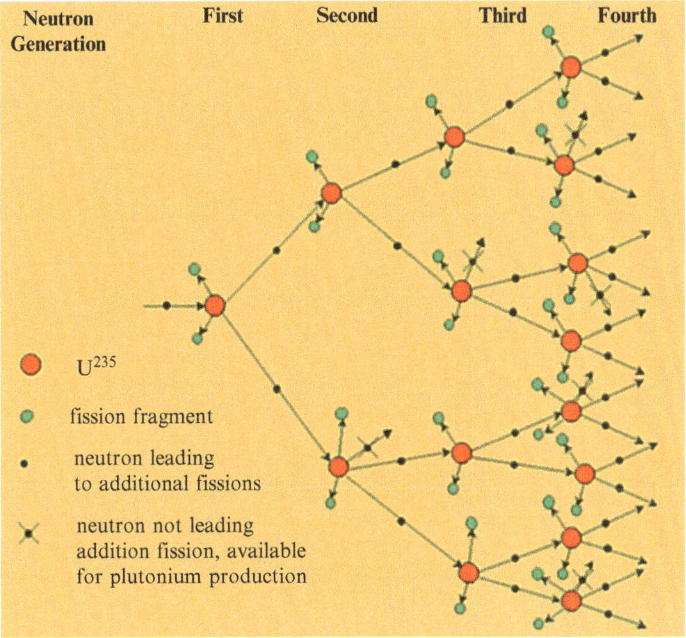

Fig. 1.1 The first generations of a nuclear chain reaction [1]

Enrico Fermi thought that he could achieve a controlled chain reaction using natural uranium. He had started this work with Leo Szilard at Columbia University but moved to the University of Chicago in early 1942.

The first nuclear reactor, called a pile, was a daring and sophisticated experiment that required nearly 50 tons of machined and shaped uranium and uranium oxide pellets along with 385 tons – the equivalent of four railroad coal hoppers – of graphite blocks, machined on site.

The pile itself was assembled in a squash court under the football field at the University of Chicago from the layered graphite blocks and uranium and uranium oxide lumps (Fermi's term) arranged roughly in a sphere with an anticipated 13-foot radius. Neutron-absorbing, cadmium-coated control rods were inserted in the pile. By slowly withdrawing the rods, neutron activity within the pile was expected to increase, and at some point, Fermi predicted there would be one neutron produced for each neutron absorbed in either producing fission or by the control rods [1]. See Fig. 1.2.

On December 2, 1942, with 57 of the anticipated 75 layers in place, Fermi began the first controlled nuclear chain reaction that occurred. At around 3:20 p.m., the reactor went critical; that is, it produced one neutron for every neutron absorbed by the uranium nuclei. Fermi allowed the reaction to continue for the next 27 min before inserting the neutron-absorbing control rods. The energy-releasing nuclear chain reaction stopped as Fermi predicted it would.

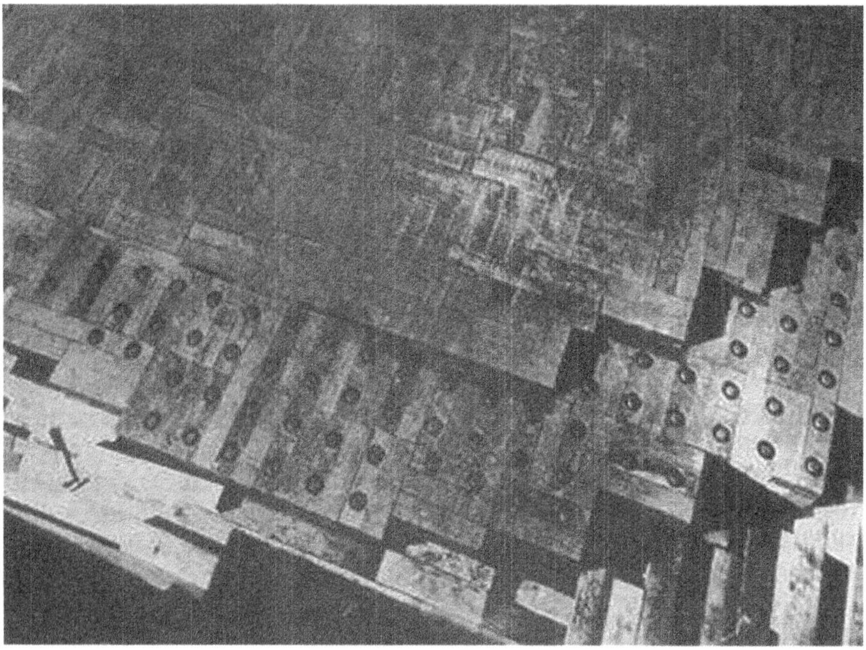

Fig. 1.2 CP-1 – graphite blocks with 3-inch diameter uranium cylinders inserted – part of a layer of CP-1, the first nuclear reactor. A layer of graphite blocks without inserted uranium is seen covering the active layer [1]

Fig. 1.3 The first controlled chain reaction, Stagg Field, Chicago, December 2, 1942. (Courtesy of the Argonne National Laboratory)

In addition to excess neutrons and energy, the pile also produced a small amount of Pu-239, the other known fissile material. See Fig. 1.3.

The achievement of the first sustained nuclear reaction was the beginning of a new age in nuclear physics and the study of the atom. Humankind could now use the tremendous potential energy contained in the nucleus of the atom. However, while a controlled chain reaction was achieved with natural uranium and could produce plutonium, it would be necessary to separate U-235 from U-238 to build a uranium bomb [1].

On December 28, 1942, upon reviewing a report from his advisors, President Franklin Roosevelt recommended building full-scale plants to produce both U-235 and Pu-239.

This changed the effort to develop nuclear weapons from experimental work in academic laboratories administered by the US Office of Scientific Research and Development to a huge effort by private industry. This work, supervised by the US Army Corps of Engineers, was codenamed the Manhattan Project. It spread throughout the entire United States, with the facilities for uranium and plutonium production being located at Oak Ridge, Tennessee, and Hanford, Washington, respectively. Work on plutonium production continued at the University of Chicago, at what became known as the Metallurgical Laboratory or Met Lab. A new laboratory at Los Alamos, New Mexico, became the focal point for development of the uranium and plutonium bombs.

1.3 Concepts in Nuclear Criticality

A nuclear reactor works on the principle of a chain reaction. An initial neutron is absorbed by a fissile nuclide and during the process of fission; additional neutrons are released to replace the neutron that was consumed. If more neutrons are produced than are consumed, then the neutron population grows. If fewer neutrons are produced than are consumed, the neutron population shrinks. The number of fissions caused by the neutron population determines the energy released.

In order to quantify this concept, let us define a multiplication factor k. We will define k as the ratio of the production to consumption of neutrons:

$$k = \frac{\text{Production}}{\text{Consumption}} \tag{1.1}$$

1.4 Fundamental of Fission Nuclear Reactors

Today many nations are considering an expanded role for nuclear power in their energy portfolios. This expansion is driven by concerns about global warming, growth in energy demand, and relative costs of alternative energy sources. In 2008, 435 nuclear reactors in 30 countries provided 16% of the world's electricity. In January 2009, 43 reactors were under construction in 11 countries, with several hundred more projected to come on line globally by 2030.

Fig. 1.4 A Nuclear power plant. (Courtesy of R2 Controls)

Concerns over energy resource availability, climate change, air quality, and energy security suggest an important role for nuclear power in future energy supplies. While the current Generation II and III nuclear power plant designs provide a secure and low-cost electricity supply in many markets, further advances in nuclear energy system design can broaden the opportunities for the use of nuclear energy. To explore these opportunities, the US Department of Energy's Office of Nuclear Energy has engaged governments, industry, and the research community worldwide in a wide-ranging discussion on the development of next-generation nuclear energy systems known as "Generation IV." See Sect. 1.12 of this chapter for more information on new generation of power plant known as Gen IV.

Nuclear reactors produce energy through a controlled fission chain reaction (see Sect. 1.1 in above: The First Chain Reaction). While most reactors generate electric power, some can also produce plutonium for weapons and reactor fuel. Power reactors use the heat from fission to produce steam, which turns turbines to generate electricity. In this respect, they are similar to plants fueled by coal and natural gas. The components common to all nuclear reactors include a fuel assembly, control rods, a coolant, a pressure vessel, a containment structure, and an external cooling facility. See Fig. 1.4.

In a nuclear reactor, neutron interacts with the nuclei of the surrounding atoms. For some nuclei (e.g., U-235), an interaction with a neutron can lead to fission: the

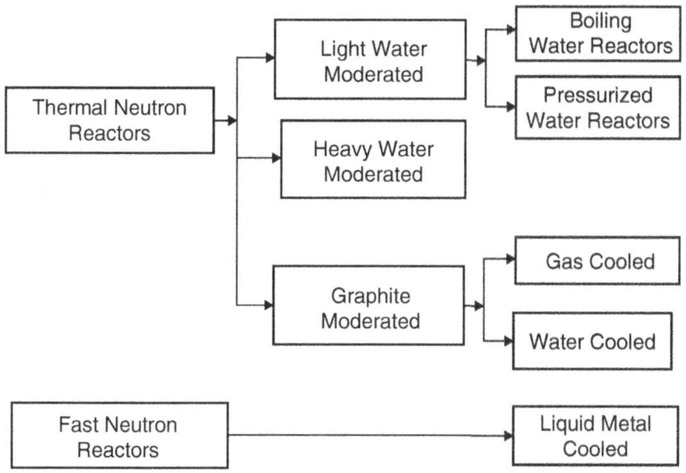

Fig. 1.5 Types of nuclear reactors. (Courtesy of Chem Cases)

nucleus is split into two parts, giving rise to two new nuclei (the so-called fission products), energy, and several new highly energetic neutrons. Other possible interactions are absorption (the neutron is removed from the system) and simple collisions, where the incident neutron transfers energy to the nucleus, either elastically (hard-sphere collision) or inelastically [7].

The speed of the neutrons in the chain reaction determines the reactor type (see Fig. 1.5). Thermal reactors use slow neutrons to maintain the reaction. These reactors require a moderator to reduce the speed of neutrons produced by fission. Fast neutron reactors, also known as fast breeder reactors (FBR), use high speed, unmoderated neutrons to sustain the chain reaction [1].

Thermal reactors operate on the principle that uranium-235 undergoes fission more readily with slow neutrons than with fast ones. Light water (H^2O), heavy water (D_2O), and carbon in the form of graphite are the most common moderators. Since slow neutron reactors are highly efficient in producing fission in uranium-235, they use fuel assemblies containing either natural uranium (0.7% U-235) or slightly enriched uranium (0.9 to 2.0% U-235) fuel. Rods composed of neutron-absorbing material such as cadmium or boron are inserted into the fuel assembly. The position of these control rods in the reactor core determines the rate of the fission chain reaction. The coolant is a liquid or gas that removes the heat from the core and produces steam to drive the turbines. In reactors using either light water or heavy water, the coolant also serves as the moderator. Reactors employing gaseous coolants (CO_2 or He) use graphite as the moderator. The pressure vessel, made of heavy-duty steel, holds the reactor core containing the fuel assembly, control rods, moderator, and coolant. The containment structure, composed of thick concrete and steel, inhibits the release of radiation in case of an accident and also secures components of the reactor from potential intruders. Finally, the most obvious components of many nuclear power plants are the cooling towers, the external components, which provide cool water for condensing the steam to water for recycling into the containment structure. Cooling towers are also employed with coal and natural gas plants.

1.5 Reactor Fundamentals

It is important to realize that while the U-235 in the fuel assembly of a thermal reactor is undergoing fission, some of the fertile U-238 present in the assembly is also absorbing neutrons to produce fissile Pu-239. Approximately one third of the energy produced by a thermal power reactor comes from fission of this plutonium. Power reactors and those used to produce plutonium for weapons operate in different ways to achieve their goals. Production reactors produce less energy and thus consume less fuel than power reactors. The removal of fuel assemblies from a production reactor is timed to maximize the amount of plutonium in the spent fuel (see Fig. 1.6). Fuel rods are removed from production reactors after only several months in order to recover the maximum amount of plutonium-239. The fuel assemblies remain in the core of power reactors for up to 3 years to maximize the energy produced. However, it is possible to recover some plutonium from the spent fuel assemblies of a power reactor.

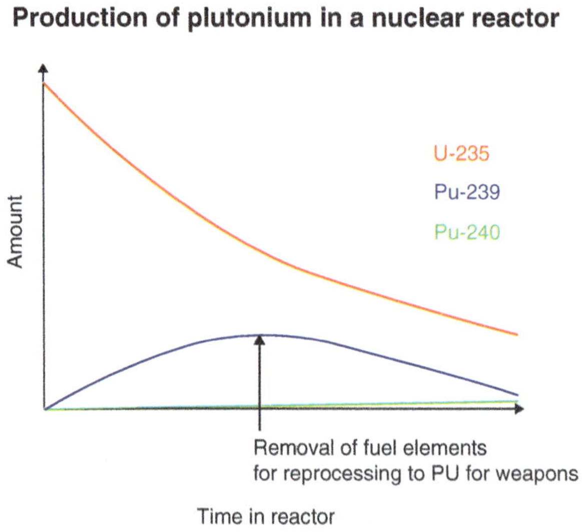

Fig. 1.6 The fate of plutonium in a thermal reactor. (Courtesy of Chem Cases)

The power output or capacity of a reactor used to generate electricity is measured in megawatts of electricity, MW(e). However, due to the inefficiency of converting heat into electricity, this represents only about one third of the total thermal energy, MW(t), produced by the reactor. Plutonium production is related to MW(t). A production reactor operating at 100 MW(t) can produce 100 grams of plutonium per day or enough for one weapon every 2 months.

Another important property of a reactor is its capacity factor. This is the ratio of its actual output of electricity for a period of time to its output if it had been operated at its full capacity. The capacity factor is affected by the time required for maintenance and repair and for the removal and replacement of fuel assemblies. The average capacity factor for US reactors has increased from 50% in the early 1970s to over 90% today. This increase in production from existing reactors has kept electricity affordable.

1.6 Thermal Reactors

Currently the majority of nuclear power plants in the world are water-moderated, thermal reactors. They are categorized as either light-water or heavy-water reactors. Light-water reactors use purified natural water (H^2O) as the coolant/moderator, while heavy-water reactors employ heavy water, deuterium oxide (D^2O). In light-water reactors, the water is either pressured to keep it in superheated form (in a pressurized water reactor, PWR) or allowed to vaporize, forming a mixture of water and steam in boiling water reactors, BWR. In a PWR (Fig. 1.10), superheated water flowing through tubes in the reactor core transfers the heat generated by fission to a heat exchanger, which produces steam in a secondary loop to generate electricity. None of the water flowing through the reactor core leaves the containment structure. In a BWR (Fig. 1.12), the water flowing through the core is converted directly to steam and leaves the containment structure to drive the turbines. Light-water reactors use low-enriched uranium as fuel. Enriched fuel is required because natural water absorbs some of the neutrons, reducing the number of nuclear fissions. All of the 103 nuclear power plants in the United States are light-water reactors; 69 are PWRs and 34 are BWRs.

1.7 Nuclear Power Plants and Their Classifications

A nuclear power plant uses controlled nuclear fission. In this section, we will explore how a nuclear power plant operates and the manner in which nuclear reactions are controlled. There are several different designs for nuclear reactors. Most of them have the same basic function, but one's implementation of this function separates it from another. There are several classification systems used to distinguish between reactor types. Below is a list of common reactor types and classification systems found throughout the world, and they are briefly explained below according to three types of classification either:

1. *Classified by moderator material* [i.e., light-water reactor, or graphite-moderated reactor, and heavy-water reactor]
2. *Classified by coolant material* [i.e., pressurized water reactor, or boiling water reactor, and gas-cooled reactor]
3. *Classified by reaction type* [i.e., fast-neutron reactor, or thermal-neutron reactor, and liquid-metal fast-breeder reactor]

1.8 Nuclear Power Plants and Their Classifications

These types of reactors and their general description are presented below.

1.8.1 Light-Water Reactors (LWR)

A light-water reactor is a type of thermal reactor that uses "light water" (plain water) as a neutron moderator or coolant instead of using deuterium oxide (^2H_2O); light-water reactors are the most commonly used among thermal reactors. Light-water reactors are contained in highly pressurized steel vessels called reactor vessels. Heat is generated by means of nuclear fission within the core of the reactor. The hundreds into a "fuel assembly," about 12 feet in length and about as thin as a pencil, group the nuclear fuel rods, each. Each fuel rod contains pellets of an oxidized form of uranium (UO_2).

A light-water fuel reactor uses ordinary water to keep the system cool. The water is circulated past the core of the reactor to absorb the generated heat. The heated water then travels away from the reactor where it leaves the system as nothing more than water vapor. This is the method used in all LWRs except the BWR for in that specific system water is boiled directly by the reactor core. See Fig. 1.7.

1.8.2 Graphite-Moderated Reactors (GMR)

A graphite-moderated reactor (GMR) is a type of reactor that is moderated with graphite. The first ongoing nuclear reaction carried out by Enrico Fermi at The University of Chicago was of this type, as well as the reactor associated with the Chernobyl accident. GMRs share a valuable property with heavy-water reactors, in that natural unenriched uranium may be used. Another highlight for the GMR is a low-power density, which is ideal if power were to suddenly stop; this would not

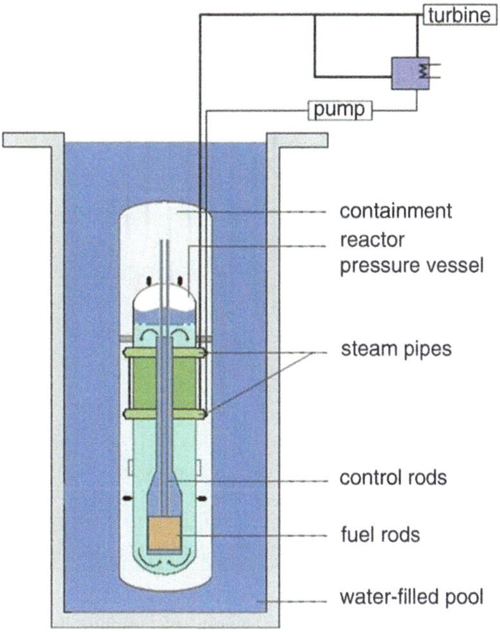

Fig. 1.7 A pumpless light-water reactor

waste as much power/fuel. The common criticisms for this design are a lack of room for steam suppression and the limited safety precautions available to the design. See Fig. 1.8.

1.8.3 Heavy-Water Reactors (HWR)

Heavy-water reactors (HWR) are a class of fission reactor that uses heavy water as a *neutron moderator*. Heavy water is deuterium oxide, D_2O. Neutrons in a nuclear reactor that use uranium must be slowed down so that they are more likely to split other atoms and get more neutrons released to split other atoms. Light water can be used, as in a light-water reactor (LWR), but since it absorbs neutrons, the uranium must be enriched for criticality to be possible. The most common pressurized heavy-water reactor (PHWR) is the CANDU reactor.

Usually the heavy water is also used as the coolant, but as example, the Lucens reactor was gas cooled. Advantage of this type of reactor is that they can operate with unenriched uranium fuel, although the opponents of heavy-water reactors suggest that such reactors pose a much greater risk of nuclear proliferation because of two characteristics:

1. They use unenriched uranium as fuel, the acquisition of which is free from supervision of international institutions on uranium enrichment.

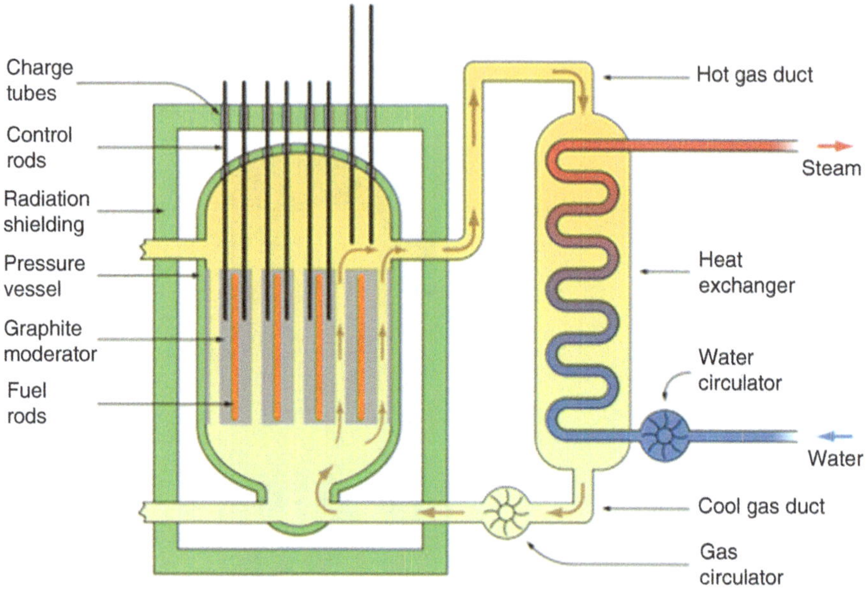

Fig. 1.8 A typical core layout of graphite-moderated reactor. (Courtesy Osterreichisches Ökologie-Institut)

2. They produce more plutonium and tritium as by-products than light-water reactors; these are hazardous radioactive substances that can be used in the production of modern nuclear weapons such as fission, boosted fission, and neutron bombs as well as the primary stages of thermonuclear weapons. For instance, India produced its plutonium for Operation Smiling Buddha, its first nuclear weapon test, by extraction from the spent fuel of a heavy-water research reactor known as "CIRUS (Canada-India Research Utility Services)." It is advocated that safeguards need to be established to prevent exploitation of heavy-water reactors in such a fashion.

In heavy-water reactors, both the moderator and coolant are heavy water (D_2O). A great disadvantage of this type comes from this fact: heavy water is one of the most expensive liquids. However, it is worth its price: this is the best moderator. Therefore, the fuel of HWRs can be slightly (1–2%) enriched or even natural uranium. Heavy water is not allowed to boil, so in the primary circuit, very high pressure, similar to that of PWRs, exists. See Fig. 1.9.

The main representative of the heavy-water type is the Canadian CANDU reactor. In these reactors, the moderator and coolant are spatially separated: the moderator is in a large tank (calandria), in which there are pressure tubes surrounding the fuel assemblies. The coolant flows in these tubes only.

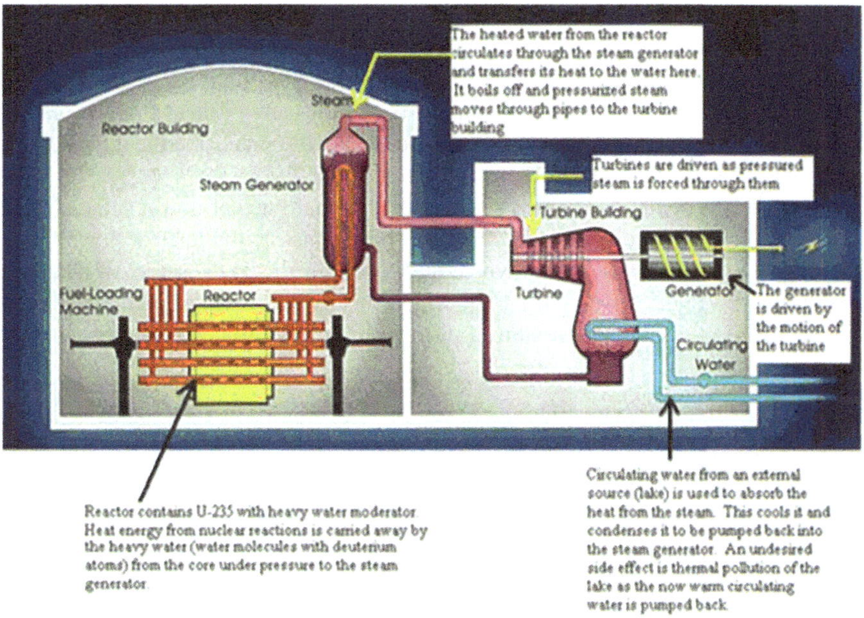

Fig. 1.9 A typical outline layout of heavy-water reactor. (Courtesy of Atomic Energy of Canada Limited)

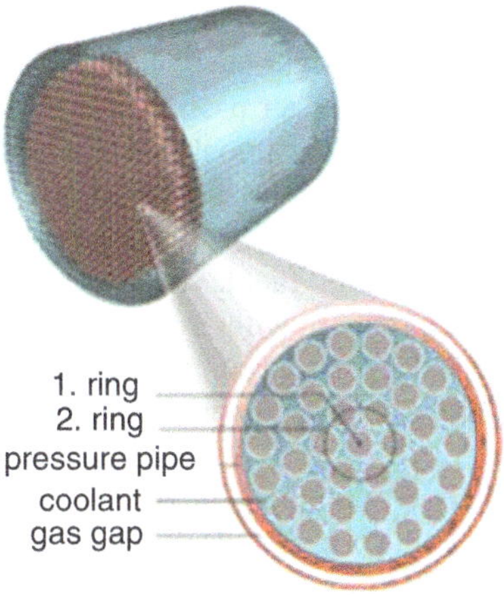

The advantage of this construction is that the whole tank need not be kept under high pressure; it is sufficient to pressurize the coolant flowing in the tubes. This arrangement is called pressurized tube reactor. Warming up of the moderator is much less than that of the coolant; it is simply lost for heat generation or steam production. The high-temperature and high-pressure coolant, similarly to PWRs, goes to the steam generator where it boils the secondary side light water. Another advantage of this type is that fuel can be replaced during operation, and thus there is no need for outages.

The other type of heavy-water reactor is the pressurized heavy-water reactor (PHWR). In this type, the moderator and coolant are the same, and the reactor pressure vessel has to stand the high pressure.

Heavy-water reactors produce cca. 6% of the total nuclear power plant (NPP), power of the world; however, 13.2% of the under-construction nuclear power plant capacity is accounted for by this type. One reason for this is the safety of the type; another is the high conversion factor, which means that during operation a large amount of fissile material is produced from U-238 by neutron capture

1.9 Classified by Coolant Material

The descriptions of these types of reactors are as follows.

1.9.1 Pressurized Water Reactors (PWR)

A pressurized water reactor (PWR) is designed by Westinghouse Bettis Atomic Power Laboratory which has used a type of light-water reactor for decades in designs for military ship applications; now the primary manufacturers are Framatome ANP and Westinghouse for present-day power plant reactors. The pressurized water reactor is unique in that although water passes through the reactor core to act as moderator and coolant, it does not flow into the turbine. Instead of the conventional flow cycle, the water passes into a pressurized primary loop; see Fig. 1.10. This step in the PWR cycle produces steam in a secondary loop that drives the turbine. Advantages of the PWR include zero fuel leaks of radioactive material into the turbine or environment and the ability to withstand higher pressures and temperatures to increase the Carnot efficiency. Disadvantages include complex reactor designs and costs. This reactor type accounts for the majority of reactors located in the United States.

Pressurized water reactor (PWR) is a type of a nuclear power reactor that uses enriched uranium as a fuel which in turn heats the light water used for producing steam. The main feature which differentiates it from a BWR nuclear reactor is that a PWR has a separate arrangement to make steam in the form of a heat exchanger.

1.9.1.1 The Arrangement of PWR

A pressurized water reactor (PWR) is a type of power plant reactor consisting of two basic circuits having light water as the working fluid. In one of the circuits, water is heated to a high temperature and kept at high pressure as well, so that it

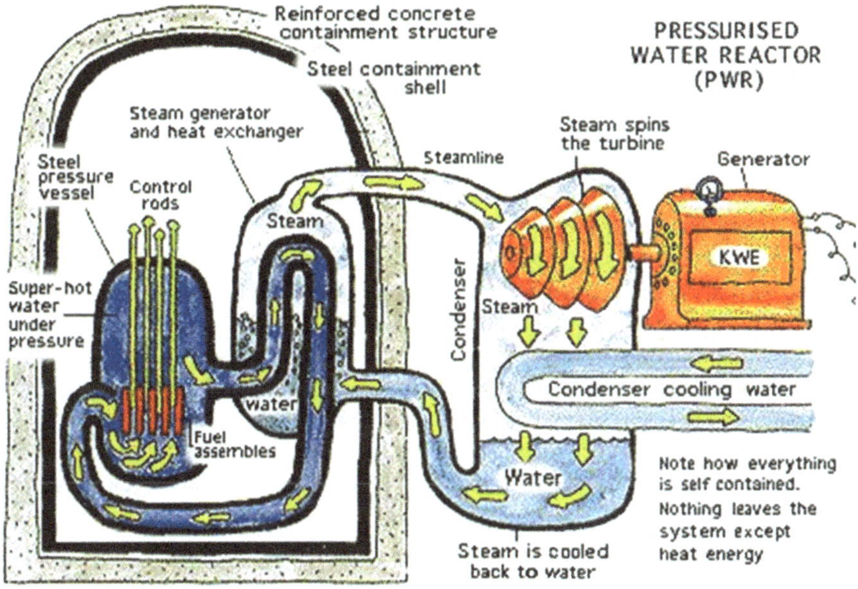

Fig. 1.10 A typical pressurized water reactor. (Courtesy of the Uranium Information Centre)

does not get converted into a gaseous state. This superheated water is used as a coolant and a moderator for the nuclear reactor core, hence the name PWR or pressurized water reactor.

The secondary circuit consists of water at high pressure in the gaseous state, i.e., steam which is used to run the turbine-alternator arrangement. The point of interaction between these two circuits is the heat exchanger or the boiler wherein heat from the superheated high-pressure water converts the water in the secondary circuit to steam.

1.9.1.2 Advantages of PWR

- Much fewer control rods are required in a PWR. In fact, for a typical 1000 MW plant, just around 5 dozen control rods are sufficient.
- Since the two circuits are independent of each other, it makes it very easy for the maintenance staff to inspect the components of the secondary circuit without having to shut down the power plant entirely.
- A PWR has got a high-power density, and this, combined with the fact that enriched uranium is used as fuel instead of normal uranium, leads to the construction of very compact core size for a given power output.
- One feature, which makes a PWR reactor very suitable for practical applications, is its positive demand coefficient, which serves to increase the output as a direct proportion to demand of power.

- The water used in the primary circuit is different from that used in the secondary circuit, and there is no intermixing between the two, except for heat transfer, which takes place in the boiler or heat exchanger. This means that the water used in the turbine side is free from radioactive steam; hence the piping on that side is not required to be clad with special shielding materials.

1.9.1.3 Drawbacks of PWR

- The primary circuit consists of high-temperature, high-pressure water which accelerates corrosion. This means that the vessel should be constructed of very strong material such as stainless steel, which adds to construction costs of PWR.
- PWR fuel charging requires the plant to be shut down, and this certainly requires a long time period of the order of at least a couple of months.
- The pressure in the secondary circuit is relatively quite low as compared to the primary circuit; hence the thermodynamic efficiency of PWR reactors is quite low of the order of 20.

1.9.1.4 Pressurizer

One important point to note here is that despite the changing loads, the pressure in the primary circuit needs to be maintained at a constant value. This is achieved by installing a device known as pressurizer in the primary circuit. It basically consists of a dome-shaped structure which has heating coils which are used to increase or decrease pressure as and when required depending on varied load conditions.

Note that in the pressurized water reactor (PWR), the water, which passes over the reactor core to act as moderator and coolant, does not flow to the turbine but is contained in a pressurized primary loop. The primary loop water produces steam in the secondary loop, which drives the turbine. The obvious advantage to this is that a fuel leak in the core would not pass any radioactive contaminants to the turbine and condenser. See Fig. 1.11.

Another advantage is that the PWR can operate at higher pressure and temperature, about 160 atmospheres and about 315 °C. This provides a higher *Carnot efficiency* than the BWR, but the reactor is more complicated and costlier to construct. Most of the US reactors are pressurized water reactors.

1.9.2 Boiling Water Reactor (BWR)

The boiling water reactor (BWR) dates back to their General Electric introduction in the 1950s. The distinguishing feature in the BWR is the boiling method for steam. In this type of reactor, water passes over the core as a coolant to expand and become steam source for a turbine placed directly above. Advantages of this design type include a simpler reactor design, a smaller reactor system, and lower costs.

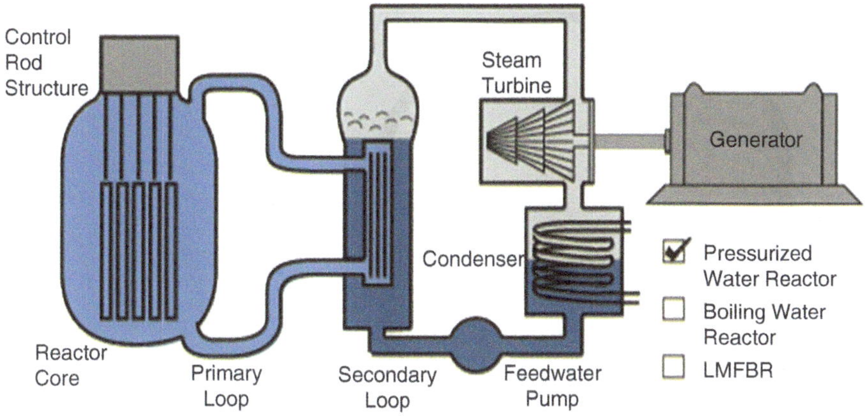

Fig. 1.11 A typical outline of pressurized water reactor

Disadvantages found are the increase of radioactive materials in the turbine and a greater chance for fuel to burn out as water quickly evaporates to expose fuel rods to an atmosphere absent of a coolant. BWRs have found fame all over the world due to the cheap simple design.

In Fig. 1.12, we see that:

1. The core inside the reactor vessel creates heat.
2. A steam-water mixture is produced when very pure water (reactor coolant) moves upward through the core, absorbing heat.
3. The steam-water mixture leaves the top of the core and enters the two stages of moisture separation where water droplets are removed before the steam is allowed to enter the steam line.
4. The steam line directs the steam to the main turbine, causing it to turn the turbine generator, which produces electricity.

Note that in the boiling water reactor (BWR), as illustrated in Fig. 1.13, the water, which passes over the reactor core to act as moderator and coolant, is also the steam source for the turbine. The disadvantage of this is that any fuel leak might make the water radioactive and that radioactivity would reach the turbine and the rest of the loop.

A typical operating pressure for such reactors is about 70 atmospheres at which pressure the water boils at about 285 °C. This operating temperature gives a *Carnot efficiency* of only 42% with a practical operating efficiency of around 32%, somewhat less than the PWR.

1.9.3 Gas-Cooled Reactors (GCR)

The gas-cooled reactor (GCR) or the gas-graphite reactors, as illustrated in Fig. 1.14, operate using graphite as moderator and some gas (mostly CO_2, lately helium) as coolant. This belongs to the oldest reactor types. The first GGR was the Calder Hall

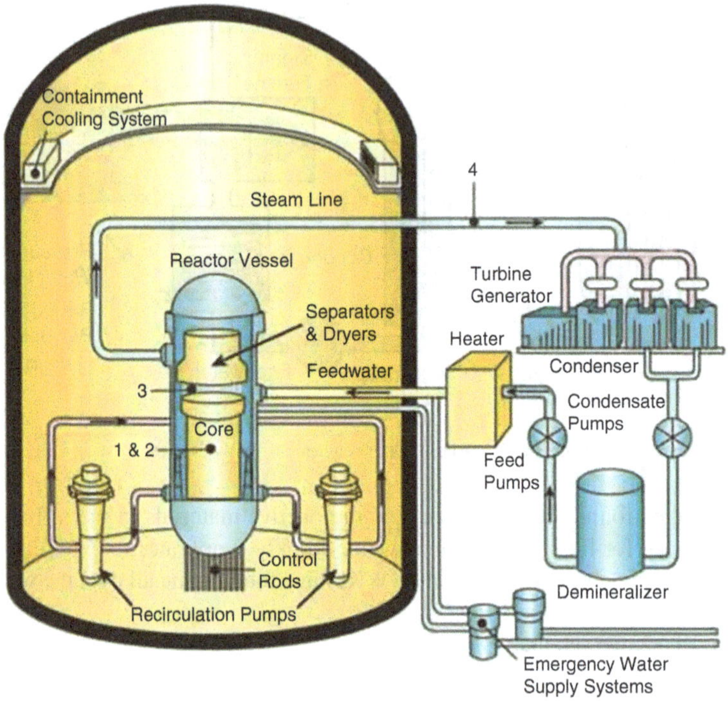

Fig. 1.12 A typical boiling water reactor. (Courtesy of the U.S. Nuclear Regulatory Commission)

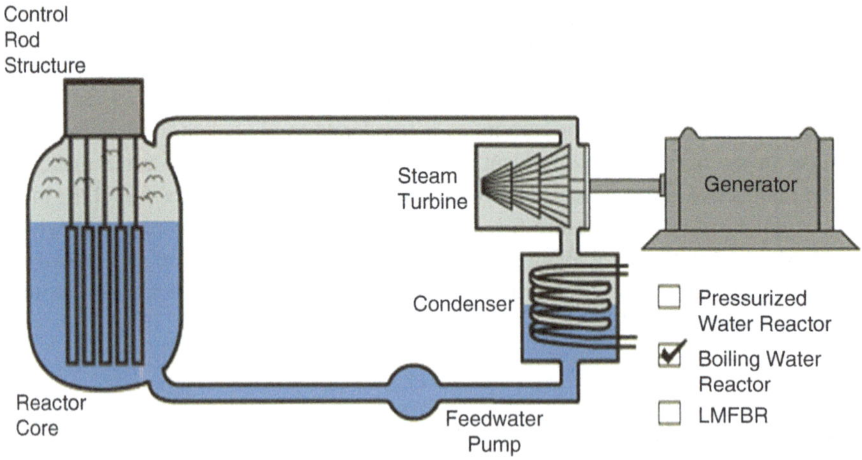

Fig. 1.13 A typical layout of boiling water reactor

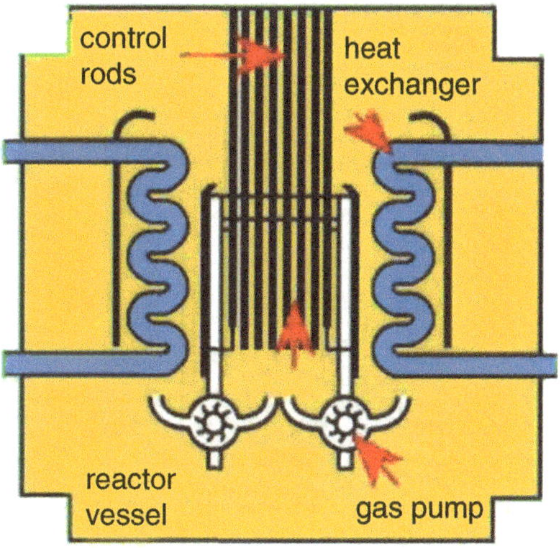

Fig. 1.14 A typical core layout of gas-cooled reactor

power plant reactor, which was built in 1955 in England. This type is called Magnox after the special magnesium alloy (Magnox), of which the fuel cladding was made. The fuel is natural uranium. These reactors account for 1.1% of the total NPP power of the world and are not built any more.

The advanced gas-cooled reactor (AGR) is a development from Magnox: the cladding is not Magnox and the fuel is slightly enriched. The moderator is also graphite and the coolant is CO_2. Contribution to total world capacity is 2.5%. This type is not manufactured any longer.

The newest gas-cooled reactor type is the high-temperature gas-cooled reactor (HTGR), which is cooled by helium and moderated by graphite. In this reactor as high as 950 °C coolant temperature can be achieved. The efficiency of a newly developed type, the gas turbine-modular helium reactor (GT-MHR), might be as high as almost 50%.

Gas-cooled reactors (GCR) and advanced gas-cooled reactors (AGR) use carbon dioxide as the coolant to carry the heat to the turbine and graphite as the moderator. Like heavy water, a graphite moderator allows natural uranium (GCR) or slightly enriched uranium (AGR) to be used as fuel. See Fig. 1.15.

1.10 Classified by Reaction Type

The descriptions of each of these reactors are given as follows.

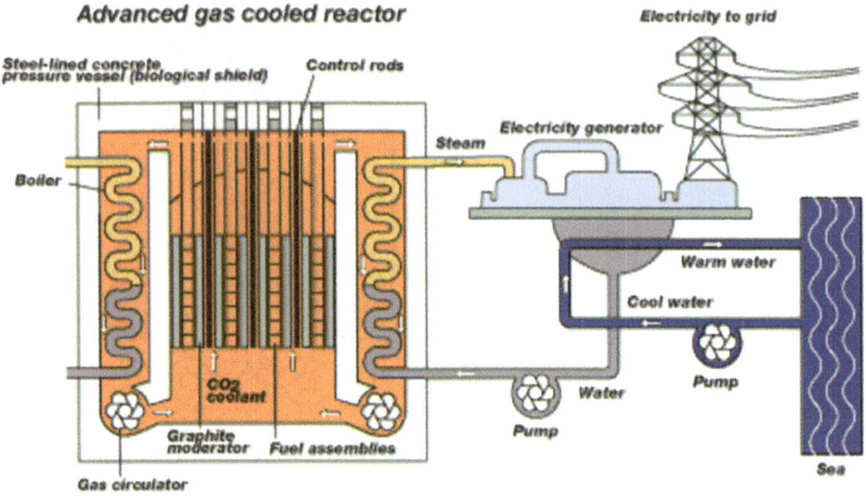

Fig. 1.15 A typical outline layout of gas-cooled reactor

1.10.1 Fast-Neutron Reactor (FNR)

Fast-neutron reactors (FNR), also known as fast-breeder reactors (FBR), use depleted nuclear waste as a form of energy. Uranium, which is composed of 0.7% uranium-235 and 99.3% uranium-238, is processed in the fast-neutron reactors into isotopes of usable plutonium of plutonium-239 and plutonium-241. Fast-neutron reactors are 60% more efficient than normal reactors; a fast-neutron reactor uses liquid metal as its coolant as opposed to water, which makes the reactor safer to use, and its fuel is metallic, which keeps the reactors under control more easily. Some cons of fast-neutron reactors though are that they are very unpredictable, making them more tedious to use. Bubbles are more present in processes, so fast-neutron reactors tend to heat up more rather than cool down, and the coolant that it requires is much more exotic, such as liquid sodium and bismuth eutectic.

Several countries have research and development programs for improved fast-breeder reactors (FBR), which are a type of fast-neutron reactors. These use the uranium-238 in reactor fuel as well as the fissile U-235 isotope used in most reactors.

Natural uranium contains about 0.7% U-235 and 99.3% U-238. In any reactor, the U-238 component is turned into several isotopes of plutonium during its operation. Two of these, Pu-239 and Pu-241, then undergo fission in the same way as U-235 to produce heat. In a fast-neutron reactor, this process is optimized so that it can "breed" fuel, often using a depleted uranium blanket around the core. FBRs can utilize uranium at least 60 times more efficiently than a normal reactor.

Fast-neutron reactors could extract much more energy from recycled nuclear fuel, minimize the risks of weapon proliferation, and markedly reduce the time nuclear waste must be isolated.

If developed sensibly, nuclear power could be truly sustainable and essentially inexhaustible and could operate without contributing to climate change. In particular, a relatively new form of nuclear technology could overcome the principal drawbacks of current methods – namely, worries about reactor accidents; the potential for diversion of nuclear fuel into highly destructive weapons; the management of dangerous, long-lived radioactive waste; and the depletion of global reserves of economically available uranium. This nuclear fuel cycle would combine two innovations: pyrometallurgical processing (a high-temperature method of recycling reactor waste into fuel) and advanced fast-neutron reactors capable of burning that fuel. With this approach, the radioactivity from the generated waste could drop to safe levels in a few hundred years, thereby eliminating the need to segregate waste for tens of thousands of years.

Fast Reactor Technology:
A Path to Long-Term Energy Sustainability
Position Statement
November 2005

"The American Nuclear Society believes that the development and deployment of advanced nuclear reactors based on fast-neutron fission technology is important to the sustainability, reliability, and security of the world's long-term energy supply. Of the known and proven energy technologies, only nuclear fission can provide the large quantities of energy required by industrial societies in a sustainable and environmentally acceptable manner."

"Natural uranium mined from the earth's crust is composed primarily of two isotopes: 99.3% is U- 238, and 0.7% is the fissile U-235. Nearly all current power reactors are of the "thermal neutron" design, and their capability to extract the potential energy in the uranium fuel is limited to less than 1% of that available. The remainder of the potential energy is left unused in the spent fuel and in the uranium, depleted in U-235 that remains from the process of enriching the natural uranium in the isotope U-235 for use in thermal reactors. With known fast reactor technology, this unutilized energy can be harvested, thereby extending by a hundred-fold the amount of energy extracted from the same amount of mined uranium."

"Fast reactors can convert U-238 into fissile material at rates faster than it is consumed making it economically feasible to utilize ores with very low uranium concentrations and potentially even uranium found in the oceans [2–4]. A suitable technology has already been proven on a small scale [5]. Used fuel from thermal reactors and the depleted uranium from the enrichment process can be utilized in fast reactors, and that energy alone would be sufficient to supply the nation's needs for several hundred years."

Fast reactors in conjunction with fuel recycling can diminish the cost and duration of storing and managing reactor waste with an offsetting increase in

(continued)

the fuel cycle cost due to reprocessing and fuel prefabrications. Virtually all long-lived heavy elements are eliminated during fast reactor operation, leaving a small amount of fission product waste that requires assured isolation from the environment for less than 500 years [5].

Although fast reactors do not eliminate the need for international proliferation safeguards, they make the task easier by segregating and consuming the plutonium as it is created. The use of onsite reprocessing makes illicit diversion from within the process highly impractical. The combination of fast reactors and reprocessing is a promising option for reasons of safety, resource utilization, and proliferation resistance [6].

"Reaping the full benefits of fast reactor technology will take a decade or more for a demonstration reactor, followed by buildup of a fleet of operating power stations. For now, and in the intermediate-term future, the looming short-term energy shortage must be met by building improved, proven thermal-reactor power plants. To assure longer-term energy sustainability and security, the American Nuclear Society sees a need for cooperative international efforts with the goal of building a fast reactor demonstration unit with onsite reprocessing of spent fuel."

1.10.2 Thermal-Neutron Reactor

Thermal reactors go through the same process as fast-neutron reactors, but in a thermal reactor, the process of obtaining plutonium is slower. These types of reactors use a neutron moderator to slow neutrons until they approach the average kinetic energy of the surrounding particles, that is, to reduce the speed of the neutrons to low-velocity thermal neutrons. The nuclear cross section of uranium-235 for slow thermal neutrons is about 1000 barns. For fast neutrons, it is in the order of 1 barn. In a thermal reactor, the neutrons that undergo the reaction process have significantly lower electron-volt energy, so the neutrons are considered to be slower. A neutron's speed will determine its chances to interact with the nucleus of an atom; the slower its speed the bigger its fission cross section becomes, and thus the higher its chance of interacting with the nucleus becomes. See Fig. 1.16.

This figure gives the value of the fission cross section for some fissile isotopes. Note that both axes are logarithmic. The thermal and fast energy regions are indicated. For thermal energies, the fission cross section equals several thousand barns; at high energies the fission cross section is of the order of 1–10 barns.

The fact that the fission cross section is rather large for low-energy neutrons has an important effect on the design of a nuclear reactor: in a reactor where the neutrons have a low energy, not much fissile material is required, because the probability

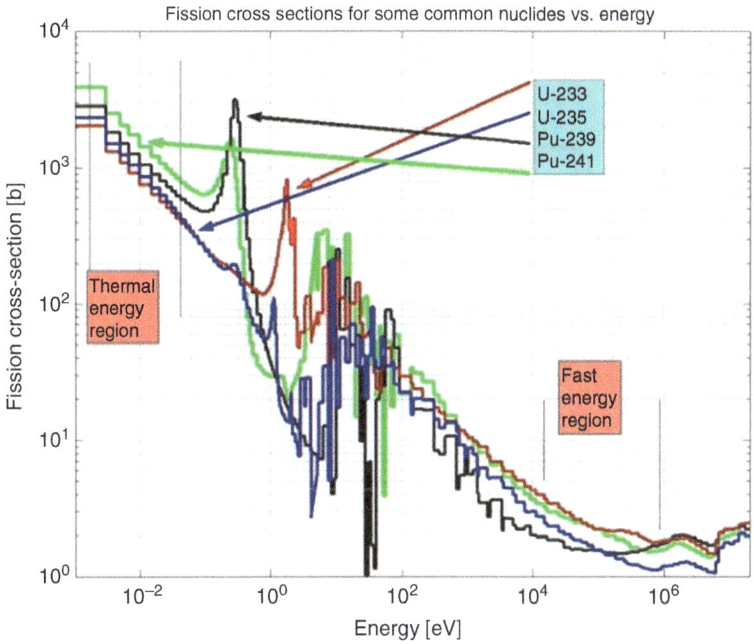

Fig. 1.16 Fission cross section for some common nuclides vs. energy. (Courtesy of TU Delft)

of an interaction is very large. The lowest energy a neutron can have in a nuclear reactor is the energy at which it is in equilibrium with its environment. The movement of the neutron is then identical to the thermal movement of the atoms that constitute the reactor. The neutrons that have slowed down from the high energy (2 MeV) where they are born to this equilibrium energy are called "thermal neutrons." The average energy of a neutron in thermal equilibrium is 0.025 eV – the neutron is slowed down over nine decades, more than a billion times. Reactors in which most fissions are induced by thermal neutrons are called thermal reactors. Thermal reactors are by far the most widely used reactors in the world today. Most reactors use water, heavy water, or graphite as moderator. The reason for the choice of thermal reactors is a simple one: a thermal reactor requires a small amount of fuel to become critical, and thus the fuel is cheap [7].

1.10.3 Liquid-Metal Fast-Breeder Reactors (LMFBR)

The plutonium-239 breeder reactor is commonly called a fast-breeder reactor, and the cooling and a liquid metal does heat transfer. The metals, which can accomplish this, are sodium and lithium, with sodium being the most abundant and most

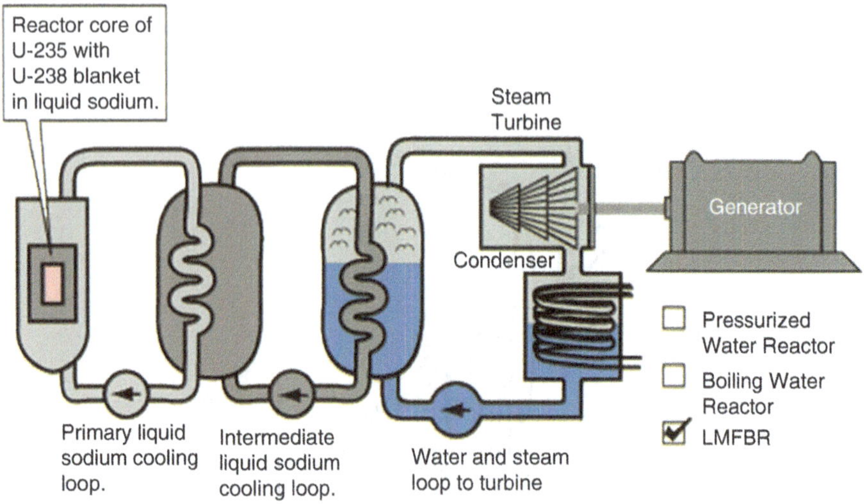

Fig. 1.17 A typical layout of liquid-metal fast-breeder reactor

commonly used. The construction of the fast breeder requires a higher enrichment of U-235 than a light-water reactor, typically 15 to 30%. The reactor fuel is surrounded by a "blanket" of non-fissionable U-238. No moderator is used in the breeder reactor since fast neutrons are more efficient in transmuting U-238 to Pu-239. At this concentration of U-235, the cross section for fission with fast neutrons is sufficient to sustain the chain reaction. Using water as coolant would slow down the neutrons, but the use of liquid sodium avoids that moderation and provides a very efficient heat-transfer medium. See Fig. 1.17.

The Super-Phenix was the first large-scale breeder reactor. It was put into service in France in 1984. It ceased operation as a commercial power plant in 1997. Such a reactor can produce about 20% more fuel than it consumes by the breeding reaction. Enough excess fuel is produced over about 20 years to fuel another such reactor. Optimum breeding allows about 75% of the energy of the natural uranium to be used compared to 1% in the standard light-water reactor. See Fig. 1.18.

Under appropriate operating conditions, the neutrons given off by fission reactions can "breed" more fuel from otherwise non-fissionable isotopes. The most common *breeding reaction* is that of plutonium-239 from non-fissile uranium-238. The term "fast breeder" refers to the types of configurations, which can actually produce more fissionable fuel than they use, such as the LMFBR. This scenario is possible because the non-fissile uranium-238 is 140 times more abundant than the fissionable U-235 and can be efficiently converted into Pu-239 by the neutrons from a fission chain reaction.

Fig. 1.18 This is a photo of a model of the containment vessel of the Super-Phenix. It is displayed at the National Museum of Nuclear Science and Technology in Albuquerque, NM

The following illustrations show the breeding reaction processes step-by-step:

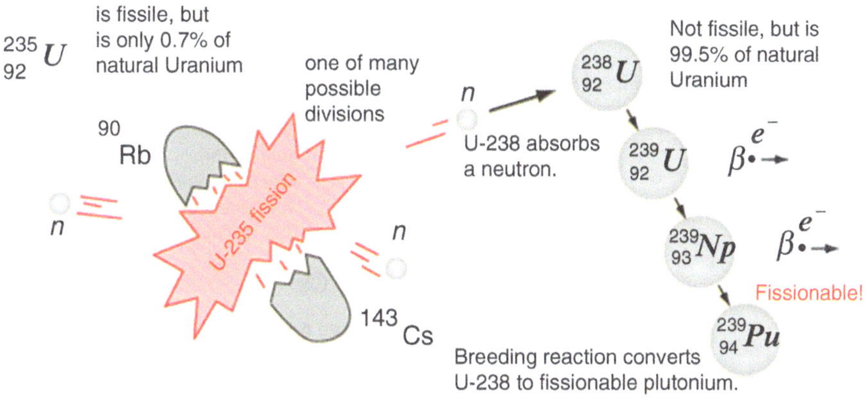

France has made the largest implementation of breeder reactors with its large Super-Phenix reactor (today is not in production line) and an intermediate Russian scale reactor (BN-600) on the Caspian Sea for electric power and desalinization [23].

Breeding plutonium-239 can be accomplished from non-fissile uranium-238 by the reaction illustrated.

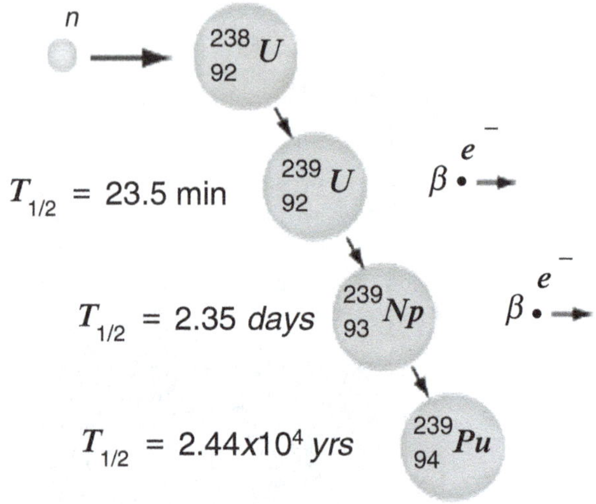

The concept of breading ratio of plutonium-239 can be defined in the following. In the breeding of plutonium fuel in breeder reactors, an important concept is the breeding ratio, the amount of fissile plutonium-239 produced compared to the amount of fissile fuel (like U-235) used to produce it. In the liquid-metal, fast-breeder reactor (LMFBR), the target breeding ratio is 1.4, but the results achieved have been about 1.2. This is based on 2.4 neutrons produced per U-235 fission, with one neutron used to sustain the reaction.

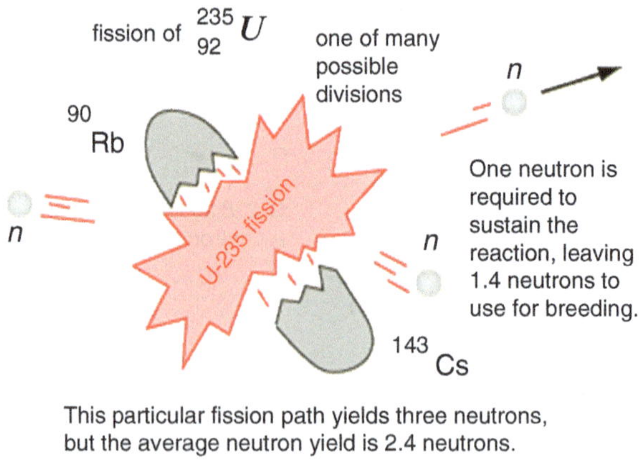

Liquid sodium is used as the coolant and heat-transfer medium in the LMFBR reactor. That immediately raised the question of safety since sodium metal is an extremely reactive chemical and burns on contact with air or water (sometimes explosively on contact with water). It is true that the liquid sodium must be protected from contact with air or water at all times, kept in a sealed system. However, it has been found that the safety issues are not significantly greater than those with high-pressure water and steam in the light-water reactors.

Sodium is a solid at room temperature but liquefies at 98 °C. It has a wide working temperature since it does not boil until 892 °C. That brackets the range of operating temperatures for the reactor so that it does not need to be pressurized as does a water-steam coolant system. It has a large *specific heat* so that it is an efficient heat-transfer fluid.

In practice, those reactors, which have used liquid-metal coolants, have been fast-neutron reactors. The liquid-metal coolant has a major advantage there because water as a coolant also moderates or slows down the neutrons. Such fast-neutron reactors require a higher degree of enrichment of the uranium fuel than do the water-moderated reactors [23].

1.11 Nuclear Fission Power Generation

Nuclear fission energy is today a competitive and mature low-carbon technology, operating at very high levels of safety. The installed nuclear electricity capacity in the Europe (EU), for example, is 132 GWe, which provides one third of the EU's generated electricity [9, 10]. Most of the current designs are light-water reactors (LWR) of the second generation, capable of providing base-load electricity often with availability factors of over 90%. There have been only a few new nuclear power plants connected to the grid in the last two decades, and as a result of decommissioning of old plants, the total number of reactors in Europe has decreased. Nevertheless, electricity supply from nuclear has remained constant, and the levelized cost has decreased owing to improved efficiency, power upgrade, and improved availability factor.

More recently, there has been a renewed interest in nuclear energy, referred to as "nuclear renaissance," mainly driven by concerns over climate change, security, and independence of supply and energy costs [8].

1.12 Generation IV Nuclear Energy Systems

Concerns over energy resource availability, climate change, air quality, and energy security suggest an important role for nuclear power in future energy supplies. While the current Generation II and III nuclear power plant designs provide a secure and low-cost electricity supply in many markets, further advances in nuclear energy system design can broaden the opportunities for the use of nuclear energy. To explore these opportunities, the US Department of Energy's Office of Nuclear Energy has engaged governments, industry, and the research community worldwide

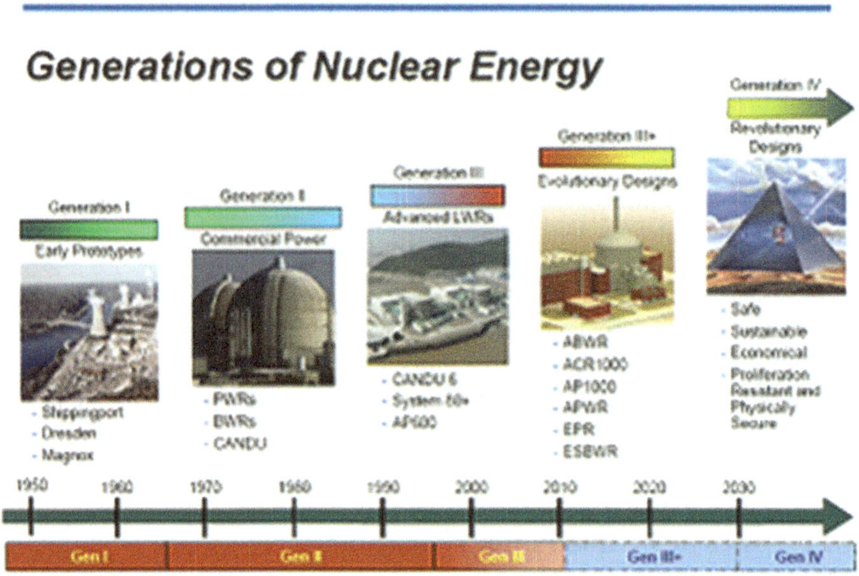

Fig. 1.19 The evolution of nuclear power

in a wide-ranging discussion on the development of next-generation nuclear energy systems known as "Generation IV." See Fig. 1.19.

The goal of the Gen IV nuclear energy systems is to address the fundamental research and development (R&D) issues necessary to establish the viability of next-generation nuclear energy system concepts to meet tomorrow's needs for clean and reliable electricity and nontraditional applications of nuclear energy. Successfully addressing the fundamental research and development (R&D) issues will allow Gen IV concepts that excel in safety, sustainability, cost-effectiveness, and proliferation risk reduction to be considered for future commercial development and deployment by the private sector.

Gen IV reactor concepts are being developed to use advanced fuels, fashioned from recycled reactor fuel and capable of high burnups. The corresponding fuel cycle strategies allow for efficient utilization of domestic uranium resources while minimizing waste. Reduction of proliferation risk and improvements in physical protection are being designed into Gen IV concepts to help thwart those who would target nuclear power plants for terrorist acts or use them improperly to develop materials for nuclear weapons. Gen IV concepts will feature advances in safety and reliability to improve public confidence in nuclear energy while providing enhanced investment protection for plant owners. Competitive life-cycle costs and acceptable financial risk are being factored into Gen IV concepts with high-efficiency electricity generation systems, modular construction, and shortened development schedules before plant startup [11]

Gen IV is also an active participant in the International Project on Innovative Nuclear Reactors and Fuel Cycles (INPRO). INPRO was established in 2001 in response to a resolution by the IAEA General Conference to help ensure that nuclear energy is available to contribute, in a sustainable manner, to meeting the energy needs of the twenty-first century and to bring together technology holders and users so that they can consider jointly the international and national actions required for achieving desired innovations in nuclear reactors and fuel cycles. INPRO provides a forum for discussion for experts and policy-makers from industrialized and developing countries on all aspects of nuclear energy planning as well as on the development and deployment of innovative nuclear energy systems in the twenty-first century.

The Generation IV International Forum (GIF) was chartered in May 2001 to lead the collaborative efforts of the world's leading nuclear technology nations to develop the next generation of nuclear energy systems. The initial efforts of GIF resulted in the identification of the six most promising reactor concepts to be investigated by this international research community and are documented in the Generation IV Technology Roadmap. Thirteen members have signed the GIF charter: Argentina, Brazil, Canada, People's Republic of China, Euratom, France, Japan, Republic of Korea, the Russian Federation, Republic of South Africa, Switzerland, the United Kingdom, and the United States. This unique international effort reached a major milestone on February 28, 2005, as five of the forum's member countries (Canada, France, Japan, the United Kingdom, and the United States) signed the world's first multilateral agreement aimed at the international development of advanced nuclear energy systems – the Framework Agreement for International Collaboration on Research and Development of Generation IV Nuclear Energy Systems. Subsequent signatories to the Framework Agreement included People's Republic of China, Euratom, Republic of Korea, Republic of South Africa, and Switzerland. The United Kingdom is a signatory of the Framework but is currently a non-active member. Argentina and Brazil have not ratified the Framework Agreement and are therefore considered non-active. The Russian Federation is working on the necessary approvals for its accession to the Framework [11]. See Fig. 1.20.

As detailed in its charter and subsequent GIF policy statements, GIF is led by the Policy Group (PG), which is responsible for the overall coordination of GIF's research and development (R&D) collaboration, policy formation, and interactions with other organizations. France currently chairs the Policy Group vice chairs from the United States and Japan. An Experts Group and the Senior Industry Advisory Panel advise the Policy Group on (R&D) strategy, priorities, and methodology and on evaluating research plans for each Generation IV system. The Framework Agreement establishes two levels of implementing arrangements in order to conduct the joint (R&D). The first level consists of a system arrangement for each Generation IV reactor concept directed by a System Steering Committee (SSC). Under each SSC, Project Arrangements are established with Project Management Boards to manage and implement the joint (R&D).

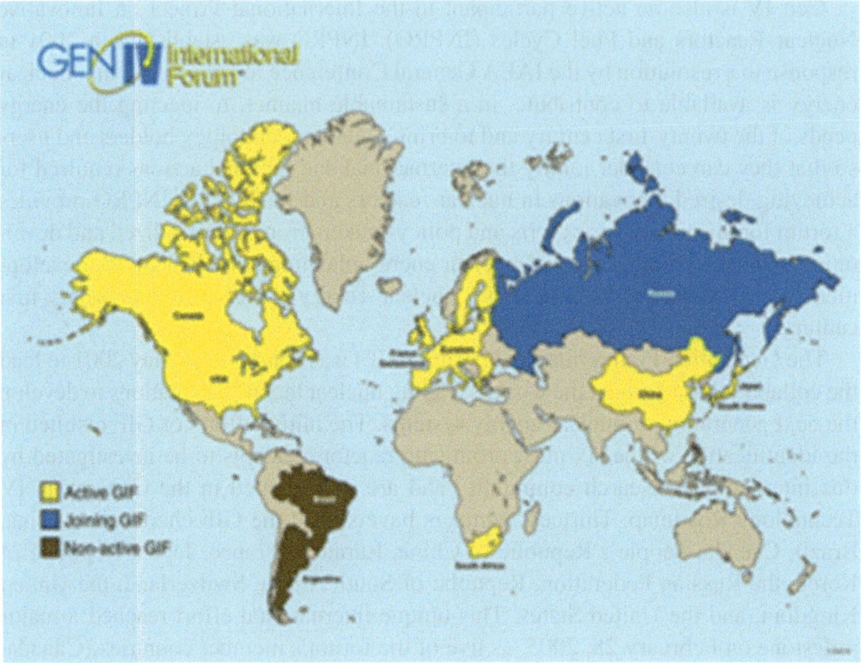

Fig. 1.20 Map of member countries

1.13 Technological State of the Art and Anticipated Developments

It has been demonstrated that Generation-II plants can be safely and economically operated for up to 60 years through the development of improved harmonized plant-life management technologies and plant license extension practices (PLIM/PLEX) and that developments in fuel technologies can still lead to improvements in reactor performance [12]. The first Generation-III reactors, which are an evolution of thermal reactors with even further improved safety characteristics and economy, are now being built. In the coming decades, nuclear electricity generation should increase or at least maintain its current level by a combination of lifetime extension and power upgrades of Generation-II reactors and new build of Generation-III reactors. Two 1.6 GWe Generation-III reactors are presently under construction in Finland and France, targeted for connection to the grid in 2012.

The Finnish reactor was a first-of-a-kind (FOAK), and the construction has suffered delays with the overnight cost increasing from 2000 to 3100 €/kWe, whereas the overnight cost for the second reactor in France is now 2400 €/kWe. In series production, the industry expects the cost to be 2000 ± 500 €/GWe, which is in line with recent international studies. An additional capacity of 100GWe of Generation-III reactors over the next 25 years is a reasonable estimate, which would require an investment in the range of 200–280 billion Euros. The capital costs represent typically 60–70% of the levelized cost for nuclear electricity, operation and maintenance 20–25%, and fuel 10–15%. The front-loaded cost profile means that the levelized cost is very sensitive to construction time and the financial schemes for the investment. Estimates in 2007 for the United Kingdom resulted in a range of 31–44 £/MWh (37–53 €/MWh).

Though uranium is relatively abundant in the Earth's crust and oceans, estimates of natural reserves are always related to the cost of mineral extraction. As the price of uranium increases on world markets, the number of economically exploitable deposits also increases. The most recent estimates [15] identified 5.5 million tons of uranium (MtU) that could be exploited below 130$/kg. The total amount of undiscovered resources (reasonably assured and speculative) available at an extraction cost below 130 $/kgU is estimated at 10.5MtU. Unconventional resources, from which uranium is extracted as a by-product only, e.g., in phosphate production, lie between 7 and 22 MtU, and reserves in seawater are estimated to be 4000 MtU. Japanese studies suggest that uranium from seawater can be extracted at 300€/kg [8]. At a conservative estimate, 25,000 tons of the uranium are required to produce the fuel to generate 1000 TWhe in an open fuel cycle. The global electricity supplied by nuclear is 2600 TWhe, which means that the conventional resources below 130$/kgU at the current rate of consumption would last for at least 85 years with the already identified resources (5.5 MtU) and 246 years, if the undiscovered are also included (5.5 + 10.5MtU). In addition to uranium, it is also possible to use thorium, which is three times more abundant in the Earth's crust, though would require different reactors and fuel cycles. Nonetheless, natural resources are plentiful and do not pose an immediate limiting factor for the development of nuclear energy.

However, in a scenario with a large expansion of nuclear energy, resources will become an issue much earlier, especially since new plants have at least a 60-year lifetime and utilities will need assurances when ordering new build that uranium supply can be maintained for the full period of operation. Eventually, known conventional reserves will all be earmarked for current plants or those under construction, and this could happen by the middle of this century. This underlines the need to develop the technology for a new generation, the so-called Generation IV, of reactors and fuel cycles that are more sustainable. In particular, fast-neutron breeder reactors could produce up to 100 times more energy from the same quantity of uranium than current designs and may significantly reduce the amount of ultimate radioactive waste for disposal.

Fast reactors convert non-fissile material (U-238) in the fuel into fissile material (Pu-239) during reactor operation so that the net amount of fissile material increases (breeding). After reprocessing of the spent fuel, the extracted fissile materials are then recycled as new fuels. Reduction of the radiotoxicity and heat load of the waste is achieved by separating some long-lived radionuclides, the minor actinides, which could then be "burned" in fast reactors or alternatively in accelerator-driven systems

(ADS), through transmutation. The fast reactor concept has been demonstrated in research programs and national prototypes in the past, but further R&D is needed to make it commercially viable and to develop the designs in compliance with true Generation IV criteria. Major issues involve new materials that can withstand higher temperatures, higher burnups and neutron doses, and corrosive coolants, reactor designs that eliminate severe accidents, and development of fuel cycles for waste minimization and elimination of proliferation risks. Fast reactors are expected to be commercially available from 2040.

So far nuclear power has primarily been used to produce electricity, but it can also be used for process heat applications [13, 14]. Currently, LWRs are already being used to a limited extent for some lower-temperature applications (200 °C), such as district heating and desalination of seawater. Existing designs of high-temperature reactors (HTR) that can reach 800 °C can be deployed in the coming decades, and very-high-temperature reactors (VHTR) that can reach gas coolant temperatures beyond 1000 °C are being studied as a Generation IV concept for later deployment. Process heat applications include petroleum refinery applications (400 °C), recovery of oil from tar sands (600–700 °C), synthetic fuel from CO2 and hydrogen (600–1000 °C), hydrogen production (600–1000 °C), and coal gasification (900–1200 °C). Small reactors that can be inherently safe and used to support specific high-energy applications and often in remote areas are another very interesting application that is receiving more attention, in particular in the IAEA INPRO initiative.

The management of radioactive waste, as either spent fuel or ultimate waste, depending on the national strategy, is a key issue for public acceptance of nuclear energy. There is scientific consensus that geological disposal is the only safe long-term solution for the management of ultimate waste. After a long period of intensive research and development coupled with in-depth political and social engagement, the world's first deep geological repositories for nuclear waste will be in operation in Sweden and Finland by 2020, with France following a few years later, demonstrating that practical solutions exist for the safe long-term management of hazardous waste from the operation of nuclear power plants. Though there will also be ultimate waste from Generation IV fast reactor fuel cycles after reprocessing, the volumes and heat loads will be greatly reduced, thereby facilitating disposal operations and optimizing use of space in available geological repositories.

1.14 Next-Generation Nuclear Plant (NGNP)

The next-generation nuclear plant (NGNP) demonstration project forms the basis for an entirely new generation of advanced nuclear plants capable of meeting the nation's emerging need for greenhouse gas-free process heat and electricity. The NGNP is based on the very-high-temperature gas-cooled reactor (VHTR) technology, which was determined to be the most promising for the United States in the medium term. The determination is documented as part of the Generation IV implementation strategy in a report submitted to Congress in 20,031 following an extensive international technical evaluation effort. The VHTR technology incorporates substantive safety and operational enhancements over existing nuclear technologies. As required by the Energy Policy Act of 2005 (EPAct), the NGNP will be a prototype nuclear power plant, built at the Idaho National Laboratory (INL). Future commercial versions of the NGNP will meet or exceed the reliability, safety, proliferation resistance, and economy of existing commercial nuclear plants [18].

It is envisioned that these advanced nuclear plants would be able to supply cost-competitive process heat that can be used to power a variety of energy-intensive industries, such as the generation of electricity, hydrogen, enhanced oil recovery, refineries, coal-to-liquids and coal-to-gas plants, chemical plants, and fertilizer plants [18].

The US Nuclear Regulatory Commission (NRC) is responsible for licensing and regulating the construction and operation of the NGNP. The EPAct authorizes the US Department of Energy (DOE) to build the NGNP at the Idaho National Laboratory and charges INL with responsibility for leading the project development. The project's completion depends on the collaborative efforts of DOE and its national laboratories, commercial industry participants, US universities, and international government agencies as well as successful licensing by the NRC. At present and pending further evaluation as the NGNP proceeds through Phase 1 in cost-shared collaboration with industry as required by the EPAct, DOE has not made a final determination on whether the license applicant will be DOE or one or more entities that reflect a partnership between DOE and private sector firms [18].

Under the provisions of Section 644 of the EPAct, the Secretary of Energy and the Chairman of the Nuclear Regulatory Commission are to jointly submit to Congress a licensing strategy for the NGNP within 3 years of the enactment of the Act on August 8, 2005. This report addresses the requirement by outlining a NGNP licensing strategy jointly developed by the NRC and DOE. The scope of the document includes all four elements of the NGNP licensing strategy described in Section 644 (b) of the EPAct:

1.0 A description of the ways in which current NRC light-water reactor (LWR) licensing requirements need to be adapted for the types of reactors considered for the project
2.0 A description of the analytical tools that the NRC will need to develop in order to independently verify the NGNP design and its safety performance
3.0 A description of other research or development activities that the NRC will need to conduct for the review of an NGNP license application
4.0 A budget estimate associated with the licensing strategy

DOE has determined that the NGNP nuclear reactor will be a very-high-temperature gas-cooled reactor (VHTR) for the production of electricity, process heat, and hydrogen. The VHTR can provide high-temperature process heat (up to 950 °C) that can be used as a substitute for the burning of fossil fuels for a wide range of commercial applications. Since the VHTR is a new and unproven reactor design, the NRC will need to adapt its licensing requirements and process, which have historically evolved around light-water reactor (LWR) designs, for licensing the NGNP nuclear reactor. Thus, Section 644 of the EPAct recognized the need for an alternative licensing strategy. This report provides the recommended NGNP licensing strategy, jointly developed by the NRC and DOE. As the technology matures, the government/industry partnership evolves, and input is provided by the general public, revisions to the strategy may be necessary and appropriate [18].

The report addresses the four elements of the licensing strategy set forth in Section 644(b) of the EPAct. These elements are summarized above [18].

1.15 Generation IV Systems

The world's population is expected to expand from 6.7 billion people today to over 9 billion people by the year 2050, all striving for a better quality of life. As the Earth's population grows, so does the demand for energy and the benefits that it brings: improved standards of living, better health and longer life expectancy, improved literacy and opportunity, and many others. Simply expanding the use of energy along the same mix of today's production options, however, does not satisfactorily address concerns over climate change and depletion of fossil resources. For the Earth to support its population while ensuring the sustainability of humanity's development, we must increase the use of energy supplies that are clean, safe, and cost-effective and which could serve for both basic electricity production and other primary energy needs. Prominent among these supplies is nuclear energy.

There is currently 370 GWe of nuclear power capacity in operation around the world, producing 3000 TWh each year – 15% of the world's electricity – the largest share provided by any non-greenhouse gas-emitting source. This reduces significantly the environmental impact of today's electricity generation and affords a greater diversity of electricity generation that enhances energy security.

For more than a decade, Generation IV International Forum (GIF) has led international collaborative efforts to develop next-generation nuclear energy systems that can help meet the world's future energy needs. Generation IV designs will use fuel more efficiently, reduce waste production, be economically competitive, and meet stringent standards of safety and proliferation resistance.

As we said the Generation IV International Forum (GIF) was initiated in May 2001 and formally chartered in mid-2001. It is an international collective representing government of 13 countries where nuclear energy is significant now and also seen as vital for the future. Most are committed to joint development of the next generation of nuclear technology. Led by the United States, Argentina, Brazil, Canada, China, France, Japan, Russia, South Korea, South Africa, Switzerland, and the United Kingdom are charter members of the GIF, along with the EU (Euratom). Most of these are party to the Framework Agreement (FA), which formally commits them to participate in the development of one or more Generation IV systems selected by GIF for further R&D. Argentina and Brazil did not sign the FA, and the United Kingdom withdrew from it; accordingly, within the GIF, these three are

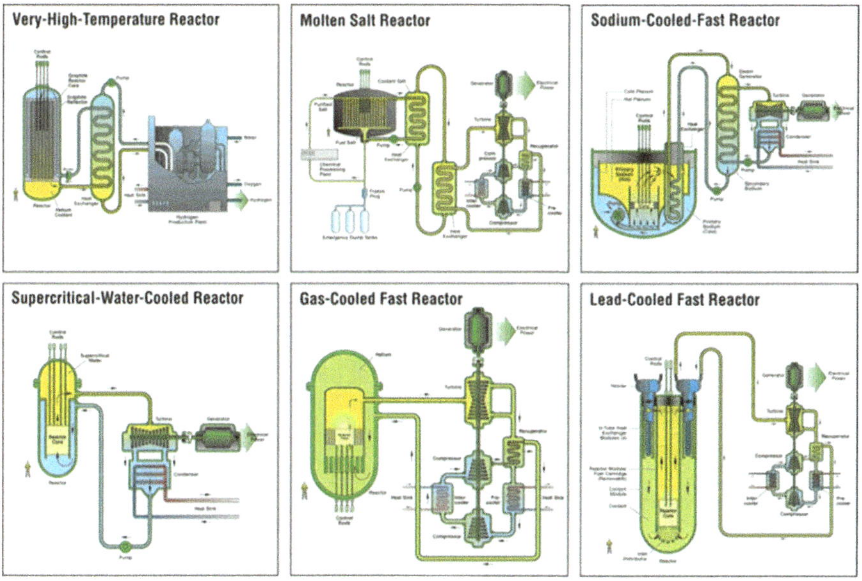

Fig. 1.21 Six reactor technologies of Generation IV. (Courtesy of the Generation IV International Forum)

designated as "inactive members." Russia formalized its accession to the FA in August 2009 as its tenth member, with Rosatom as implementing agent. In 2011 the 13 members decided to modify and extend the GIF charter indefinitely.

With these goals in mind, some 100 experts evaluated 130 reactor concepts before GIF selected 6 reactor technologies for further research and development.

These include:

1. The **v**ery-**h**igh-**t**emperature **r**eactor **(VHTR)**
2. The **m**olten **s**alt **r**eactor **(MSR)**
3. The **s**odium-cooled **f**ast **r**eactor **(SFR)**
4. The **s**uper**c**ritical **w**ater-cooled **r**eactor **(SCWR)**
5. The **g**as-cooled **f**ast **r**eactor **(GFR)**
6. The **l**ead-cooled **f**ast **r**eactor **(LFR)**

Figure 1.21 is an illustration of the six types of reactors that are considered as part of Generation IV power plant. More details of each of these reactors are provided in later sections.

1.15.1 Very-High-Temperature Reactor (VHTR)

Among the six candidates of the Gen IV nuclear systems in the technical roadmap of Generation IV International Forum (GIF), the very-high-temperature reactor (VHTR) is primarily dedicated to the cogeneration of electricity and hydrogen, the

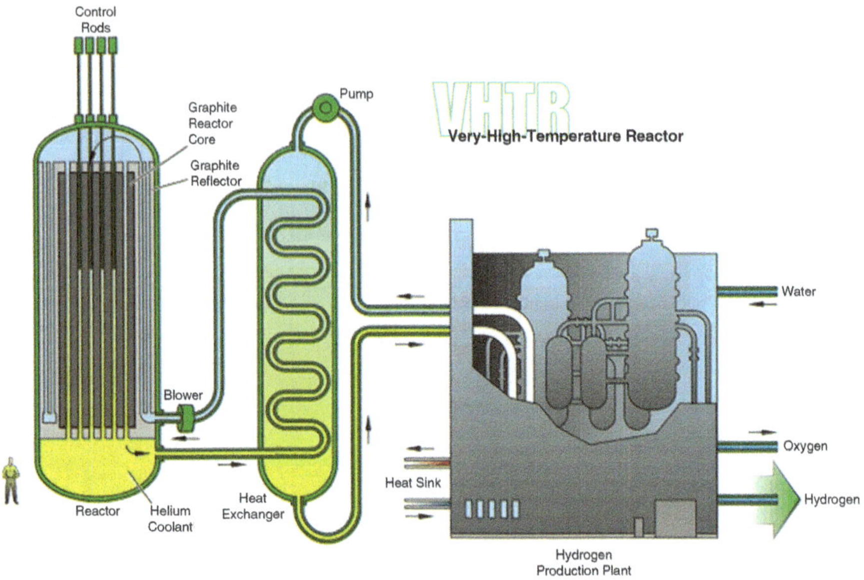

Fig. 1.22 Very-high-temperature reactor. (Courtesy of the Generation IV International Forum)

latter being extracted from water by using thermochemical, electrochemical, or hybrid processes. Its high outlet temperature makes it attractive also for the chemical, oil, and iron industries. Original target of outlet temperature of 1000 °C from VHTR can support the efficient production of hydrogen by thermochemical processes. The technical basis for VHTR is the TRISO-coated particle fuel, the graphite as the core structure, helium coolant, as well as the dedicated core layout and lower-power density to the removal of decay heat in a natural way. The VHTR has potential for inherent safety, high thermal efficiency, process heat application capability, low operation and maintenance costs, and modular construction (Fig. 1.22).

The VHTR is a next step in the evolutionary development of high-temperature gas-cooled reactors. It is a graphite-moderated, helium-cooled reactor with thermal-neutron spectrum. It can supply nuclear heat and electricity over a range of core outlet temperatures between 700 and 950 °C, or more than 1000 °C in the future. The reactor core type of the VHTR can be a prismatic block core such as the Japanese HTTR, or a pebble-bed core such as the Chinese HTR-10. For electricity generation, a helium gas turbine system can be directly set in the primary coolant loop, which is called a direct cycle, or at the lower end of the outlet temperature range, a steam generator can be used with a conventional Rankine cycle. For nuclear heat applications such as process heat for refineries, petrochemistry, metallurgy, and hydrogen production, the heat application process is generally coupled with the reactor through an intermediate heat exchanger (IHX), the so-called indirect cycle. The VHTR can produce hydrogen from only heat and water by using thermochemical processes (such as the sulfur-iodine (S-I) process or the hybrid sulfur process) and high-temperature steam electrolysis (HTSE) or from heat, water, and natural gas by applying the steam reformer technology.

While the original approach for VHTR at the start of the Generation IV program focused on very high outlet temperatures and hydrogen production, current market assessments have indicated that electricity production and industrial processes based on high-temperature steam that require modest outlet temperatures (700–850 °C) have the greatest potential for application in the next decade. This also reduces technical risk associated with higher outlet temperatures. As a result, over the past decade, the focus has moved from higher outlet temperature designs such as GT-MHR and PBMR to lower outlet temperature designs such as HTR-PM in China and the NGNP in the United States.

The VHTR has two typical reactor configurations, namely:

I. The pebble-bed type
II. The prismatic block types

Although the shapes of the fuel element for two configurations are different, the technical basis for both configuration is the same, such as the TRISO-coated particle fuel in the graphite matrix, full ceramic (graphite) core structure, helium coolant, and low-power density.

This will allow achieving high outlet temperature and the retention of fission production inside the coated particle under normal operation condition and accident condition. The VHTR can support alternative fuel cycles such as U-Pu, Pu, MOX, and U-Th.

1.15.2 Molten Salt Reactor (MSR)

The MSR is distinguished by its core, in which the fuel is dissolved in molten fluoride salt. The technology was first studied more than 50 years ago. Modern interest is on fast reactor concepts as a long-term alternative to solid-fueled fast-neutron reactors. The onsite fuel-reprocessing unit using pyrochemistry allows breeding plutonium or uranium-233 from thorium. R&D progresses toward resolving feasibility issues and assessing safety and performance of the design concepts. Key feasibility issues focus on a dedicated safety approach and the development of salt redox potential measurement and control tools in order to limit corrosion rate of structural materials. Further work on the batch-wise online salt processing is required. Much work is needed on molten salt technology and related equipment. See Fig. 1.23.

Molten salt reactor (MSR) technology was partly developed, including two demonstration reactors, in the 1950s and 1960s in the United States (Oak Ridge National Laboratory). The MSR demonstrations were thermal-neutron-spectrum graphite-moderated concepts. Since 2005, R&D has focused on the development of fast-spectrum MSR concepts (MSFR) combining the generic assets of fast-neutron reactors (extended resource utilization, waste minimization) with those relating to molten salt fluorides as fluid fuel and coolant (low pressure and high boiling temperature, optical transparency).

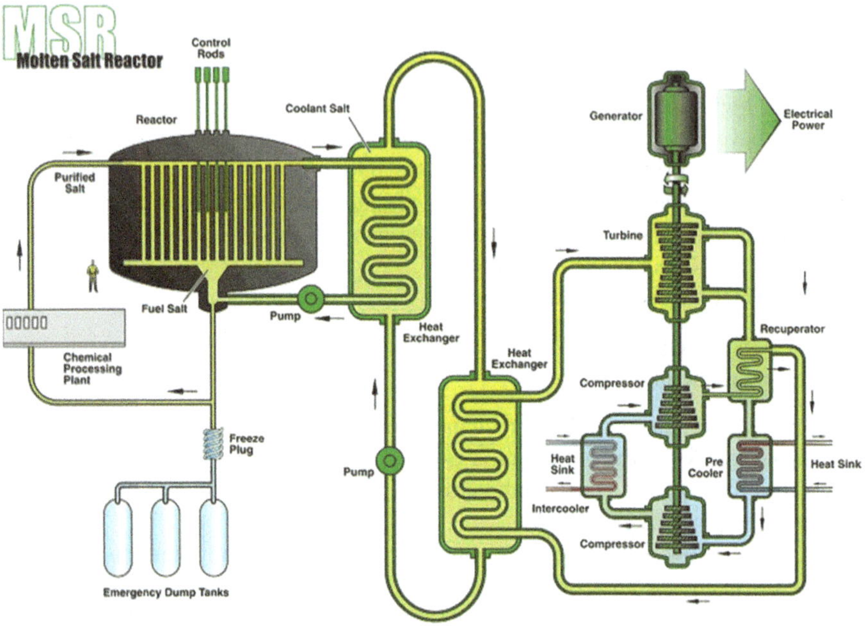

Fig. 1.23 Molten salt reactor. (Courtesy of the Generation IV International Forum)

In contrast to most other molten salt reactors previously studied, the MSFR does not include any solid moderator (usually graphite) in the core. This design choice is motivated by the study of parameters such as feedback coefficient, breeding ratio, graphite lifespan, and 233 U initial inventory. MSFR exhibits large negative temperature and void reactivity coefficients, a unique safety characteristic not found in solid-fuel fast reactors.

Compared with solid-fuel reactors, MSFR systems have lower fissile inventories, no radiation damage constraint on attainable fuel burnup, no requirement to fabricate and handle solid fuel, and a homogeneous isotopic composition of fuel in the reactor. These and other characteristics give MSFRs potentially unique capabilities for actinide burning and extending fuel resources.

MSR developments in Russia on the molten salt actinide recycler and transmuter (MOSART) aim to be used as efficient burners of transuranic (TRU) waste from spent UOX and MOX light-water reactor (LWR) fuel without any uranium and thorium support and also with it. Other advanced reactor concepts are being studied, which use the liquid-salt technology, as a primary coolant for fluoride salt-cooled high-temperature reactors (FHRs), and coated particle fuels similar to high-temperature gas-cooled reactors.

More generally, there has been a significant renewal of interest in the use of liquid salt as a coolant for nuclear and nonnuclear applications. These salts could facilitate heat transfer for nuclear hydrogen production concepts, concentrated solar electricity generation, oil refineries, and shale oil processing facilities among other applications.

1.15.3 Sodium-Cooled Fast Reactor (SFR)

The sodium-cooled fast reactor (SFR) uses liquid sodium as the reactor coolant, allowing high-power density with low coolant volume fraction and operation at low pressure. While the oxygen-free environment prevents corrosion, sodium reacts chemically with air and water and requires a sealed coolant system. See Fig. 1.24.

Plant size options under considerations are ranging from small, 50 to 300 MWe, modular reactors to larger plants up to 1500 MWe. The outlet temperature is 500–550 °C for the options, which allows the use of the materials developed and proven in prior fast reactor programs.

The SFR closed fuel cycle enables regeneration of fissile fuel and facilitates management of minor actinides. However, this requires that recycle fuels be developed and qualified for use. Important safety features of the Generation IV system include a long thermal response time, a reasonable margin to coolant boiling, a primary system that operates near atmospheric pressure, and an intermediate sodium system between the radioactive sodium in the primary system and the power conversion system. Water/steam, supercritical carbon dioxide, or nitrogen can be considered as working fluids for the power conversion system to achieve high performance in terms of thermal efficiency, safety, and reliability. With innovations to reduce capital cost, the SFR is aimed to be economically competitive in future electricity markets. In addition, the fast-neutron spectrum greatly extends the uranium

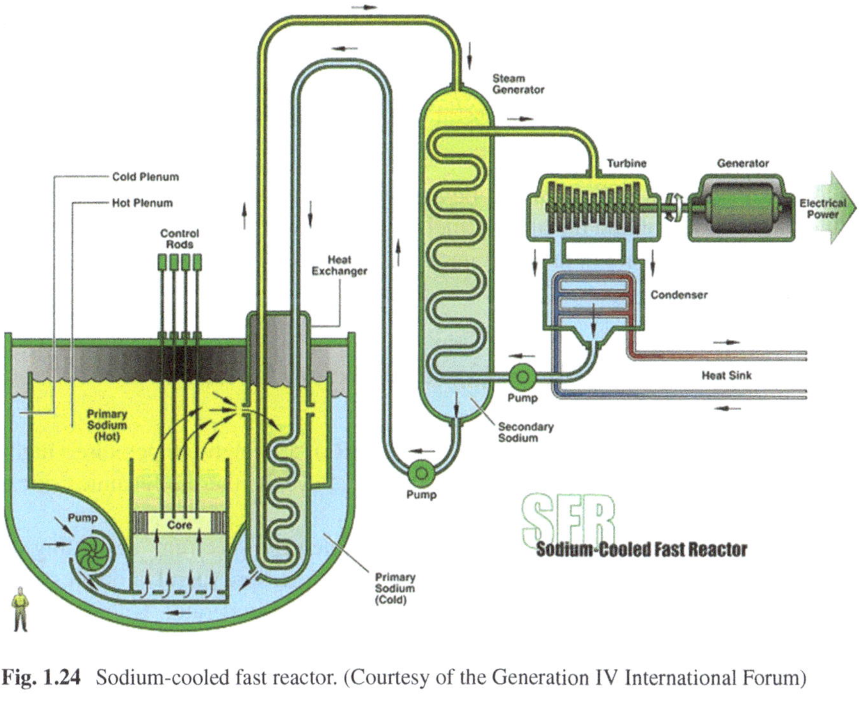

Fig. 1.24 Sodium-cooled fast reactor. (Courtesy of the Generation IV International Forum)

resources compared to thermal reactors. The SFR is considered to be the nearest-term deployable system for actinide management.

Much of the basic technology for the SFR has been established in the former fast reactor programs and is being confirmed by the Phenix end-of-life tests in France, the restart of Monju in Japan, and the lifetime extension of BN-600 in Russia. New programs involving SFR technology include the Chinese experimental fast reactor (CEFR) which was connected to the grid in July 2011, and India's prototype fast-breeder reactor (PFBR) which is currently planned to go critical in 2013.

The SFR is an attractive energy source for nations that desire to make the best use of limited nuclear fuel resources and manage nuclear waste by closing the fuel cycle.

Fast reactors hold a unique role in the actinide management mission because they operate with high-energy neutrons that are more effective at fissioning actinides. The main characteristics of the SFR for actinide management mission are:

- Consumption of transuranics in a closed fuel cycle, thus reducing the radiotoxicity and heat load, which facilitates waste disposal and geologic isolation
- Enhanced utilization of uranium resources through efficient management of fissile materials and multi-recycle

High level of safety achieved through inherent and passive means also allows accommodation of transients and bounding events with significant safety margins.

The reactor unit can be arranged in a pool layout or a compact loop layout. Three options are considered:

- A large-sized (600–1500 MWe) loop-type reactor with mixed uranium-plutonium oxide fuel and potentially minor actinides, supported by a fuel cycle based upon advanced aqueous processing at a central location serving a number of reactors
- An intermediate-to-large-sized (300–1500 MWe) pool-type reactor with oxide or metal fuel

A small-sized (50–150 MWe) modular-type reactor with uranium-plutonium-minor-actinide-zirconium metal alloy fuel, supported by a fuel cycle based on pyrometallurgical processing in facilities integrated with the reactor.

1.15.4 Supercritical Water-Cooled Reactor (SCWR)

The supercritical water-cooled reactors (SCWRs) are high-temperature, high-pressure, light-water-cooled reactors that operate above the thermodynamic critical point of water (374 °C, 22.1 MPa). See Fig. 1.25.

The reactor core may have a thermal- or a fast-neutron spectrum, depending on the core design. The concept may be based on current pressure vessel or on pressure-tube reactors and thus use light water or heavy water as moderator. Unlike current water-cooled reactors, the coolant will experience a significantly higher enthalpy

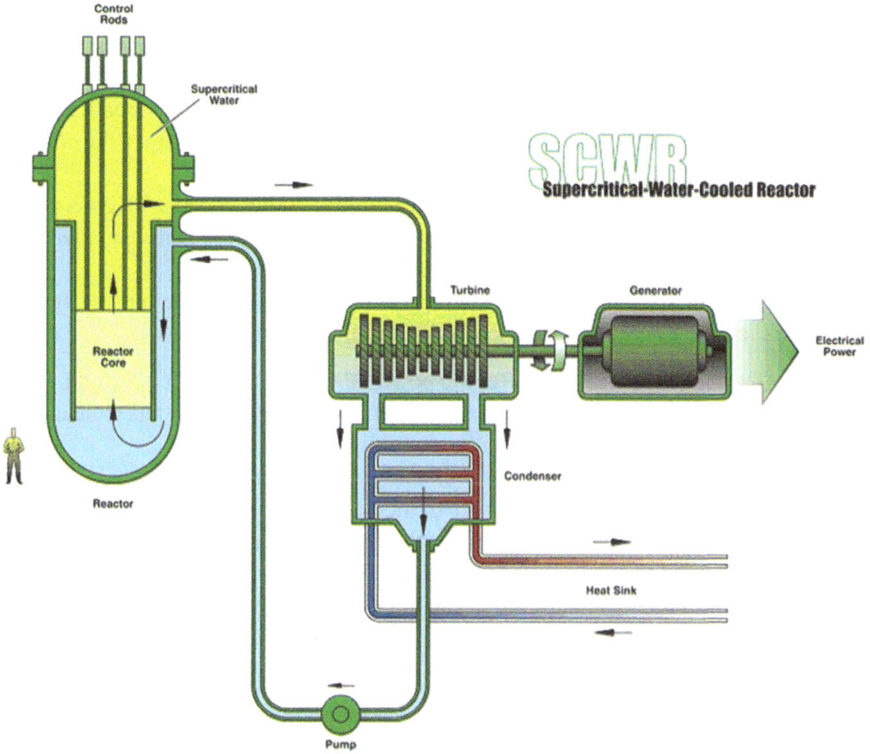

Fig. 1.25 Supercritical water-cooled reactor. (Courtesy of the Generation IV International Forum)

rise in the core, which reduces the core mass flow for a given thermal power and increases the core outlet enthalpy to superheated conditions. For both pressure-vessel and pressure-tube designs, a once through steam cycle has been envisaged, omitting any coolant recirculation inside the reactor. As in a boiling water reactor, the superheated steam will be supplied directly to the high-pressure steam turbine, and the feed water from the steam cycle will be supplied back to the core. Thus, the SCWR concepts combine the design and operation experiences gained from hundreds of water-cooled reactors with those experiences from hundreds of fossil-fired power plants operated with supercritical water (SCW). In contrast to some of the other Generation IV nuclear systems, the SCWR can be developed incrementally step-by-step from current water-cooled reactors.

A. *Advantage and Challenges*

Such SCWR designs have unique features that offer many advantages compared to state-of the-art water-cooled reactors:

- SCWRs offer increases in thermal efficiency relative to current-generation water-cooled reactors. The efficiency of a SCWR can approach 44% or more, compared to 34–36% for current reactors.

- Reactor coolant pumps are not required. The only pumps driving the coolant under normal operating conditions are the feed water pumps and the condensate extraction pumps.
- The steam generators used in pressurized water reactors and the steam separators and dryers used in boiling water reactors can be omitted since the coolant is superheated in the core.
- Containment, designed with pressure suppression pools and with emergency cooling and residual heat removal systems, can be significantly smaller than those of current water-cooled reactors can.
- The higher steam enthalpy allows to decrease the size of the turbine system and thus to lower the capital costs of the conventional island.

These general features offer the potential of lower capital costs for a given electric power of the plant and of better fuel utilization, and thus a clear economic advantage compared with current light water reactors.

However, there are several technological challenges associated with the development of the SCWR, and particularly the need to validate transient heat-transfer models (for describing the depressurization from supercritical to subcritical conditions), qualification of materials (namely, advanced steels for cladding), and demonstration of the passive safety systems.

B. *GIF Progress up to 2012*

Preconceptual core design studies for a core outlet temperature of more than 500 °C have been performed in Japan, assuming either a thermal-neutron spectrum or a fast-neutron spectrum. Both options are based on a coolant heat-up in two steps with intermediate mixing underneath the core. Additional moderator for a thermal-neutron spectrum is provided by feed water inside water rods. The fast-spectrum option uses zirconium-hydride (ZrH2) layers to minimize hardening of the neutron spectrum in case of core voiding. A preconceptual design of safety systems for both options has been studied with transient analyses.

A preconceptual plant design with 1700 MW net electric power based on a pressure-vessel-type reactor has been studied by Yamada et al. and has been assessed with respect to efficiency, safety, and cost. The study confirms the target net efficiency of 44% and estimates a cost reduction potential of 30% compared with current pressurized water reactors. Safety features are expected to be similar to advanced boiling water reactors.

A preconceptual design of a pressure-vessel-type reactor with a 500 °C core outlet temperature and 1000 MW electric power has been developed in Europe, as summarized by Schulenberg and Starflinger. The core design is based on coolant heat-up in three steps. Additional moderator for the thermal-neutron spectrum is provided in water rods and in gaps between assembly boxes. The design of the nuclear island and of the balance of the plant confirms results obtained in Japan, namely, an efficiency improvement up to 43.5% and a cost reduction potential of 20 to 30% compared with latest boiling water reactors. Safety features as defined by the stringent European Utility Requirements are expected to be met.

Canada is developing a pressure-tube-type SCWR concept with a 625 °C core outlet temperature at the pressure of 25 MPa. The concept is designed to generate 1200 MW electric power (a 300 MW concept is also being considered). It has a modular fuel channel configuration with separate coolant and moderator. A high-efficiency fuel channel is incorporated to house the fuel assembly. The heavy-water moderator directly contacts the pressure tube and is contained inside a low-pressure calandria vessel. In addition to providing moderation during normal operation, it is designed to remove decay heat from the high-efficiency fuel channel during long-term cooling using a passive moderator cooling system. A mixture of thorium oxide and plutonium is introduced as the reference fuel, which aligns with the GIF position paper on thorium fuel. The safety system design of the Canadian SCWR is similar to that of the ESBWR. However, the introduction of the passive moderator cooling system coupled with the high-efficiency channel could reduce significantly the core damage frequency during postulated severe accidents such as large-break loss-of-coolant or station blackout events.

Preconceptual designs of three variants of pressure-vessel supercritical reactors with thermal, mixed, and fast-neutron spectrum have been developed in Russia, which joined the SCWR System Arrangement in 2011.

Outside of the GIF framework, two conceptual SCWR designs with thermal- and mixed-neutron spectrum cores have been established by some research institutes in China. This is done under the framework of the Chinese national R&D projects from 2007 to 2012, covering some basic research projects on materials and thermo-hydraulics, the core/fuel design, the main system design (including the conventional part), safety systems design, reactor structure design, and fuel assembly structure design. The related feasibility studies have also been completed and show that the design concept has promising prospects in terms of the overall performance, integration of design, and component structure feasibility and manufacturability.

Prediction of heat transfer in SCW can be based on data from fossil-fired power plants as discussed by Pioro et al. Computational tools for more complex geometries like fuel assemblies are available but still need to be validated with bundle experiments. System codes for transient safety analyses have been upgraded to include SCW, including depressurization transients to subcritical conditions. Flow stability in the core has been studied numerically. As in boiling water reactors, flow stability can be ensured using suitable inlet orifices in fuel assemblies.

A number of candidate cladding materials have been tested in capsules, autoclaves, and recirculating loops up to 700 °C at a pressure of 25 MPa. Stainless steels with more than 20% chromium (Cr) are expected to have the required corrosion resistance up to a peak cladding temperature of 650 °C. More work is needed to develop alloys suitable for use at the design peak cladding temperatures of 850 °C for the Canadian SCWR concept. Further work is also needed to better identify the coolant conditions that lead to stress corrosion cracking. It has been shown that the creep resistance of existing alloys can be improved by adding small amounts of elements, such as zirconium (Zr), as reported by Kaneda et al. In the longer term, the steel experimental oxide dispersion strengthened (ODS) alloys offer an even higher potential, whereas nickel-base alloys are being considered for use in ultra-supercritical fossil-fired plants and are less favorable for use in SCWRs due to their high neutron absorption and associated swelling and embrittlement.

Key water chemistry issues have been identified by Guzonas et al.; predicting and controlling water radiolysis and corrosion product transport (including fission products) remain the major R&D areas. In this regard, the operating experience using nuclear steam reheat at the Beloyarsk nuclear power plant in Russia is extremely valuable.

1.15.5 Gas-Cooled Fast Reactor (GFR)

The gas-cooled reactor (GFR) system is a high-temperature helium-cooled fast-spectrum reactor with a closed fuel cycle. It combines the advantages of fast-spectrum systems for long-term sustainability of uranium resources and waste minimization (through fuel multiple reprocessing and fission of long-lived actinides), with those of high-temperature systems (high thermal cycle efficiency and industrial use of the generated heat, for hydrogen production for example). See Fig. 1.26.

The GFR uses the same fuel recycling processes as the SFR and the same reactor technology as the VHTR. Therefore, its development approach is to rely, in so far as feasible, on technologies developed for the VHTR for structures, materials, components, and power conversion system. Nevertheless, it calls for specific R&D beyond the current and foreseen work on the VHTR system, mainly on core design and safety approach.

The reference design for GFR is based around a 2400 MWth reactor core contained within a steel pressure vessel. The core consists of an assembly of hexagonal fuel elements, each consisting of ceramic-clad, mixed-carbide-fueled pins contained within a ceramic hex-tube. The favored material at the moment for the pin clad and hex-tubes is silicon carbide fiber-reinforced silicon carbide. The figure below shows the reactor core located within its fabricated steel pressure vessel surrounded by main heat exchangers and decay heat removal loops. The whole of the primary circuit is contained within a secondary pressure boundary, the guard containment.

As it is illustrated in Fig. 1.27, the coolant is helium, and the core outlet temperature will be of the order of 850 °C. A heat exchanger transfers the heat from the primary helium coolant to a secondary gas cycle containing a helium-nitrogen mixture, which in turn drives a closed cycle gas turbine. The waste heat from the gas turbine exhaust is used to raise steam in a steam generator, which is then used to

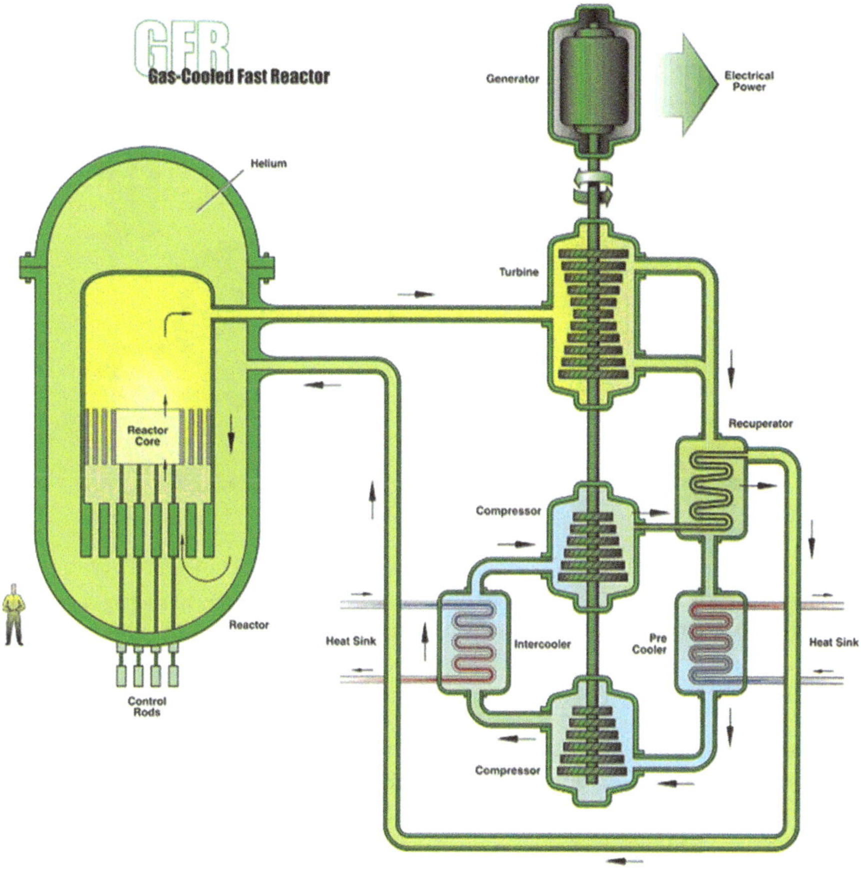

Fig. 1.26 Gas-cooled fast reactor. (Courtesy of the Generation IV International Forum)

drive a steam turbine. Such a combined cycle is common practice in natural gas-fired power plant and so represents an established technology, with the only difference in the GFR case being the use of a closed cycle gas turbine.

1.15.6 Lead-Cooled Fast Reactor (LFR)

The lead-cooled fast reactor (LFR) features a fast-neutron spectrum, high-temperature operation, and cooling by molten lead or lead-bismuth eutectic (LBE), low-pressure, chemically inert liquids with very good thermodynamic properties. It would have multiple applications including production of electricity, hydrogen, and

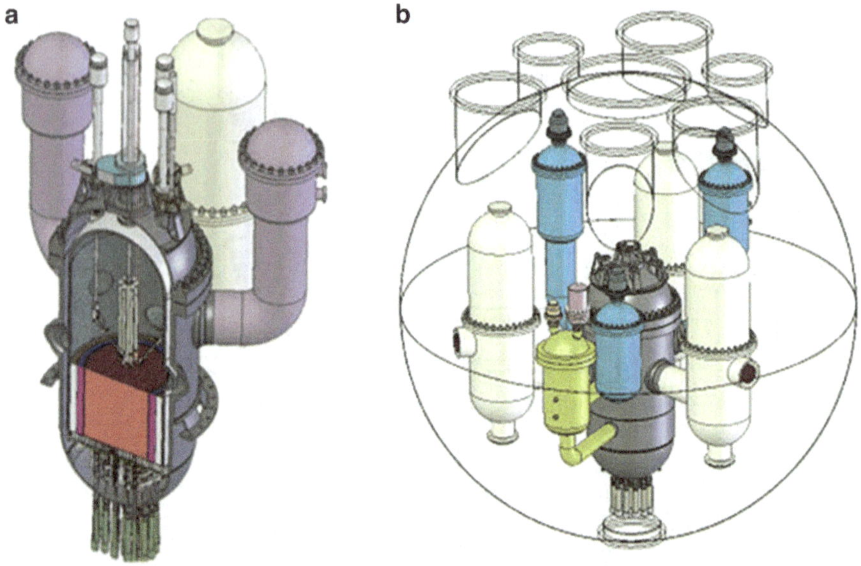

Fig. 1.27 (**a**) GFR-reactor, decay heat loops, main heat exchangers, and fuel-handling equipment. (**b**) GFR spherical guard vessel. (Courtesy of the Generation IV International Forum)

process heat. System concepts represented in plans of the Generation IV International Forum (GIF) System Research Plan (SRP) are based on Europe's ELFR lead-cooled system, Russia's BREST-OD-300, and the SSTAR system concept designed in the United States. See Fig. 1.28.

The LFR has excellent materials management capabilities since it operates in the fast-neutron spectrum and uses a closed fuel cycle for efficient conversion of fertile uranium. It can also be used as a burner to consume actinides from spent LWR fuel and as a burner/breeder with thorium matrices. An important feature of the LFR is the enhanced safety that results from the choice of molten lead as a chemically inert and low-pressure coolant. In terms of sustainability, lead is abundant and hence available, even in case of deployment of a large number of reactors. More importantly, as with other fast systems, fuel sustainability is greatly enhanced by the conversion capabilities of the LFR fuel cycle. Because they incorporate a liquid coolant with a very high margin to boiling and benign interaction with air or water, LFR concepts offer substantial potential in terms of safety, design simplification, proliferation resistance, and the resulting economic performance. An important factor is the potential for benign end state to severe accidents.

The LFR has development needs in the areas of fuels, materials performance, and corrosion control. During the next 5 years, progress is expected on materials, system design, and operating parameters. Significant test and demonstration activities are underway and planned during this period.

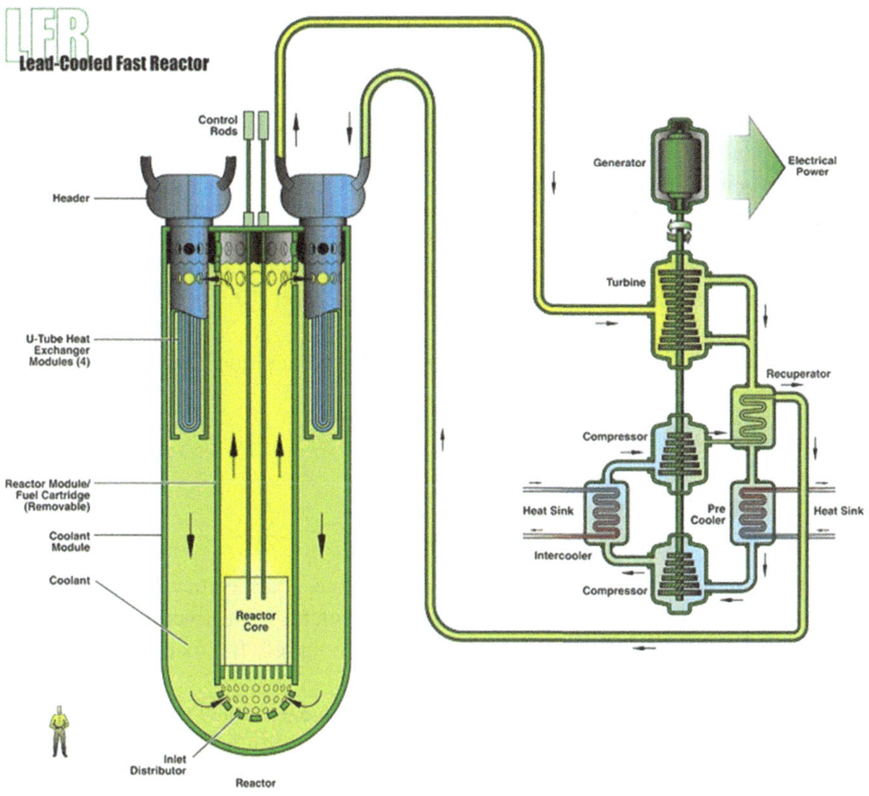

Fig. 1.28 Lead-cooled fast reactor. (Courtesy of the Generation IV International Forum)

1.16 Next-Generation of Nuclear Power Reactors for Power Production

Experts are projecting worldwide electricity consumption will increase substantially in the coming decades, especially in the development world, accompanying economic growth and social progress that have direct impact on rising electricity prices, and have focused fresh attention on nuclear power plants. New, safer, and more economical nuclear reactors could not only satisfy many of our future energy needs but could combat global warming as well. Today's existing nuclear power plants on line in the United States provide fifth of the nation's total electrical output.

Taking into account the expected increase in energy demand worldwide and the growing awareness about global warming, climate change issues, and sustainable development, nuclear energy will be needed to meet future global energy demand.

Nuclear power plant technology has evolved as distinct design generations as we mentioned in the previous section and briefly summarized here again as follows:

- First generation: prototypes, and first realizations (~1950–1970)
- Second generation: current operating plants (~1970–2030)
- Third generation: deployable improvements to current reactors (~2000 and on)
- Fourth generation: advanced and new reactor systems (2030 and beyond)

The **G**eneration IV **I**nternational **F**orum, or **GIF**, was chartered in July 2001 to lead the collaborative efforts of the world's leading nuclear technology nations to develop next-generation nuclear energy systems to meet the world's future energy needs.

Eight technology goals have been defined for Generation IV systems in four broad areas:

1. Sustainability
2. Economics
3. Safety and reliability
4. Proliferation resistance and physical protection

A large number of countries share these ambitious goals as they aim at responding to economic, environmental, and social requirements of the twenty-first century. They establish a framework and identify concrete targets for focusing GIF R&D efforts.

Eight technology goals have been defined for Generation IV systems in four broad areas: sustainability, economics, safety and reliability, and proliferation resistance and physical protection.

1.17 Goals for Generation IV Nuclear Energy Systems

The next generation ("Generation IV") of nuclear energy systems is intended to meet the below goals (while being at least as effective as the "third" generation in terms of economic competitiveness, safety, and reliability) in order to provide a sustainable development of nuclear energy.

In principle, the Generation IV systems should be marketable or deployable from 2030 onward. The systems should also offer a true potential for new applications compatible with an expanded use of nuclear energy, in particular in the fields of hydrogen or synthetic hydrocarbon production, seawater desalination, and process heat production.

It has been recognized that these objectives, widely and officially shared by a large number of countries, should be at the basis of an internationally shared R&D program, which allows keeping open and consolidating the technical options, and avoiding any early or premature down selection.

In fact, because the next-generation nuclear energy systems will address needed areas of improvement and offer great potential, many countries share a common interest in advanced R&D that will support their development. The international research community should explore such development benefits from the identification of promising research areas and collaborative efforts. The collaboration on R&D by many nations on the development of advanced next-generation nuclear energy systems will in principle aid the progress toward the realization of such systems, by leveraging resources, providing synergistic opportunities, avoiding unnecessary duplication, and enhancing collaboration.

Sustainability – 1	Generation IV nuclear energy systems will provide sustainable energy generation that meets clean air objectives and provides long-term availability of systems and effective fuel utilization for worldwide energy production
Sustainability – 2	Generation IV nuclear energy systems will minimize and manage their nuclear waste and notably reduce the long-term stewardship burden, thereby improving protection for the public health and the environment
Economics – 1	Generation IV nuclear energy systems will have a clear life-cycle cost advantage over other energy sources
Economics – 2	Generation IV nuclear energy systems will have a level of financial risk comparable to other energy projects
Safety and reliability – 1	Generation IV nuclear energy systems operations will excel in safety and reliability
Safety and reliability – 2	Generation IV nuclear systems will have a very low likelihood and degree of reactor core damage
Safety and reliability – 3	Generation IV nuclear energy systems will eliminate the need for offsite emergency response
Proliferation resistance and physical protection	Generation IV nuclear energy systems will increase the assurance that they are very unattractive and the least desirable route for diversion or theft of weapon-usable materials and provide increased physical protection against acts of terrorism

As it is illustrated by Fig. 1.19, we are able to see the evolution of nuclear power plant life cycles.

In 2009, the Experts Group published an outlook on Generation IV R&D, to provide a view of what GIF members hope to achieve collectively in the period 2010–2014. All Generation IV systems have features aiming at performance improvement, new applications of nuclear energy, and/or more sustainable approaches to the management of nuclear materials. High-temperature systems offer the possibility of efficient process heat applications and eventually hydrogen production. Enhanced sustainability is achieved primarily through adoption of a closed fuel cycle with reprocessing and recycling of plutonium, uranium, and minor actinides using fast reactors; this approach provides significant reduction in waste generation and uranium resource requirements (Table 1.1).

The following section of this chapter summarizes the main characteristics of the six Generation IV systems.

Table 1.1 Summary of the main characteristics of the six Generation IV systems

System	Neutron spectrum	Coolant	Temp. °C	Fuel cycle	Size (MWe)
VHTR (very-high-temperature reactor)	Thermal	Helium	900–1000	Open	250–300
SFR (Sodium-cooled fast reactor)	Fast	Sodium	550	Closed	30–150 300–1500 1000–2000
SCWR (Supercritical water-cooled reactor)	Thermal/fast	Water	510–625	Open/closed	300–700 1000–2000
GFR (Gas-cooled fast reactor)	Fast	Helium	850	Closed	1200
LFR (Lead-cooled fast reactor)	Fast	Lead	480–800	Closed	20–180 300–1200 600–1000
MSR (Molten salt reactor)	Epithermal	Fluoride salt	700–800	Closed	1000

1.18 Why We Need to Consider the Future Role of Nuclear Power Now

The following reasonings are some arguments that show why we need to consider the future role in design of new nuclear power plant:

1.0 Nuclear power has been part of the global energy need mix for the past five decades. Currently it provides about 18% of the electricity we use in our homes and workplaces. For example, in the United Kingdom, about one third of our emissions of carbon dioxide come from electricity generation [19]. The vast majority of those emissions come from coal and gas power plants.

2.0 Energy companies will need to invest in around 30–35GW of new electricity generating capacity – as coal and nuclear plants retire – over the next two decades, with around two thirds needed by 2020. This is equivalent to about one third of our existing capacity. The world needs a clear and stable regulatory framework to reduce uncertainty for business to help ensure sufficient and timely investment in technologies that contribute to our energy goals.

3.0 Of the capacity that is likely to close over the two decades, two thirds is from carbon-intensive fossil fuel generation, and about 10GW is nuclear and therefore low carbon. So companies' decisions on the type of power stations they invest in to replace this capacity will have significant implications for the level of carbon emissions. As an illustration, if our existing nuclear power stations were all replaced with fossil fuel-fired power stations, our emissions would be between eight and sixteen MtC (million tons of carbon) a year higher as a result (depending on the mix of gas and coal-fired power stations). This would be equivalent to about 30–60% of the total carbon savings we project to achieve under our central scenario from all the measures we are bringing forward in the Energy White Paper [20]. Our gas demand would also be higher, at a time when we are becoming more dependent on imported sources of fossil fuels.

4.0 Electricity demand in the United States is expected to grow significantly in the future. Over the past decade, Americans used 17% more electricity, but domestic capacity rose only 2.3% (National Energy Policy, May 2001). Unless the United States significantly increases its generating capacity, the country will face an energy shortage that is projected to adversely affect our economy, our standard of living, and our national security. Coupled with this challenge is the need to improve our environment.

5.0 New nuclear power stations have long lead times. This time is necessary to secure the relevant regulatory and development consents, which must be obtained before construction can begin, and there is also a long construction period compared to other generating technologies. Our conservative assumption is that for the first new nuclear plant, the pre-construction period would last around 8 years (to secure the necessary consents) and the construction period would last around 5 years. For subsequent plants, this is assumed to be 5 and 5 years, respectively. New nuclear power stations are therefore unlikely to make a significant contribution to the need for new capacity before 2020.

6.0 Even with our expectations that the share of renewable will grow, it is likely that fossil fuel generation will meet some of this need. However, beyond that date there are still significant amounts of new capacity needed; for example, in 2023 one third or 3GW of our nuclear capacity will still be operational, based on published lifetimes. Given the likely increase in fossil fuel generation before this date, it is important that much of this capacity is replaced with low-carbon technologies. New nuclear power stations could make an important contribution, as outlined in this consultation document, to meeting our needs for low-carbon electricity generation and energy security in this period and beyond 2050. Because of the lead times, without clarity now we will foreclose the opportunity for nuclear power.

7.0 The existing approach on new nuclear build was set out in 200,311: "Nuclear power is currently an important source of carbon-free electricity. However, its current economics make it an unattractive option for new, carbon-free generating capacity and there are also important issues of nuclear waste to be resolved. These issues include our legacy waste and continued waste arising from other sources. This white paper does not contain specific proposals for building new nuclear power stations. However, we do not rule out the possibility that at some point in the future new nuclear build might be necessary if we are to meet our carbon targets. Before any decision to proceed with the building of new nuclear power stations, there will need to be the fullest public consultation and the publication of a further white paper setting out our proposals."

8.0 Since 2003 there have been a number of developments, which have led the government to consider afresh the potential contribution of new nuclear power stations. Firstly, there has been significant progress in tackling the legacy waste issue:

- We have technical solutions for waste disposal that scientific consensus and experience from abroad suggest could accommodate all types of wastes from existing and new nuclear power stations.
- There is now an implementing body (the Nuclear Decommissioning Authority), with expertise in this area, and the government is reconstituting the Committee on Radioactive Waste Management (CoRWM) in order to provide continued independent scrutiny and advice.
- A framework for implementing long-term waste disposal in a geological repository will be consulted on in the coming months.

9.0 The government has also made progress in considering the issue of waste management in relation to potential new nuclear power stations:

- This consultation provides the opportunity to discuss the ethical, intergenerational, and public acceptability issues associated with a decision to allow the private sector to invest in new nuclear power stations and generate new nuclear waste.
- The government is developing specific proposals to protect the taxpayer. Under these proposals, private sector developers would meet the full decommissioning costs and full share of waste management costs. The proposals would be implemented in the event that we conclude that energy companies should be allowed to invest in new nuclear power stations. They would need to be in place before proposals for new power stations could go ahead.

10.0 Secondly, the high-level economic analysis of nuclear power, prepared for the Energy Review, concluded that under likely scenarios for gas and carbon prices and taking prudent estimates of nuclear costs, nuclear power would offer general economic benefit to the United Kingdom in terms of reduced carbon emissions and security of supply benefits [21]. Therefore, the government believes that it has a potential contribution to make, alongside other low-carbon generating technologies.

11.0 Thirdly, some energy companies have expressed a strong interest in investing in new nuclear power stations. They assess that new nuclear power stations could be an economically attractive low-carbon investment, which could help diversify their generation portfolios. Their renewed interest reflects assessments that with carbon being priced to reflect its impacts and gas prices likely to be higher than previously expected, the economics of new nuclear power stations are becoming more favorable.

12.0 Nuclear power stations have long lead times. If they are to be an option to replace the capacity closing over the next two decades, and in particular after 2020, a decision on whether allowing energy companies the option of investing in new nuclear power stations would be in the public interest needs to be

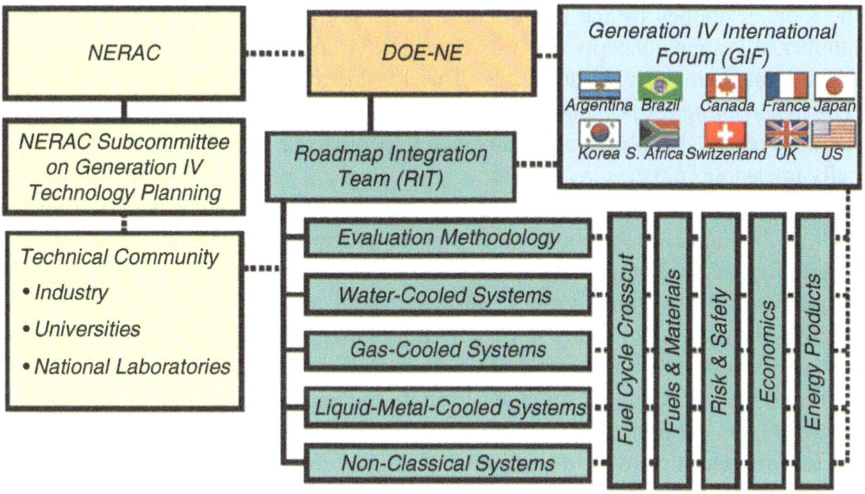

Fig. 1.29 The roadmap organization

> taken now. Energy companies would need to begin their initial preparations in the near future in order to have a reasonable prospect of building new generation in this period. Not taking the public interest decision now would foreclose the option of new nuclear being one of our options for tackling climate change and achieving energy security.

1.19 The Generation IV Roadmap Project

As the Generation IV goals were being finalized, preparations were made to develop the Generation IV technology roadmap. The organization of the roadmap is shown in Fig. 1.29. The Roadmap Integration Team (RIT) is the executive group. Groups of international experts were organized to undertake identification and evaluation of candidate systems, and to define R&D to support them [22].

In a first step, an Evaluation Methodology Group was formed to develop a process to systematically evaluate the potential of proposed Generation IV nuclear energy systems to meet the Generation IV goals. A discussion of the Evaluation Methodology Group's evaluation methodology is included in this report. At the same time, a solicitation was issued worldwide, requesting that concept proponents submit information on nuclear energy systems that they believe could meet some or all of the Generation IV goals. Nearly 100 concepts and ideas were received from researchers in a dozen countries [22].

Technical working groups (TWGs) were formed – covering nuclear energy systems employing water-cooled, gas-cooled, liquid-metal-cooled, and nonclassical reactor concepts – to review the proposed systems and evaluate their potential using the tools developed by the Evaluation Methodology Group. Because of the large number of system concepts submitted, the TWGs collected their concepts into sets of concepts with similar attributes. The TWGs conducted an initial screening, *termed screening for potential*, to eliminate those concepts or concept sets that did not have reasonable potential for advancing the goals or were too distant or technically infeasible [22].

A Fuel Cycle Crosscut Group (FCCG) was also formed at a very early stage to explore the impact of the choice of fuel cycle on major elements of sustainability – especially waste management and fuel utilization. Their members were equally drawn from the working groups, allowing them to compare their insights and findings directly. Later, other Crosscut Groups were formed covering economics, risk and safety, fuels and materials, and energy products. The Crosscut Groups reviewed the TWG reports for consistency in the technical evaluations and subject treatment and continued to make recommendations regarding the scope and priority for crosscutting R&D in their subject areas. Finally, the TWGs and Crosscut Groups worked together to report on the R&D needs and priorities of the most promising concepts [22].

The international experts that contributed to this roadmap represented all ten GIF countries, the Organization for Economic Cooperation and Development Nuclear Energy Agency, the European Commission, and the International Atomic Energy Agency [22].

At least in the United Sates under the Department of Energy (DOE), with Idaho SMR project, with a roadmap, federal agencies may wish to follow, when making power purchase decisions that may involve a small modular reactor, the key steps in the decision process may be summarized in Fig. 1.30. SMRs are designed to provide valuable resilience services as a secure, reliable, and flexible source of primary and backup power. SMRs, coupled with transmission hardening, could provide highly reliable, non-intermittent, clean, and carbon-free power. SMRs can also easily store 2 years' worth of fuel on-site. Certain SMR designs allow for output to be varied over days, hours, or minutes, thereby enabling the SMR to adjust to be in line with changing load demands.

However, implementing SMR projects around the country is difficult due to the FOAK technology, construction challenges, and licensing requirements. SMRs introduce significant expenses and risks that may be challenging for a project to bear without any financial support from the intended end user. The recommendations offered in this chapter, along with the potential solutions described elsewhere in this report, may be considered to assist with overcoming these challenges and advance the deployment of SMRs in the United States while improving grid resiliency at the Oak Ridge Reservation.

More specifically, this chapter identifies how the federal government can assist with making the financing and development of SMRs easier – both in its role as a customer and as a governing body, which permits federal agencies to enter into agreements for a term of up to 30 years to purchase power produced by small modular reactors.

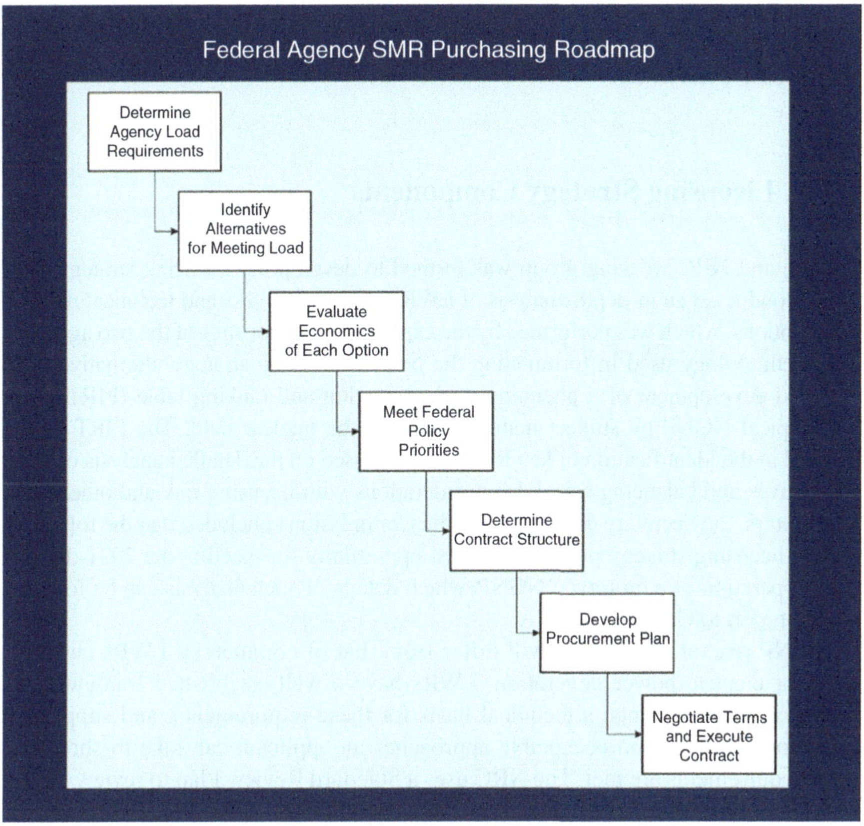

Fig. 1.30 Federal agency SMR purchasing roadmap

Congress may wish to consider enacting legislation which would permit federal agencies to share in the risks associated with the construction of SMRs. Leveraging the federal government's strong credit standing as a purchaser of the power and its continual need for baseload power is important in the development of SMRs. Federal agency purchasers can help to set the market and offer more certainty to other initial buyers. While there are a range of legal authorities, federal agencies may use to purchase power, most often GSA's 40 U.S.C. § 501 is used, limiting PPA terms to 10 years. This 10-year limitation impacts a party's ability to take advantage of government purchasing for financing.

Federal agencies should be able to purchase power produced by an SMR for a term of up to 30 years. Currently, only DOD (pursuant to 10 U.S.C. § 2922a) has the authority to purchase power for a term of up to 30 years in limited circumstances. By creating an authority that permits federal agencies to purchase power for up to 30 years, SMR developers will be able to use traditional financing to repay a financed project or a long-term bond for up to 30-year term, making the financing more affordable. Depending upon the size of the federal agency's off-take, as compared to the size of the power source being funded, this discrepancy may make it difficult for financing.

1.20 Licensing Strategy Components

A DOE and NRC working group was formed to develop the licensing strategy. This group conducted an in-depth analysis of LWR licensing process and technical requirement options, which was performed by the experienced senior staff of the two agencies. The methodology used in formulating the NGNP licensing strategy alternatives also included development of a phenomena identification and ranking table (PIRT) for a prototypical NGNP by subject matter experts in the nuclear field. The PIRT results assisted in the identification of key R&D needs. Based on the detailed analysis of these alternatives and balancing schedule considerations with licensing risk and other pertinent factors, the Secretary of Energy and the Commission concluded that the following NGNP licensing strategy provides the best opportunity for meeting the 2021 date for initial operation of a prototype NGNP, which details of such analysis can be found in NGNP report to Congress [18].

NGNP reactor technology will differ from that of commercial LWRs currently used for electric power generation. LWRs have a well-established framework of regulatory requirements, a technical basis for these requirements, and supporting regulatory guidance on acceptable approaches an applicant can take to show that NRC requirements are met. The NRC uses a Standard Review Plan to review licensing applications for these reactor designs. Additionally, the NRC has a well-established set of validated analytical codes and methods and a well-established infrastructure for conducting safety research needed to support its independent safety review of an LWR plant design and the technical adequacy of a licensing application.

New nuclear power plants can be licensed under either of two existing regulatory approaches. The first approach is the traditional "two-step" process described in Title 10, Part 50, "Domestic Licensing of Production and Utilization Facilities," of the *Code of Federal Regulations* (10 CFR Part 50), which requires both a construction permit (CP) and a separate operating license (OL). The second approach is the new "one-step" licensing process described in 10 CFR Part 52, "Licenses, Certifications, and Approvals for Nuclear Power Plants," which incorporates a combined construction and operating license (COL). Both of these processes allow a deterministic or risk-informed performance-based approach to technical requirements.

Many of the regulatory requirements and supporting review guidance for LWRs are technology-neutral; that is, they are applicable to non-LWR designs as well as LWR designs. However, certain LWR requirements may not apply to the unique aspects of a VHTR design. Accordingly, in developing the NGNP licensing strategy, the NRC and DOE considered the various options available to the NRC staff for adapting current NRC LWR licensing requirements for the NGNP VHTR. These options related to legal, process, technical, research, and regulatory infrastructure matters and included an examination of historical licensing activities. These considerations led to selection of a licensing strategy that would comply best with the considerations identified in the EPAct.

The licensing strategy outlined in this report is composed of two distinct aspects. The first aspect is a recommended approach for how the NRC will adapt the current LWR technical requirements to apply to a VHTR. The second aspect is a recommended licensing process alternative that identifies which of the procedural alternatives in the NRC regulations would be best for licensing the NGNP. To arrive at these recommendations, NRC and DOE evaluated a number of options and alternatives.

1.21 Market and Industry Status and Potentials

Europe plays a leading role in the development of nuclear energy and has 35% of the globally installed capacity. The reactors in Europe have been in operation for 27 years on average. Current plans in most EU member countries are to extend their lifetime on a case-by-case basis beyond 40 years, and even beyond 60 years in some cases, in combination with power upgrades. The first two Generation-III reactors, European pressurized-water reactor (EPR), are currently being constructed.

The global growth of the nuclear energy can be measured by the increasing number of reactors [9, 10] (three more in 2005 and 2006, seven in 2007, and ten in 2008), but with a strong concentration in Asia. Nevertheless, a number of these reactors are of European design. There are presently four reactors under construction in Europe: the EPRs in Finland and France and two smaller reactors of Generation-II type (VVER 440) in Slovakia and with plans to build new reactors in France, Romania, Bulgaria, and Lithuania. Perhaps more importantly the United Kingdom has taken concrete steps toward new build with bidding beginning in 2009 from leading utilities, and Italy has declared that it intends to start a nuclear program with a target to produce 25% of the electricity by 2030. The estimated maximum potential installed capacities of nuclear fission power for the EU-27 (150 GWe by 2020 and 200 GWe by 2030) appear more realistic than the baseline (115 GWe in 2020 and 100 GWe in 2030).

Programs to build fast reactor and high-temperature reactor demonstrators are being implemented in Russia and several Asian countries. Although these are not Generation IV designs, transfer of knowledge and experience from operation will contribute significantly to future Generation IV development. In Europe, a concerted effort is proposed in the form of a European Industrial Initiative in sustainable nuclear fission as part of the Community's SET-Plan. The EII has singled out the sodium-cooled fast reactor (SFR) as its primary system with the basic design selected by 2012 and construction of a prototype of 250–600 MWe that is connected to the grid and operational by 2020.

In parallel, a gas- or lead-cooled fast reactor (GFR/LFR) will also be investigated. The selection of the alternative fast reactor technology is scheduled for 2012 on the basis of a current program of preconceptual design research. The reactor will be a 50–100 MWth demonstrator reactor that should also be in operation by 2020. The SFR prototype and LFR/GFR demonstrator will be complemented by a fuel fabrication workshop that should serve both systems, and by a range of new or refurbished supporting experimental facilities for qualification of safety systems, components, materials, and codes. A commercial deployment for a SFR is expected from 2040 and for the alternative design a decade later.

High-temperature reactors dedicated to cogeneration of process heat for the production of synthetic fuels or industrial energy products could be available to meet market needs by 2025, which would trigger requirements to construct "first-of-a-kind" demonstrators in the next few years. Indeed, such programs are currently being set up in some countries (the United States, Japan, South Africa, and China). The key aspect is the demonstration of the coupling with the conventional industrial plant. Supercritical water reactors and molten salt reactors, as well as accelerator-driven subcritical systems dedicated to transmutation of nuclear waste, are currently being assessed in terms of feasibility and performance, though possible industrial applications have yet to be clearly identified.

1.22 Barriers

The high capital cost of nuclear energy in combination with uncertain long-term conditions constitutes a financial risk for utilities and investors. The lack of widespread support in the EU member states may undermine the strength of EU industry for the development of new technologies. Harmonized regulations, codes, and standards at the EU level would strengthen the competitiveness of Europe's nuclear sector and promote deployment of Generation-III technology in the near term. The industry, infrastructures, and services that support nuclear power have shrunk significantly during the last decades. This situation in Europe is not unique, but it may pose a bottleneck for the deployment of reactors in the relatively near future. One example is large forgings needed for pressure-vessel heads. World capacity is limited, and even at the present new build construction rate, there is a waiting list for delivery of these components.

Public acceptance remains an important issue, but even though opinion is not very favorable in a number of member states, there are signs that the mood is changing. Nevertheless, concerted efforts are still required, based on objective and open dialogue among all stakeholders. International cooperation currently exists at the level of research, and this is being facilitated in the area of Generation IV systems by the Generation IV International Forum (GIF). However, EU industry is facing stiff competition, especially in Asia where strong corporate support for R&D is putting industry in a better position to gain leadership in the near future. Another significant potential barrier for nuclear fission is the shortage of qualified engineers and scientists as a result of the lack of interest in nuclear careers during the 1990s and the reduced availability of specialist courses at universities. Preservation of nuclear knowledge remains a major issue, especially since most of the current generations of nuclear experts are nearing retirement.

1.23 Needs

The high initial capital investments and sensitive nature of the technology involved means that renewed deployment of currently available nuclear technology can only take place in a stable (or, at least, predictable) regulatory, economic, and political environment. In June 2009, the EU established a common binding framework on nuclear safety with the adoption of the Council Directive establishing a community framework for the safety of nuclear installations [16, 17]. This is the first binding EU legislation in this field.

In order to retain its leading position and to overcome bottlenecks in the supply chain, Europe also needs to reinvigorate the industrial supply chains supporting the nuclear sector. Apart from this overriding requirement for a clear European strategy on nuclear energy, a new research and innovation system is needed that can assure additional funding, especially for the development of Generation IV technology. In this context the Sustainable Nuclear Energy Technology Platform [11] plays a key role. The timescales involved, and the fact that key political and strategic decisions are yet to be taken regarding this technology, mean that a significant part of this additional funding must be public. The launch of the European Sustainable Nuclear Industrial Initiative under the Community's SET-Plan, bringing together key industrial and R&D organizations, would be a very significant step toward the construction and operation of the necessary demonstrators and prototypes.

High-temperature reactors based on existing technology can also be deployed in the near future with the aim of demonstrating the cogeneration of process heat and the coupling with industrial processes. This would need to be built and funded through a European or International Consortium, which should also include the process heat end-user industries. In the meantime, an enhanced research effort is needed to ensure Europe's leadership in sustainable nuclear energy technologies that include continuous innovation in LWRs, qualification and development of materials, and closed fuel cycle with U-Pu multi-recycling and (very) high-temperature reactors and related fuel technology.

Breakthroughs are especially sought in the fields of materials to enhance safety, nuclear fuels, and fuel cycle processes. Additionally, there is a need for harmonization of European standards and a strategic planning of national and European research infrastructures for use by the European research community. The implementation of geological disposal of high-level waste is also being pursued as part of national waste management programs, though some countries are not as advanced as others. The new Implementing Geological Disposal Technology Platform, launched in November 2009, is coordinating the remaining necessary applied research in Europe leading up to the start of operation of the first geological repositories for high-level and long-lived waste around 2020 and will facilitate progress in and technology transfer with other national programs.

More effort is needed to inform and interact with the public and other stakeholders, and the education and training of a new generation of nuclear scientists and engineers and transfer of knowledge from the generation that designed and built reactors in the 1970s and 1980s needs urgent attention. The European Nuclear Energy Forum (ENEF) provides a unique platform for a broad open discussion on the role nuclear power plays today and could play in the low-carbon economy of the future. ENEF analyzes and discusses the opportunities (competitiveness, financing, grid, etc.) and risks (safety, waste) and need for education and training associated with the use of nuclear power and proposes effective ways to foster communication with and participation of the public.

1.24 Synergies with Other Sectors

Nuclear energy provides a very stable base-load electricity supply and can therefore work in synergy with renewable energies that are more intermittent. Nuclear energy should also contribute significantly to a low-carbon transport sector as high-temperature application can provide synthetic fuel and hydrogen, while generated electricity could provide a large share of the energy for electrical cars. Interactions are anticipated with activities in "hydrogen energy and fuel cells" through the potential of nuclear hydrogen production and with "grids" from the characteristics of nuclear electricity generation. With respect to basic materials research, there should be synergies with other applications, such as "biofuels" and "clean coal," where materials are subjected to extreme environments. In addition, the opportunities for important common research with the fusion program, especially in the area of materials, need to be fully exploited. The European Energy Research Alliance under the SET-Plan is also expected to provide opportunities for synergies and collaborative work in the area of nuclear materials. In general, crosscutting research would benefit from more clearly defined channels of interaction, responsibilities, and increased flexibility regarding funding and programming.

References

1. F. Settle web Site http://www.chemcases.com/nuclear/index.html.
2. B. L. Cohen "Breeder Reactors: A Renewable Energy Source," American Journal of Physics, 51, 1 (Jan. 1983).
3. A. Weinberg "The Second Fifty Years of Nuclear Fission," *Proc. Special Symposium: 50 Years of Nuclear Fission in Review*, Ontario, Canada, June 5, 1989, Canadian Nuclear Society.
4. N. Seko, "Aquaculture of Uranium in Seawater by a Fabric-Adsorbent Submerged System," *Nuclear Technology*, **144**, 274 (Nov. 2003).

5. "Some Physics of Uranium," available on the Internet at http://www.worldnuclear.org/education/phys.html (December 2005).
6. W. H. Hannum, "The Technology of the Integral Fast Reactor and Its Associated Fuel Cycle," *Progress in Nuclear Energy*, **31**, 1 (1997).
7. TUDelft, http://www.iri.tudelft.nl/~rooijen/spectra_en.html.
8. http://setis.ec.europa.eu/newsroom-items-folder/nuclear-fission-power-generation
9. World Nuclear Association website http://www.world-nuclear.org/
10. IAEA Power Reactor Information System website http://www.iaea.org/programmes/a2/
11. http://www.ne.doe.gov/geniv/neGenIV1.html
12. Sustainable Nuclear Energy Technology Platform, Strategic Research Agenda, May 2009, www.SNETP.eu
13. Nuclear Energy Outlook 2008, OECD/NEA Report No. 6348, 2008, Nuclear Energy Agency, Paris.
14. 2009 Update of the MIT 2003 Future of Nuclear Power, An Interdisciplinary MIT Study, 2003, Massachusetts Institute of Technology, Cambridge USA, May 2009 http://web.mit.edu/nuclearpower/pdf/nuclearpower-update2009.pdf
15. Uranium 2007: Resources, Production and Demand. OECD Nuclear Energy Agency and the International Atomic Energy Agency, OECD 2008 NEA N 6345.
16. Proposal for a COUNCIL DIRECTIVE (EURATOM) setting up a Community framework for nuclear safety COM (2008) 790/3, November 2008
17. COUNCIL OF THE EUROPEAN UNION Legislative Acts and Other Instruments 10667/09, June 2009.
18. http://www.ne.doe.gov/pdfFiles/NGNP_reporttoCongress.pdf
19. Updated Emissions Projections, July 2006, DTI, http://www.dti.gov.uk/files/file31861.pdf
20. DTI: Energy White Paper, Meeting the Energy Challenge, http://www.dti.gov.uk/energy/whitepaper
21. DTI Analysis, http://www.dti.gov.uk/energy/whitepaper/consultations/nuclearpower2007
22. http://www.ne.doe.gov/geniv/documents/gen_iv_roadmap.pdf
23. http://hyperphysics.phy-astr.gsu.edu/hbase/nucene/fasbre.html

Chapter 2
Nuclear Power Structure from Past to Present

Abstract In the years just before and during World War II, nuclear research focused mainly on the development of defense weapons. Later, scientists concentrated on peaceful applications of nuclear technology. An important use of nuclear energy is the generation of electricity. After years of research, scientists have successfully applied nuclear technology to many other scientific, medical, and industrial purposes. Governments have been deeply involved in the development of nuclear energy. Some of them initiated and led the development of nuclear energy since its military beginnings in World War II, because of its strategic nature and the scope of its risks and benefits. Governments later supported the development of civilian nuclear energy, primarily for the generation of electricity. In the postwar period, governments played an increasing overall role in the economies of the industrial countries. Science and technology were essential instruments of government action, and nuclear energy was a highly visible symbol of their successful application.

2.1 Introduction

The concept of the atom has existed for many centuries. But we only recently began to understand the enormous power contained in the tiny mass: the history of our discoveries about atoms. We begin with the ideas of the Greek philosophers. Then we follow the path to the early scientists who discovered radioactivity. Finally, we reach modern-day use of atoms as a valuable source of energy.

Ancient Greek philosophers first developed the idea that all matter is composed of invisible particles called *atoms*. The word atom comes from the Greek word, *atoms*, meaning indivisible. Scientists in the eighteenth and nineteenth centuries revised the concept based on their experiments. By 1900, physicists knew the atom contains large quantities of energy. British physicist Ernest Rutherford was called the father of nuclear science because of his contribution to the theory of atomic structure. In 1904 he wrote:

> *If it were ever possible to control at will the rate of disintegration of the radio elements, an enormous amount of energy could be obtained from a small amount of matter.*

B. Zohuri, *Small Modular Reactors as Renewable Energy Sources*,
https://doi.org/10.1007/978-3-319-92594-3_2

Albert Einstein developed his theory of the relationship between mass and energy 1 year later. The mathematical formula is $E = mc^2$, or "energy equals mass times the speed of light squared." It took almost 35 years for someone to prove Einstein's theory [1].

As part of history of nuclear energy and early discoveries of this fission drive energy, we can state that no scientific progress ever really starts. Rather, it builds on the work of countless other discoveries. Since we have to start somewhere, this story will start in Germany, in 1895, where a fellow named Roentgen was experimenting with cathode rays in a glass tube that he had sucked the air out of. At one point, he had the device covered but noticed that the photographic plates off to the side were lighting up when the device was energized. He realized that he was looking at a new kind of ray and called it what any reasonable physicist would call an unknown: the X-ray. He systematically studied these rays and took the first X-ray photo of his wife's hand 2 weeks later, thereby becoming the father of modern medical diagnostics.

Soon after in France, in 1896, a guy named Becquerel noticed that if he left uranium salts sitting on photographic plates, they would expose even though no cathode ray tube was energized. The energy must have been coming from inside the salts themselves. Marie Curie and her husband Pierre studied the phenomenon and isolated two new elements that exhibited this spontaneous energy production: polonium and radium. They named the phenomenon radioactivity.

In England, Ernest Rutherford starts studying radioactivity and discovers that there are two types of rays that come out that are different from X-rays. He calls them alpha- and beta- radiation. He later discovers the shocking fact that the vast majority of the mass of atoms is concentrated in their centers and thus discovers the atomic nucleus. He is widely regarded today as the father of nuclear physics. He later discovers gamma radiation. In 1920, he theorizes the existence of a neutral particle in the nucleus called a neutron, though there is no evidence that neutrons exist yet.

In 1932, Chadwick reads some published results from the Curie's kid, Irene Joliot-Curie, which says gamma radiation was found to knock protons out of wax. Disbelieving, he suspects they are seeing Rutherford's neutrons and does experiments to prove this, thus discovering the neutron.

In 1934, physicist Enrico Fermi (Fig. 2.1) conducted experiments in Rome that showed neutrons could split many kinds of atoms. The results surprised even Fermi himself. When he bombarded uranium with neutrons, he did not get the elements he expected. The elements were much lighter than uranium.

He led the team of scientists who created the first self-sustaining nuclear chain reaction at University of Chicago, where he built the first fissionable nuclear reactor.

In the fall of 1938, German scientists Otto Hahn and Fritz Strassman fired neutrons from a source containing the elements *radium* and *beryllium* into uranium (atomic number 92). They were surprised to find lighter elements, such as *barium* (atomic number 56), in the leftover materials [1].

Fig. 2.1 Enrico Fermi, an Italian physicist

These elements had about half the atomic mass of uranium. In previous experiments, the leftover materials were only slightly lighter than uranium.

Although they are tiny, atoms have a large amount of energy holding their nuclei together. Certain *isotopes* of some elements can be split and will release part of their energy as heat. This splitting is called *fission*. The heat released in fission can be used to help generate electricity in power plants.

Uranium-235 (*U-235*) is one of the isotopes that fissions easily. During fission, U-235 atoms absorb loose neutrons. This causes U-235 to become unstable and split into two light atoms called *fission products*.

The combined mass of the fission products is less than that of the original U-235. The reduction occurs because some of the matter changes into energy. The energy is released as heat. Two or three neutrons are released along with the heat. These neutrons may hit other atoms, causing more fission. This scenario and process of fission product chain reaction is illustrated in Fig. 2.2.

Some governments initiated and led the development of nuclear energy since its military beginnings in World War II, because of its strategic nature and the scope of its risks and benefits. They later sponsored and managed the development of civilian nuclear energy, primarily for the generation of electricity. As a unique new source of energy with seemingly unlimited promise, nuclear energy enjoyed a very high priority in the economic and energy policies of the leading industrial countries for many decades.

A series of fissions is called a *chain reaction*. If enough uranium is brought together under the right conditions, a continuous chain reaction occurs. This is

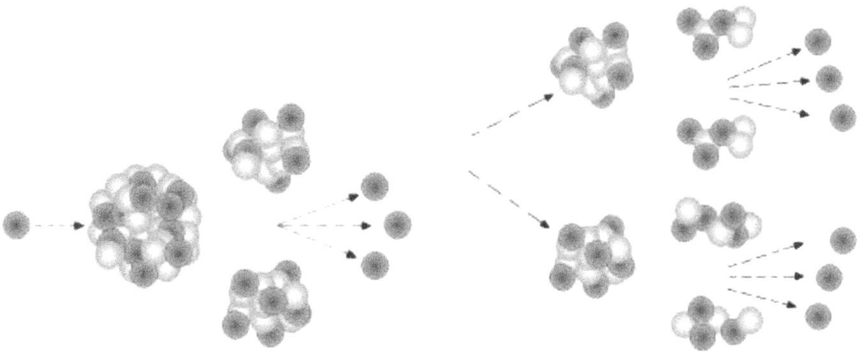

Fig. 2.2 Illustration of fission process chain reaction

called a *self-sustaining chain reaction*. A self-sustaining chain reaction creates a great deal of heat, which can be used to help generate electricity.

2.2 The First Self-Sustaining Chain Reaction

In 1939, Bohr came to America. He shared with Einstein the Hahn-Strassman-Meitner discoveries. Bohr also met Fermi at a conference on theoretical physics in Washington, D.C. They discussed the exciting possibility of a self-sustaining chain reaction. In such a process, atoms could be split to release large amounts of energy.

Scientists throughout the world began to believe a self-sustaining chain reaction might be possible. It would happen if enough uranium could be brought together under proper conditions. The amount of uranium needed to make a self-sustaining chain reaction is called a *critical mass*.

Fermi and his associate, Leo Szilard, suggested a possible design for a uranium chain reactor in 1941. Their model consisted of uranium placed in a stack of graphite to make a cube-like frame of fissionable material (see Fig. 2.3).

With neutrons around, everyone's shooting them at various nuclides. Soon enough, Hahn and Strassman shoot them at uranium atoms and see some strange behavior which Lise Meitner and her nephew Frisch identify as the splitting of the atom, releasing much energy. They name it fission, after binary fission in biology.

Szilard recognizes fission as a potential way to form a chain reaction (which he had been considering for a long time). He and Fermi do some neutron multiplication studies and see that it is indeed possible. They go home, knowing that the world is about to change forever.

Szilard, Wigner, and Teller write a letter to President Roosevelt, warning of nuclear weapons, and have Einstein sign it and send it (he was more famous). Roosevelt authorizes a small study into uranium. In 1942, Fermi successfully created the first man-made nuclear chain reaction in a squash court under the stadium

Fig. 2.3 Picture of Fermi and Szilard

at the University of Chicago. The Manhattan project kicked into full gear. Two types of bombs were pursued simultaneously, one made with enriched uranium and the other made with plutonium. Giant secret cities were built very quickly. The one in Oak Ridge, TN, had a reactor that created the first gram-quantities of plutonium for study, but its main task was to enrich uranium. The one in Hanford, WA, is the site of plutonium production reactors (the first high-power nuclear reactors) and plutonium extraction chemistry plants. Another, in Los Alamos, NM, is the site where the technology that turns weapon materials into weapons is developed. Both paths to the bomb are successful. The more uncertain design, the plutonium implosion device (like Fat Man) is successfully tested at the Trinity site in New Mexico in July 1945.

The scientific community in the United States, including those Europeans now calling the United States their safe home, pursued the idea that uranium fission and the production of excess neutrons could be the source of extraordinary new weapons. They knew Lise Meitner's interpretation, in Sweden, of Hahn's experiments would likely be known in Germany. Clearly there might now be a race commencing for the development and production of a new, super weapon based on the fission of $^{235}U_{92}$ or $^{239}Pu_{94}$ (see Fig. 2.4).

The decision is made to drop Little Boy and Fat Man on Hiroshima and Nagasaki, Japan, on August 6 and 9, 1945. The cities are devastated, with up to 250,000 people dead. Japan surrenders unconditionally 6 days later, on August 15, 1945. This is the first time the public realizes that the United States has been developing bombs [2].

Fig. 2.4 Lise Meitner and Otto R. Frisch

2.3 Nuclear Energy Expands and Stagnates for Peace Usages

As we know, historically, the first nuclear reactor was only the beginning. Most early atomic research focused on developing an effective weapon for use in World War II. The work was done under the code name *Manhattan Project.* However, during pre- and post-World War II time frame, some scientists worked on making *breeder reactors*, which would produce fissionable material in the chain reaction. Therefore, they would create more fissionable material than they would use. Enrico Fermi led a group of scientists in initiating the first self-sustaining nuclear chain reaction. The historic event, which occurred on December 2, 1942, in Chicago, is recreated in this painting as it is shown in Fig. 2.5.

In the process of fission energy that was expanding in a peaceful application, an experimental liquid-metal cooled reactor in Idaho called EBR-I was attached to a generator in 1951, producing the first nuclear-generated electricity. But before civilian power plants came to be, Admiral Rickover pushed to use reactors to power submarines, since they wouldn't need to refuel or to use oxygen for combustion. The USS Nautilus launched in 1954 was the first nuclear-powered submarine. Soon after, the Soviet Union opens the first nonmilitary, electricity-producing reactor [3]. Based on the submarine reactor design, the Shippingport reactor opens in 1957 as the first commercial reactor in the United States.

Through the 1960s and 1970s, lots of nuclear reactors are built for making electricity, using designs very similar to those made for the submarines. They work

Fig. 2.5 Enrico Fermi and group of scientists in Chicago

well and produce cheap, emission-free electricity with a very low mining and transportation footprint. A nuclear-powered future is envisioned by many. In 1974, France decided to make a major push for nuclear energy and ended up with 75% of their electricity coming from nuclear reactors. The United States built 104 reactors and got about 20% of its electricity from them. Eventually, labor shortages and construction delays started bringing the cost of nuclear reactors up, slowing their growth.

The 1979 Three Mile Island accident and the 1986 Chernobyl accident further slowed the deployment of nuclear reactors. Tighter regulations brought costs higher. The 1986 passive safety tests at EBR-II prove that advanced reactor designs (besides the ones originally used to make submarines) can be substantially safer. These tests have major failure occur with no control rods inserted, and the reactors shut themselves down automatically.

In 1994, the Megatons to Megawatts treaty with Russia is signed to down-blend nuclear warheads into reactor fuel. Eventually, 10% of US electricity comes from dismantled nuclear weapons.

In the late 1990s and 1900s, the phenomenal safety record of the US commercial reactor fleet (0 deaths) and smooth operation of reactors combined with ongoing worries of global climate change due to carbon emissions bring about substantial talk of a "nuclear renaissance," where new builds might start up substantially again. Meanwhile, strong interest in Asia strengthens, and ambitious plans to build large fleets are made to satisfy growing energy needs without adding more fossil fuel.

On March 2011, a large earthquake and tsunami inundate the reactors at Fukushima Daiichi. Backup diesel generators fail, and the decay heat cannot be cooled. Fuel melts, and hydrogen builds up and explodes (outside of containment). Radiation is released, but much of it goes out to the sea instead of into populated area. No people are expected to die from radiation dose.

2.4 Government and Nuclear Energy

After the war, the United States government encouraged the development of nuclear energy for peaceful civilian purposes. The Congress created the Atomic Energy Commission (AEC) in 1946. The AEC authorized the construction of Experimental Breeder Reactor I at a site in Idaho. The reactor generated the first electricity from nuclear energy on December 20, 1951. Figure 2.6 is an illustration of an early Experimental Breeding Reactor I which generated electricity to light four 200 watt bulbs on December 20, 1951. This milestone symbolized the beginning of the nuclear power industry.

A major goal of nuclear research in the mid-1950s was to show that nuclear energy could produce electricity for commercial use. The first commercial electricity-generating plant powered by nuclear energy was located in Shippingport, Pennsylvania. It reached its full design power in 1957. Light-water reactors like Shippingport use ordinary water to cool the reactor core during the chain reaction. They were the best designs then available for nuclear power plants.

Private industry became more and more involved in developing light-water reactors after Shippingport became operational. Federal nuclear energy programs shifted their focus to developing other reactor technologies.

The nuclear power industry in the United States grew rapidly in the 1960s. Utility companies saw this new form of electricity production as economical, environmentally clean, and safe. In the 1970s and 1980s, however, growth slowed. Demand for

Fig. 2.6 The Experimental Breeder Reactor I

electricity decreased, and concern grew over nuclear issues, such as reactor safety, waste disposal, and other environmental considerations.

Still, the United States had twice as many operating nuclear power plants as any other country in 1991. This was more than one-fourth of the world's operating plants. Nuclear energy supplied almost 22% of the electricity produced in the United States.

Regulation of nuclear safety and security remains a core function of the government. It should guarantee the existence of an independent, competent regulator with adequate resources and authority. The emphasis now is on the safety culture of organizations, beginning at the most senior levels. This brings in the need to ensure good governance. Nuclear regulation should be in line with modern regulatory practice across the government, allowing nuclear energy to compete fairly. Governments looking for a future contribution from nuclear energy should ensure that regulation is prepared to deal with issues of decommissioning, refurbishment, uprating, life extension, and new reactor designs.

In the 1980s and 1990s, problems with exclusive government ownership and control of production equipment appeared. Governments came under pressure to cut expenditures and diminish their direct involvement in the economy. Expanding international trade forced all industries to be more competitive. Markets were championed as an alternative to government direction and regulation. Simultaneously, environmental protection and the concept of sustainable development increased in importance in policy-making, while the need to ensure security of energy supplies persisted or even increased.

At the end of 1991, 31 other countries also had nuclear power plants in commercial operation or under construction. That is an impressive worldwide commitment to nuclear power technology.

During the 1990s, the United States faces several major energy issues and has developed several major goals for nuclear power, which are:

- To maintain exacting safety and design standards
- To reduce economic risk
- To reduce regulatory risk
- To establish an effective high-level nuclear waste disposal program

Several of these nuclear power goals were addressed in the Energy Policy Act of 1992, which was signed into law in October of that year.

The United States is working to achieve these goals in a number of ways. For instance, the US Department of Energy has undertaken a number of joint efforts with the nuclear industry to develop the next generation of nuclear power plants. These plants are being designed to be safer and more efficient.

There is also an effort under way to make nuclear plants easier to build by standardizing the design and simplifying the licensing requirements, without lessening safety standards.

In the area of waste management, engineers are developing new methods and places to store the radioactive waste produced by nuclear plants and other nuclear

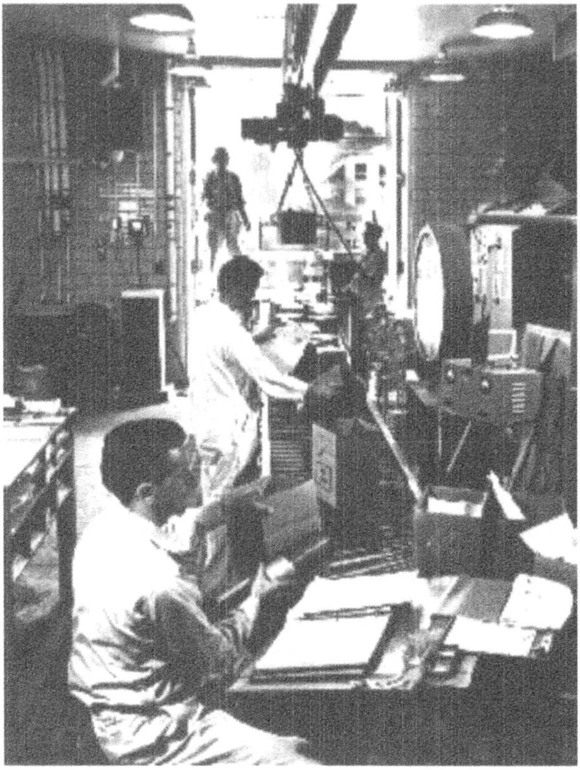

Fig. 2.7 In Oak Ridge, Tennessee, workers package isotopes

processes. Their goal is to keep the waste away from the environment and people for very long periods of time.

Scientists are also studying the power of nuclear fusion. Fusion occurs when atoms join – or fuse – rather than split. Fusion is the energy that powers the sun. On Earth, the most promising fusion fuel is deuterium, a form of hydrogen. It comes from water and is plentiful. It is also likely to create less radioactive waste than fission. However, scientists are still unable to produce useful amounts of power from fusion and are continuing their research. See Fig. 2.7, where in Oak Ridge, Tennessee, workers package isotopes, which are commonly used in science, industry, and medicine.

In the current era of privatization and competitive markets, government still has an essential role in energy, electricity, and nuclear energy. While, in some countries, it may not exercise as much direct control through ownership and economic regulation as in the past, it still has the basic responsibility for creating policy frameworks within which market forces can function and public policy goals can be achieved. Thus, with fewer direct instruments, governments will need alternative policy measures.

2.5 Going Forward with Nuclear Energy

Research in other nuclear areas is also continuing in the 1990s. Nuclear technology plays an important role in medicine, industry, science, food and agriculture, as well as power generation. For example, doctors use radioisotopes to identify and investigate the causes of disease. They also use them to enhance traditional medical treatments. In industry, radioisotopes are used for measuring microscopic thicknesses, detecting irregularities in metal casings, and testing welds. Archaeologists use nuclear techniques to date prehistoric objects accurately and to locate structural defects in statues and buildings. Nuclear irradiation is used in preserving food. It causes less vitamin loss than canning, freezing, or drying.

Nuclear research has benefited mankind in many ways. But today, the nuclear industry faces huge, very complex issues. How can we minimize the risk? What do we do with the waste? The future will depend on advanced engineering, scientific research, and the involvement of an enlightened citizenry.

In March 2013, famous climate scientist James Hansen co-publishes a paper from NASA computing that even with worst case estimates of nuclear accidents and nuclear waste, nuclear energy as a whole has saved 1.8 million lives and counting by offsetting the air pollution-related deaths that come from fossil fuel plants.

September 2013, Voyager 1 enters interstellar space, 36 years after its launch. It is powered by a plutonium-238 radio isotopic thermal generator. Figure 2.8 is an illustration of NASA Voyager at age 40, still reaching for the stars.

The Voyager 1 and now Voyager 2 spacecraft explored Jupiter, Saturn, Uranus, and Neptune before starting their journey toward interstellar space.

2.6 Small Modular Reactors

In Chap. 1 of this book, pretty much covered are different generations of nuclear power reactors, including the ones that are in operation in the United States at present time under Generation III that is known as GEN-III configuration, as well as conceptual design of six types of Generation IV, which we know them as GEN-IV, where some of the conceptual design of small modular reactor configuration falls in these types of reactors. These generations are considered as the new generation of nuclear power plants, and they are also called International Forum (GIF).

Small- and medium-sized or modular reactors are an option to fulfil the need for flexible power generation for a wider range of users and applications. Small modular reactors, deployable either as single- or multi-module plant, offer the possibility to combine nuclear with alternative energy sources, including renewables, and they are flexible and affordable power generation, and most importantly they are very safe to operate (see next section of this chapter), while their efficiencies of their thermal output are under investigation by this author [4] as well as other scientists (Zohuri et al. [5]).

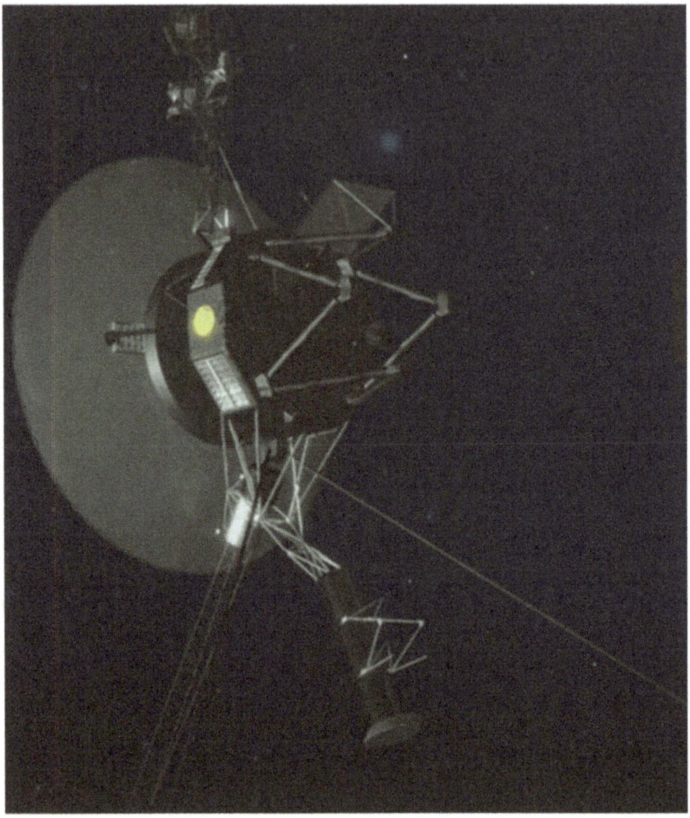

Fig. 2.8 Voyager I image

Global interest in small- and medium-sized modular reactors has been increasing due to their ability to meet the need for flexible power generation for a wider range of users and applications and replace aging fossil fuel-fired power plants. They also display an enhanced safety performance through inherent and passive safety features, offer better upfront capital cost affordability, and are suitable for cogeneration and non-electric applications. In addition, they offer options for remote regions with less developed infrastructures and the possibility for synergetic hybrid energy systems that combine nuclear and alternate energy sources, including renewables.

Many member states are focusing on the development of small modular reactors, which are defined as advanced reactors that produce electricity of up to 300 MW(e) per module. These reactors have advanced engineered features, are deployable either as a single- or multi-module plant, and are designed to be built in factories and shipped to utilities for installation as demand arises.

There are about 50 SMR designs and concepts globally. Most of them are in various developmental stages, and some are claimed as being near-term deployable. There are currently four SMRs in advanced stages of construction in Argentina,

China, and Russia, and several existing and newcomer nuclear energy countries are conducting SMR research and development.

Dealing with the new generation of these SMRs, the IAEA is coordinating the efforts of its member states to develop SMRs of various types by taking a systematic approach to the identification and development of key enabling technologies, with the goal to achieve competitiveness and reliable performance of such reactors. The Agency also helps them address common infrastructure issues that could facilitate the SMRs' deployment.

2.7 Small Modular Reactors: Safety, Security, and Cost Concerns

Aftermath of the major accidents at Three Mile Island in 1979 and Chernobyl in 1986 and then, recent devastated Japan's Fukushima nuclear power plant frailer in Japan in March of 2011, pretty much nuclear power fell out of favor, and some countries applied the brakes to their nuclear programs. Concerns about climate change and air pollution, as well as growing demand for electricity, led many governments to reconsider their aversion to nuclear power, which emits little carbon dioxide and had built up an impressive safety and reliability record. Some countries reversed their phaseouts of nuclear power, some extended the lifetimes of existing reactors, and many developed plans for new ones.

Despite all these given concerns and issues in respect to the nuclear energy, still we are facing the fact of why we still need nuclear power as clean source of energy, particularly when we deal with renewable source of energy arguments [6].

Today, roughly 60 nuclear plants are under construction worldwide, which will add about 60,000 megawatts of generating capacity – equivalent to a sixth of the world's current nuclear power capacity; however, this movement has been lost after March of 2001 and Japan's Fukushima nuclear power episode.

Nuclear power's track record of providing clean and reliable electricity compares favorably with other energy sources. Low natural gas prices, mostly the result of newly accessible shale gas, have brightened the prospects that efficient gas-burning power plants could cut emissions of carbon dioxide and other pollutants relatively quickly by displacing old, inefficient coal plants, but the historical volatility of natural gas prices has made utility companies wary of putting all their eggs in that basket. Besides, in the long run, burning natural gas would still release too much carbon dioxide. Wind and solar power are becoming increasingly widespread, but their intermittent and variable supply makes them poorly suited for large-scale use in the absence of an affordable way to store electricity. Hydropower, meanwhile, has very limited prospects for expansion in the United States because of environmental concerns and the small number of potential sites [7].

Part of any nuclear power plant safety that one should consider as part of design and operation of such source of energy is the reactor stability. Understanding time-dependent behaviors of nuclear reactors and the methods of their control is essential to the operation and safety of nuclear power plants. This chapter provides researchers and engineers in nuclear engineering very general yet comprehensive information on the fundamental theory of nuclear reactor kinetics and control and the state-of-the-art practice in actual plants, as well as the idea of how to bridge the two. The dynamics and stability of engineering equipment that affects their economical and operation from safety and reliable operation point of view. In this chapter, we will talk about the existing knowledge that is today's practice for design of reactor power plants and their stabilities as well as available techniques to designers, although stable power processes are never guaranteed. An assortment of unstable behaviors wrecks power apparatus, including mechanical vibration, malfunctioning control apparatus, unstable fluid flow, unstable boiling of liquids, or combinations thereof. Failures and weaknesses of safety management systems are the underlying causes of most accidents [8].

The safety and capital cost challenges involved with traditional nuclear power plants may be considerable, but a new class of reactors in the development stage holds promise for addressing them. These reactors, called small modular reactors (SMRs), produce anywhere from 10 to 300 megawatts, rather than the 1000 megawatts produced by a typical reactor. An entire reactor, or at least most of it, can be built in a factory and shipped to a site for assembly, where several reactors can be installed together to compose a larger nuclear power station. SMRs have attractive safety features, too. Their design often incorporates natural cooling features that can continue to function in the absence of external power, and the underground placement of the reactors and the spent-fuel storage pools are more secure.

As concisely stated by the IAEA in TECDOC-1524, reproduced in the bullet points shown below, small modular reactors (SMRs) are highlighted as a viable alternative to the NPPs that have been used as desalination plant energy sources for the following reasons:

- SMRs have lower investment costs.
- Almost all SMR concepts appear to show increased availability ($90%).
- Because of inherent safety features, most SMRs have good potential for location near population centers, hence lowering the transport costs.

Since small modular reactors (SMRs) are smaller than conventional nuclear plants, the construction costs for individual projects are more manageable, and thus the financing terms may be more favorable. And because they are factory assembled, the on-site construction time is shorter. The utility company can build up its nuclear power capacity step by step, adding additional reactors as needed, which means that it can generate revenue from electricity sales sooner. This helps not only the plant owner but also customers, who are increasingly being asked to pay higher rates today to fund tomorrow's plants [7].

Generation IV reactors all have a combination of novel features, such as modular construction, rail transportability, and seismic isolation techniques that are not present in current Generation II reactor designs and started to first appear on some Generation III and III+ reactor designs.

The trend to modularization was adopted by the Westinghouse AP1000, in which factory-prefabricated modules, consisting of steel plates and additional structural elements such as tie rods and stiffening elements, are delivered to the site and assembled into larger, crane-liftable modules, which are then installed and filled with concrete. This type of construction technique is different to the traditional stick build, in which all construction works are performed on-site from the ground up.

The purpose of modular construction is to cut down on construction time and cost by transferring part of the construction process to a factory setting in which highly repetitive tasks can be automated or performed in a controlled environment. By the time the order book reaches "nth-of-a-kind," where tooling and supply chains are in place, quality issues are ironed out, and vendors and contractors are far along the assembly and construction learning curve. Modular construction thus lends itself better to SMRs.

2.7.1 Safety Concepts of the MSR

In the current light-water reactor (LWR) licensing, there are two major guidelines for safety. One is used as a guide for safety design of the reactor at a conceptual design stage, which is called "General Design Criteria (GDC)." And another is used for accident analysis, which defines the events to be studied in licensing and provides the criteria of the analyzed results. Both guidelines have not been issued for the MSR. As for the former one, GDC for fluoride salt-cooled high-temperature reactor (FHR) is now being established in the framework of the American Nuclear Society [9].

The SMR has very high safety, because of the following unique features, including practically no possibility of a severe accident (Furukawa et al. [10]).

1. Its primary loop and a secondary loop are operated at a very low pressure (about 0.5 MPa), which essentially eliminates accidents such as system rupture due to high pressure.
2. The molten salt is chemically inert, that is, it does not react violently with air or water and is not flammable. The corrosion of Hastelloy N can be minimized by the appropriate chemical control and maintenance of the molten salt.
3. Pressure increase in a primary loop is incredible because boiling temperature of the fuel salt is very high (about 1400 °C) compared with the operating temperature (about 700 °C).
4. Since there is no water within a containment, there is no possibility of high pressure by steam generation and no possibility of hydrogen explosion at any accidental conditions.
5. The fuel salt is drained to a drain tank through a freeze valve, if required. In case of a rupture in the primary loop, the spilled fuel salt is drained to an emergency drain tank without passing a freeze valve.

6. The fuel salt can keep criticality only where graphite exists in an appropriate fraction. In an accident, the fuel salt, which is drained to a drain tank, cannot cause a re-criticality accident. (As for the MSR without graphite moderator, this drain tank is appropriately designed to prevent re-criticality.)
7. The MSR has a large negative reactivity coefficient of a fuel salt temperature that can suppress an abnormal change of the reactor power. Although a temperature coefficient of graphite is positive, it does not affect the safety, because the heat capacity of graphite is large enough.
8. Since gaseous FPs can be removed by separating from the fuel salt, the danger due to the release of radioactivity from the core at accidental conditions can be minimized.
9. Since fuel composition can be adjusted when necessary, an excess reactivity and a reactivity margin to be compensated by control rods are small. Therefore, reactivity requirements for control rods are also small.
10. The delayed neutron fraction of 233 U is lower than that of 235 U, and some delayed neutrons are generated outside the core. However, safe control of the reactor is possible because of a large negative reactivity coefficient with fuel salt temperature and small reactivity insertion.
11. Since there is no airflow and no heat source within a core when fuel salt is drained at an accidental condition, graphite fire does not occur.

With the US federal budget under tremendous pressure, it is hard to imagine taxpayers' funding demonstrations of a new nuclear technology. But if the United States takes a hiatus from creating new clean energy options – be it SMRs, renewable energy, advanced batteries, or carbon capture and sequestration – Americans will look back in 10 years with regret. There will be fewer economically viable options for meeting the United States' energy and environmental needs, and the country will be less competitive in the global technology market.

Small modular reactors are unlikely to solve the economic and safety problems faced by nuclear power. According to the US Department of Energy (DOE) and some members of the nuclear industry, the next big thing in nuclear energy will be a small thing: the "small modular reactor" (SMR).

SMRs – "small" because they generate a maximum of about 30% as much power as typical current reactors and "modular" because they can be assembled in factories and shipped to power plant sites – have been getting a lot of positive attention recently, as the nuclear power industry has struggled to remain economically viable in an era of flat demand and increasing competition from natural gas and other energy alternatives.

SMRs have been touted as both safer and more cost-effective than older, larger nuclear reactor designs. Proponents have even suggested that SMRs are so safe that some current Nuclear Regulatory Commission (NRC) regulations can be relaxed for them, arguing that they need fewer operators and safety officers, less robust containment structures, and less elaborate evacuation plans. Are these claims justified?

2.7.2 Economies of Scale and Catch

SMR-based power plants can be built with a smaller capital investment than plants based on larger reactors. Proponents suggest that this will remove financial barriers that have slowed the growth of nuclear power in the recent years.

However, there's a catch: "affordable" doesn't necessarily mean "cost-effective." Economies of scale dictate that, all other things being equal, larger reactors will generate cheaper power. SMR proponents suggest that mass production of modular reactors could offset economies of scale, but a 2011 study concluded that SMRs would still be more expensive than current reactors.

Even if SMRs could eventually be more cost-effective than larger reactors due to mass production, this advantage will only come into play when many SMRs are in operation. But utilities are unlikely to invest in SMRs until they can produce competitively priced electric power. This Catch-22 has led some observers to conclude that the technology will require significant government financial help to get off the ground.

2.7.3 Are Small Modular Reactors Safer?

One of the chief selling points for SMRs is that they are supposed to be safer than current reactor designs. However, their safety advantages are not as straightforward as some proponents suggest.

- SMRs use passive cooling systems that do not depend on the availability of electric power. This would be a genuine advantage under many accident scenarios, but not all. Passive systems are not infallible, and credible designs should include reliable active backup cooling systems. But this would add to cost.
- SMRs feature smaller, less robust containment systems than current reactors. This can have negative safety consequences, including a greater probability of damage from hydrogen explosions. SMR designs include measures to prevent hydrogen from reaching explosive concentrations, but they are not as reliable as a more robust containment – which, again, would add to cost.
- Some proponents have suggested siting SMRs underground as a safety measure. However, underground siting is a double-edged sword – it reduces risk in some situations (such as earthquake) and increases it in others (such as flooding). It can also make emergency intervention more difficult. And it too increases cost.
- Proponents also point out that smaller reactors are inherently less dangerous than larger ones. While this is true, it is misleading, because small reactors generate less power than large ones, and therefore, more of them are required to meet the same energy needs. Multiple SMRs may actually present a higher risk than a single large reactor, especially if plant owners try to cut costs by reducing support staff or safety equipment per reactor.

2.7.4 Shrinking Evacuation Zones

Because of SMRs' alleged safety advantages, proponents have called for shrinking the size of the emergency planning zone (EPZ) surrounding an SMR plant from the current standard of 10 miles to as little as 1000 feet, making it easier to site the plants near population centers and in convenient locations such as former coal plants and military bases.

However, the lessons of Fukushima, in which radiation levels high enough to trigger evacuation or long-term settlement were measured at as much as 20 to 30 miles from the accident, suggest that these proposals, which are based on assumptions and models that have yet to be tested in practice, may be overoptimistic.

2.7.5 Safety Conclusions of Nuclear Power Plants

- Unless a number of optimistic assumptions are realized, SMRs are not likely to be a viable solution to the economic and safety problems faced by nuclear power.
- While some SMR proponents are worried that the United States is lagging in the creation of an SMR export market, cutting corners on safety is a shortsighted strategy.
- Since safety and security improvements are critical to establishing the viability of nuclear power as an energy source for the future, the nuclear industry and the DOE should focus on developing safer reactor designs rather than weakening regulations.
- Congress should direct the DOE to spend taxpayer money only on support of technologies that have the potential to provide significantly greater levels of safety and security than currently operating reactors.
- The DOE should not be promoting the idea that SMRs do not require 10-mile emergency planning nor should it be encouraging the NRC to weaken its other requirements just to facilitate SMR licensing and deployment.

As part of discussion of Reactor Safety Study (RSS) and responses for such important factor in respect to design of Generation IV rectors for future usage of producing electricity to meet the of such energy, we need to have some basic understanding of how fissionable nuclear reactors are fundamentally working. One of the fundamental issues that we need to pay attention to, when it comes to operate fissionable nuclear power, is physics of reactor kinetics.

As part of reactor kinetics, we should look at the fission reactivity due to neutron interaction with uranium (^{235}U) or plutonium (^{239}P) as example of fuel, and that is if the multiplication exceeds unity by more than a small amount, the reactor power will build up at a rapid rate of interaction. For constant power production, the effective multiplication factor k_{eff} must be kept at unity. Thus, a key quantity is the difference $k_{eff} - 1$. This quantity is usually expressed in terms of the *reactivity* ρ, where by definition is given as

$$\rho = \frac{k_{\text{eff}} - 1}{k_{\text{eff}}} \tag{2.1}$$

From Eq. 2.1, we can conclude that

$$k_{\text{eff}} = \frac{1}{1 - \rho} \tag{2.2}$$

Off course, a runaway chain reaction is one in which k_{eff} rises appreciably above unity, or equivalently, the reactivity ρ is appreciably greater than zero. A major aspect of reactor safety is avoidance of such an excursion.

To put the matter of effective multiplication factor in different perspectives, which describes all the possible events in the life of a neutron and effectively describes the state of finite multiply system, will be defined as follows:

$$k_{\text{eff}} = \frac{\text{neutron production from fission in one neutron generation}}{\text{neutron absorption and leakage in the preceding neutron generation}} \tag{2.3}$$

in which in Eq. 2.3 the required condition for a *stable, self-sustained fission chain reaction* in a multiplying system (in a nuclear reactor) is that *exactly every fission initiates another fission*. The minimum condition is for each nucleus undergoing fission to produce, on the average, at least one neutron that causes fission of another nucleus. Also, the number of fissions occurring per unit time (the reaction rate) within the system must be constant.

This condition can be expressed conveniently in terms of the multiplication factor. The effective multiplication factor is the ratio of the neutrons produced by fission in one neutron generation to the number of neutrons lost through absorption in the preceding neutron generation. This can be expressed mathematically as shown in Eq. 2.3 in above.

It is obvious *the effective multiplication factor* in a multiplying system is a measure of the change in the fission neutron population from one neutron generation to the subsequent generation.

- $k_{\text{eff}} < 1$. If the multiplication factor for a multiplying system is *less than 1.0*, then the *number of neutrons is decreasing* in time (with the mean generation time), and the chain reaction will never be self-sustaining. This condition is known as *the subcritical state.*
- $k_{\text{eff}} = 1$. If the multiplication factor for a multiplying system is *equal to 1.0*, then there is *no change in neutron population* in time, and the chain reaction will be *self-sustaining*. This condition is known as *the critical state.*
- $k_{\text{eff}} > 1$. If the multiplication factor for a multiplying system is *greater than 1.0*, then the multiplying system produces *more neutrons* than are needed to be self-sustaining. The number of neutrons is exponentially increasing in time (with the mean generation time). This condition is known as *the supercritical state.*

In the spite of possible misapprehension, it is worth nothing noting that a bomb-like nuclear explosion cannot occur in a nuclear reactor. In a bomb, a critical mass of almost pure fissile material (^{235}U or ^{239}Pu) is brought together violently and compressed by the force of a chemical explosion, and the chain reaction develops fully within one-millionth of a second – quickly enough for much of the fuel to fission before the mass is disassembled – considering the critical mass for nuclear weapons with and without reflectors [25].

In this situation, a spherical form of fissionable material will have a mass that is less than the critical mass if its radius is small compared to the mean-free path for fission, λ. For 1-MeV neutrons in ^{235}U, $\lambda = (M/\rho N_A \sigma_f) = 17$ cm for numerical values of the fission cross section σ_f and the uranium density ρ, there exist tables in public domain that can be looked up [25].

Although λ sets a crude scale for the dimensions, it does not alone determine the critical radius R_c. For a reasonable estimate of R_c, it is necessary to treat the geometry in greater detail and take into account two crucial nuclear parameters, which are beyond the scope of this book for the time being.

To continue with our safety discussion here, we state that the presence of the non-fissile material has two consequences that are pertinent to the issue of explosions:

1. The multiplication factor k_{eff} in a reactor is close to unity, whereas in a bomb, it approaches 2.
2. The average time between fission generations (the mean neutron lifetime, l, can be expressed as the sum of terms for the prompt and delayed components as $l = (1 - \beta)l_p + \beta\tau_e$, where l_p and τ_e are mean lifetimes for prompt and delayed neutrons, respectively, and β is the delayed neutron fraction and ^{235}U is about 0.65%) is greater in a reactor than in a bomb, because the most frequent neutron reactions in a reactor are elastic or inelastic scattering, not fission. Note that the fission cross section for neutrons colliding with ^{238}U is small for neutron energies below 2 MeV and is negligible below 1 MeV.

As a result, the chain reaction builds up much more slowly in a reactor than in a bomb. Overall, the first "line of defense" against an explosion in a reactor is the negative feedback that prevents criticality accidents. This should suffice. However, if there are mistakes in the design or operation of the reactor and the chain reaction reaches too high a power level, there is time for the ultimate "negative feedback" to come into play – the partial disassembly of the reactor core, which stops the chain reaction after only a relatively small amount of energy has been produced (i.e., only a small fraction of the nuclei have fissioned). This is what happened in the Chernobyl accident, where most of the energy of the explosion came from chemical reactions, including steam interacting with hot metal. Such an accident can be very serious, but the consequences are not on the scale of the consequences of a nuclear explosion.

As part of general requirements for means of achieving reactor safety, underlying the approach to safety, for any sort of equipment, are high standards in design, construction, and the reliability of components. In nuclear reactors, concern about possible accidents has led to particularly intense efforts to achieve high standards. Individual components of the reactor and associated equipment must be of a codified high quality. As described in an OECD report,

> In the early years of water reactor development in the USA, a tremendous effort was put into development of very detailed codes and standards for nuclear plants, and these were widely adopted by other countries where nuclear plants were initially built under US licenses. [3, p. 62]

The efforts of the United States have since been supplemented by parallel efforts by other countries and the International Atomic Energy Agency (IAEA). In parallel, a nuclear reactor safety philosophy has developed which includes a number of special features.

In the summary, we need to mention that the results of the RSS included estimates of the probability distributions for a variety of forms of harm: early fatalities, early illness, latent cancer fatalities, thyroid nodules, genetic effects, property damage, and magnitude of the area in which relocation and decontamination would be required. These results were presented in the form of graphs of the probability of occurrence as a function of the magnitude of the harm.

To provide perspective, the RSS also compared the risks from reactor accidents to those from other sorts of accidents or natural mishaps. For these other accidents, there are few data on latent effects.

In Fig. 2.9, the annual risks from 100 reactors, as estimated in the RSS, are compared with the annual risks from other causes, such as airplane accidents and dam failures. For example, Fig. 2.9 indicates an average of 1 airplane accident causing 100 or more fatalities every 3 years, whereas a nuclear reactor accident with this early toll was predicted to occur only once every 80,000 years.

Based on the safety arguments in above and the conclusion that we have seen, the question that comes to our mind is why we need nuclear power plant, which is subject of the following section here in this chapter.

2.8 Why We Need Nuclear Power Plants

Generally speaking, originally the nuclear power plants, perceived as a cheap and potential source of peaceful power post-World War II (WW II), and the commercial use of energy driven by nuclear have been controversial for decades. Naturally, environmentalists constantly are worried about the dangers of nuclear power plants, and their radioactive waste poses a threat to nearby communities as time grew, and plant constructions in the United States virtually begin to die after the early 1980 during Carter administration, who was particularly opposing any research around liquid-metal fast-breeder reactor (LMFBR) and their commercial implementations within the United States (i.e., Clinch River Project in Oak Ridge, TN), yet French government moved on to their commercialization of this type of reactor that is known as Phenix II. The disaster at Chernobyl in former Soviet Union (FSU)

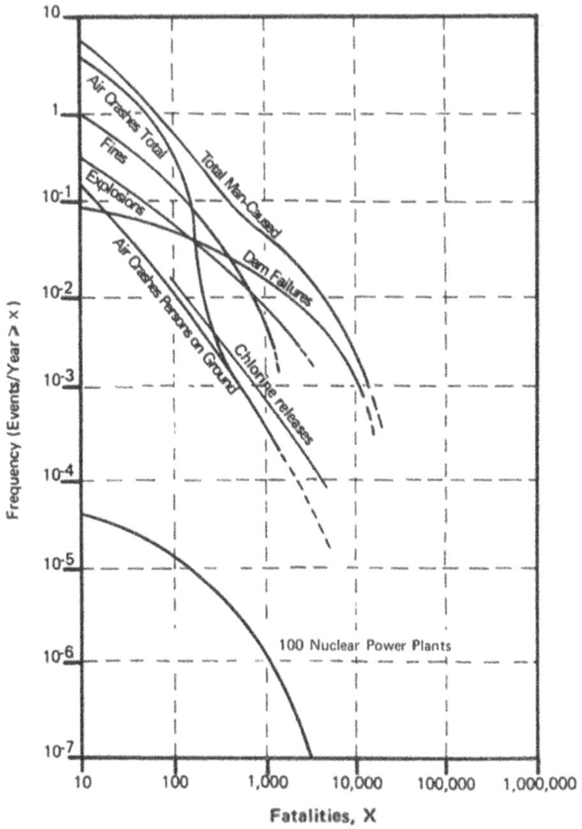

Fig. 2.9 RSS comparison of annual probabilities of accident 100 nuclear reactors and man-caused

followed by Three Mile Island episode in the United States enforced a lot of negative images toward nuclear power plant and their industries involving their design and commercialization. Yet in the decade prior to the Japanese nuclear crises of Fukushima Daiichi of 2011, sentiment about nuclear power underwent a marked change. The alarming acceleration of global warming due to the burning of fossil fuels and concern about dependence of foreign fuel have forced policy-makers, climate scientist, and energy experts to look once again at nuclear power as ultimate source of energy and encouraging the nuclear engineers and scientist to look into new generation of these plants with better life cycle and safety, while they are more efficient and cost-effective for their operating owners to build [4].

Such enforcement also building momentum based on the population growth globally and consequently increases on demand for the electricity that let us start thinking as a way of finding a new pat to renewable source of energy. The major growth in the electricity production industry in the last 30 years has centered on the expansion of natural gas power plants based on gas turbine cycles. The most popular extension of the simple Brayton gas turbine has been the combined cycle power plant with the Air-Brayton cycle serving as the topping cycle and the Steam-Rankine cycle serving as the bottoming cycle for new generation of nuclear power plants that are known as GEN-IV.

The Air-Brayton cycle is an open-air cycle, and the Steam-Rankine cycle is a closed cycle. The Air-Brayton cycle for a natural gas-driven power plant must be an open cycle, where the air is drawn in from the environment and exhausted with the products of combustion to the environment. This technique is suggested as an innovative approach to GEN-IV nuclear power plants in the form and type of small modular reactors (SMRs). The hot exhaust from the Air-Brayton cycle passes through a Heat Recovery Steam Generator (HSRG) prior to exhausting to the environment in a combined cycle. The HRSG serves the same purpose as a boiler for the conventional Steam-Rankine cycle [4].

In 2007 gas turbine-combined cycle plants had a total capacity of 800 GW and represented 20% of the installed capacity worldwide. They have far exceeded the installed capacity of nuclear plants, though in the late 1990s, they had less than 5% of the installed capacity worldwide. There are number of reasons for this. First, natural gas is abundant and cheap. Second, combined cycle plants achieve the greatest efficiency of any thermal plant. And third, they require the least amount of waste heat/cooling water of any thermal plant.

A typical gas turbine plant consists of a compressor, combustion chamber, turbine, and an electrical generator. A combined cycle plant takes the exhaust from the turbine and runs it through a Heat Recovery Steam Generator (HRSG) before exhausting to the local environment. The HRSG serves the function of the boiler for a typical closed-cycle steam plant. The steam plant consists of a steam turbine, a condenser, a water pump, an evaporator (boiler), and an electrical generator. In a combined cycle plant, the gas turbine and steam turbine can be on the same shaft to eliminate the need for one of the electrical generators. However, the two-shafts two-generator systems provide a great deal more flexibility at a slightly higher cost. In addition to the closed loop for the steam, an open loop circulating water system is required to extract the waste heat from the condenser.

The waste heat extracted by this "circulating" water system is significantly less per megawatt for a combined cycle system as the open Brayton cycle exhausts its waste heat directly to the air.

The layout for the components of a typical combined cycle power plant is given below in Fig. 2.10.

GE currently markets a system that will produce 61% efficiency at design power better than 60% efficiency down to 87% of design power [11] for gas turbine-combined cycle plants.

An approximate efficiency can be calculated for a combined cycle power plant by the following simple argument [12].

$$\text{Brayton cycle efficiency} = \frac{W_B}{Q_{\text{in}}} = \eta_B$$

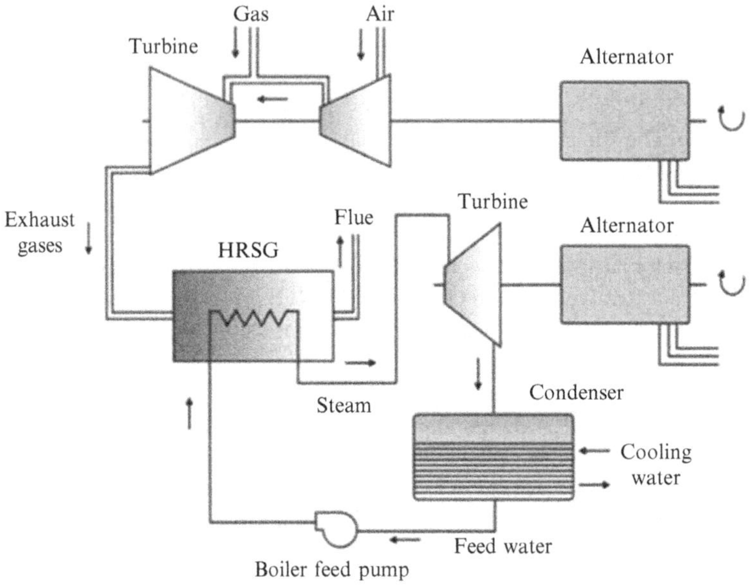

Fig. 2.10 Typical gas turbine-combined cycle power plant

$$\text{Heat to Rankine cycle} = Q_R = (1-\eta_B)Q_{\text{in}}$$

$$\text{Rankine cycle efficiency} = \frac{W_R}{Q_R} = \eta_R$$

$$\text{Overall efficiency} = \frac{W_B + W_R}{Q_{\text{in}}} = \eta_T = \frac{\eta_B Q_{\text{in}} + \eta_R Q_R}{Q_{\text{in}}}$$

$$= \frac{\eta_B Q_{\text{in}} + \eta_R (1-\eta_B) Q_{\text{in}}}{Q_{\text{in}}} = \eta_B + \eta_R - \eta_B \eta_R$$

$$\eta_T = \eta_B + \eta_R - \eta_B \eta_R$$

This efficiency must be corrected for pressure losses and assumes that all the heat in the Brayton exhaust is used in the Heat Recovery Steam Generator (HSRG). For a combustion gas turbine, this is not usually possible if condensation of the water in the exhaust products is to be avoided. The detailed models developed in this effort give a more accurate answer.

For the nuclear reactor system, the heat transfer is in the opposite direction. All reactor components and fluids in the primary and secondary loops must be at a higher temperature than the peak temperature of the gas exiting the heat exchanger. This severely restricts the peak temperature that can be achieved for the air entering the turbine. However, all is not lost.

In a typical combustion system, there are pressure losses approaching 5% of the total pressure to complete the combustion process [13]. Heat exchangers can be built with significantly lower-pressure drops than 5% approaching 1% [14]. Therefore, the most straightforward method to overcome this severe temperature limitation is to borrow a technique from steam power plants and implement multiple reheat cycles. That is, the first heat exchanger heats the air to its peak temperature. Then the air is expanded through the first turbine. The air is then reheated to the same peak temperature and expanded through the second turbine. Based on the relative pressure losses that appear possible, up to five turbines might be considered. All five turbines will be driving the same compressor. Multiple compressors on concentric shafts driven by different sets of turbines might be possible, but that has not been considered here.

For a nuclear system to take advantage of combined cycle technology, there are many numbers of changes to the plant components that must be made. The most significant of course is that the combustion chamber must be replaced by a heat exchanger in which the working fluid from the nuclear reactor secondary loop is used to heat the air. The normal Brayton cycle is an internal combustion one where the working fluid is heated by the combustion of the fuel with the air in the combustion chamber. The walls of the combustion chamber can be cooled, and peak temperatures in the working fluid can be significantly above the temperature that the walls of the chamber can tolerate for any length of time.

2.8.1 Methodology of Combined Cycle

The approach taken in the combined cycle (CC) code developed for this effort is to model the thermodynamics of the components making up the power conversion systems as real components with non-ideal efficiencies. Pressure drops are included for every component except the connected piping. The compressor design is modeled with a small stage polytropic efficiency to take into account state-of-the-art designs. The gas turbines are likewise modeled with a polytropic efficiency. The steam turbines are modeled with a simple overall thermal efficiency. Pressure drops in each of the heat exchangers are included. The input files specify the pressure drops, and the heat exchangers are designed to meet these specifications if possible [15].

Some scientists are calling the nuclear power plants' source of energy as 100% renewable energy, and of course environmentalists arguably are saying that is wrong approach, just because in the core of these plants, there exists uranium or plutonium as fuel when we are talking about fission-type nuclear power plants that exist in grid today and produce electricity to the net. However, on the other side of the spectrum, researchers and scientist at national laboratories and universities around the globe that are working toward fusion program to achieve a breakeven passionately argue that nuclear power plants of fusion type are totally clean so long as the source of energy comes in the form of two hydrogen isotopes such as deuterium (D) and tritium (T) as sources of fusion reaction and driving energy from it.

This is a dream that is too far away from reality of today's need and demand for electricity yet is not out of scope of the near future. Physics of plasma for driving energy via inertial confinement fusion (ICF) [16] or magnetic confinement fusion (MCF) [17] are in agreement with such innovative approaches.

2.8.2 Why We Still Need Nuclear Power

"Nuclear power's track record of providing clean and reliable electricity compares favorably with other energy sources. Low natural gas prices, mostly the result of newly accessible shale gas, have brightened the prospects that efficient gas-burning power plants could cut emissions of carbon dioxide and other pollutants relatively quickly by displacing old, inefficient coal plants, but the historical volatility of natural gas prices has made utility companies wary of putting all their eggs in that basket. Besides, in the long run, burning natural gas would still release too much carbon dioxide. Wind and solar power are becoming increasingly widespread, but their intermittent and variable supply makes them poorly suited for large-scale use in the absence of an affordable way to store electricity. Hydropower, meanwhile, has very limited prospects for expansion in the United States because of environmental concerns and the small number of potential sites."

"The United States must take a number of decisions to maintain and advance the option of nuclear energy. The NRC's initial reaction to the safety lessons of Fukushima must be translated into action; the public needs to be convinced that nuclear power is safe. Washington should stick to its plan of offering limited assistance for building several new nuclear reactors in this decade, sharing the lessons learned across the industry. It should step up its support for new technology, such as SMRs and advanced computer modeling tools. And when it comes to waste management, the government needs to overhaul the current system and get serious about long-term storage. Local concerns about nuclear waste facilities are not going to magically disappear; they need to be addressed with a more adaptive, collaborative, and transparent waste program."

These are not easy steps, and none of them will happen overnight. But each is needed to reduce uncertainty for the public, the energy companies, and the investors. A more productive approach to developing nuclear power – and confronting the mounting risks of climate change – is long overdue. Further delay will only raise the stakes.

2.8.3 Is Nuclear Energy Renewable Source of Energy

Assuming for the time being we are taking fission reaction as foundation for present (GEN-III) and future (GEN-IV) nuclear power reactors, as nuclear energy sources to somewhat degree, we can argue it is a clean source of energy.

Although nuclear energy is considered clean energy, its inclusion in the renewable energy list is a subject of major debate. To understand the debate, we need to understand the definition of renewable energy and nuclear energy first. However, until we manage through future technology of these fission reactors to manage to bring down the price electricity per kilowatt hours driven by fusion energy down to the point of those by gas or fossil fuels, there is no chance to push these reactors beyond GEN-III.

However, efforts toward reduction price of electricity driven by nuclear fission power plants, especially using some innovative design of GEN-IV plants with high-temperature baseline in conjunction with some thermodynamics cycles such as Brayton and Rankine, are on the way by so many universities and national laboratories such as Idaho National Laboratory and universities such as MIT, UC Berkeley and University of New Mexico, as well as this author.

Renewable energy is defined as an energy source/fuel type that can regenerate and can replenish itself indefinitely. The five renewable sources used most often are biomass, wind, solar, hydro, and geothermal.

Nuclear energy on the other hand is a result of heat generated through the fission process of atoms. All power plants convert heat into electricity using steam. At nuclear power plants, the heat to make the steam is created when atoms split apart – called fission. The fission releases energy in the form of heat and neutrons. The released neutrons then go on to hit other neutrons and repeat the process, hence generating more heat. In most cases the fuel used for nuclear fission is uranium.

One question we can raise here in order to further understand whether or not we need present nuclear technology as a source of energy is that:

What is the difference between clean energy and renewable energy? Put another way, why is nuclear power in the doghouse when it comes to revamping the nation's energy mix?

The issue has come to the forefront the time during the debate over the Waxman-Markey energy and climate bill and its provisions for a national renewable energy mandate.

To simply put it, Republicans have tried and failed several times to pass amendments that would christen nuclear power as a "low emissions" power source eligible for all the same government incentives and mandates as wind power and solar power.

Many environmental groups are fundamentally opposed to the notion that nuclear power is a renewable form of energy on the grounds that it produces harmful waste byproducts and relies on extractive industries to procure fuel like uranium.

Even so, the nuclear industry and pronuclear officials from countries including France have been trying to brand the technology as renewable on the grounds that it produces little or no greenhouse gases. Branding nuclear energy as renewable could also enable nuclear operators to benefit from some of the same subsidies and friendly policies offered to clean energies like wind, solar, and biomass.

2.8.4 Argument for Nuclear as Renewable Energy

Most supporters of nuclear energy point out the low carbon emission aspect of nuclear energy as its major characteristic to be defined as renewable energy. According to nuclear power opponents, if the goal to build a renewable energy infrastructure is to lower carbon emission, then there is no reason for not including nuclear energy in that list [18]

But one of the most interesting arguments for including nuclear energy in the renewable energy portfolio came from Bernard L. Cohen, former professor at the University of Pittsburg. Professor Cohen defined the term "indefinite" (time span required for an energy source to be sustainable enough to be called renewable energy) in numbers by using the expected relationship between the sun (source of solar energy) and the Earth. According to Professor Cohen, if the uranium deposit could be proved to last as long as the relationship between the Earth and Sun is supposed to last (5 billion years), then nuclear energy should be included in the renewable energy portfolio [19].

In his paper Professor Cohen claims that using breeder reactors (nuclear reactor able to generate more fissile material than it consumes), it is possible to fuel the Earth with nuclear energy indefinitely. Although the amount of uranium deposit available could only supply nuclear energy for about 1000 years, Professor Cohen believes actual amount of uranium deposit available is way more than what is considered extractable right now. In his arguments he includes uranium that could be extracted at a higher cost, uranium from the seawater, and also uranium from eroding Earth crust by river water. All of those possible uranium resources if used in a breeder reactor would be enough to fuel the Earth for another 5 billion years and hence renders nuclear energy as renewable energy.

2.8.5 Argument Against Nuclear Energy as Renewable Energy

One of the biggest arguments against including nuclear energy in the list of renewables is the fact that uranium deposit on Earth is finite, unlike solar and wind. To be counted as renewable, the energy source (fuel) should be sustainable for an indefinite period of time, according to the definition of renewable energy.

Another major argument proposed by the opponents including nuclear energy as renewable energy is the harmful nuclear waste from nuclear power reactors. The nuclear waste is considered as a radioactive pollutant that goes against the notion of a renewable energy source. Yucca Mountain is one of the examples used quite often to prove this point. Most of the opponents in the United States also point at the fact that while most renewable energy source could render the US energy independent, uranium would still keep the country energy dependent as the United States would still have to import uranium.

2.8.6 Today's Safety of Nuclear Power Plant

Aftermath of the major accidents at Three Mile Island in 1979 and Chernobyl in 1986 and then, recent devastated Japan's Fukushima nuclear power plant frailer in Japan in March of 2011, pretty much nuclear power fell out of favor, and some countries applied the brakes to their nuclear programs. Concerns about climate change and air pollution, as well as growing demand for electricity, led many governments to reconsider their aversion to nuclear power, which emits little carbon dioxide and had built up an impressive safety and reliability record. Some countries reversed their phaseouts of nuclear power, some extended the lifetimes of existing reactors, and many developed plans for new ones.

Despite all these given concerns and issues in respect to the nuclear energy, still we are facing the fact of why we still need nuclear power as clean source of energy, particularly when we deal with renewable source of energy arguments [20].

Today, roughly 60 nuclear plants are under construction worldwide, which will add about 60,000 megawatts of generating capacity – equivalent to a sixth of the world's current nuclear power capacity; however, this movement has been lost after March of 2001 and Japan's Fukushima nuclear power episode.

Nuclear power's track record of providing clean and reliable electricity compares favorably with other energy sources. Low natural gas prices, mostly the result of newly accessible shale gas, have brightened the prospects that efficient gas-burning power plants could cut emissions of carbon dioxide and other pollutants relatively quickly by displacing old, inefficient coal plants, but the historical volatility of natural gas prices has made utility companies wary of putting all their eggs in that basket. Besides, in the long run, burning natural gas would still release too much carbon dioxide. Wind and solar power are becoming increasingly widespread, but their intermittent and variable supply makes them poorly suited for large-scale use in the absence of an affordable way to store electricity. Hydropower, meanwhile, has very limited prospects for expansion in the United States because of environmental concerns and the small number of potential sites [20].

Part of any nuclear power plant safety that one should consider as part of design and operation of such source of energy is the reactor stability. Understanding time-dependent behaviors of nuclear reactors and the methods of their control is essential to the operation and safety of nuclear power plants. This chapter provides researchers and engineers in nuclear engineering very general yet comprehensive information on the fundamental theory of nuclear reactor kinetics and control and the state-of-the-art practice in actual plants, as well as the idea of how to bridge the two. The dynamics and stability of engineering equipment that affects their economical and operation from safety and reliable operation point of view. In this chapter, we will talk about the existing knowledge that is today's practice for design of reactor power plants and their stabilities as well as available techniques to designers, although stable power processes are never guaranteed. An assortment of unstable behaviors wrecks power apparatus, including mechanical vibration, malfunctioning control apparatus, unstable fluid flow, unstable boiling of liquids, or combinations thereof. Failures and weaknesses of safety management systems are the underlying causes of most accidents [21].

The safety and capital cost challenges involved with traditional nuclear power plants may be considerable, but a new class of reactors in the development stage holds promise for addressing them. These reactors, called small modular reactors (SMRs), produce anywhere from 10 to 300 megawatts, rather than the 1000 megawatts produced by a typical reactor. An entire reactor, or at least most of it, can be built in a factory and shipped to a site for assembly, where several reactors can be installed together to compose a larger nuclear power station. SMRs have attractive safety features, too. Their design often incorporates natural cooling features that can continue to function in the absence of external power, and the underground placement of the reactors and the spent-fuel storage pools are more secure.

Since small modular reactors (SMRs) are smaller than conventional nuclear plants, the construction costs for individual projects are more manageable, and thus the financing terms may be more favorable. And because they are factory-assembled, the on-site construction time is shorter. The utility company can build up its nuclear power capacity step by step, adding additional reactors as needed, which means that it can generate revenue from electricity sales sooner. This helps not only the plant owner but also customers, who are increasingly being asked to pay higher rates today to fund tomorrow's plants [22]

With the US federal budget under tremendous pressure, it is hard to imagine taxpayers funding demonstrations of a new nuclear technology. But if the United States takes a hiatus from creating new clean energy options – be it SMRs, renewable energy, advanced batteries, or carbon capture and sequestration – Americans will look back in 10 years with regret. There will be fewer economically viable options for meeting the United States' energy and environmental needs, and the country will be less competitive in the global technology market.

2.8.7 Summary

It seems like at the heart of debate lies the confusion over the exact definition of renewable energy and the requirements that need to be met in order to be one. The recent statement by Helene Pelosi, the interim director general of International Renewable Energy Agency (IRENA), saying IRENA will not support nuclear energy programs because it is a long, complicated process and it produces waste and is relatively risky proves that their decision has nothing to do with having a sustainable supply of fuel [23]. And if that's the case, then nuclear proponents would have to figure out a way to deal with the nuclear waste management issue and other political implications of nuclear power before they can ask IRENA to reconsider including nuclear energy in the renewable energy list [24].

One more strong argument against fission nuclear power plants as source of renewable energy comes from Dr. James Singmaster in August 3, 2009, and has been republished here as follows:

> *The basic problem of the climate crisis is the ever-expanding overload of heat energy in the closed biosphere of earth. Temperatures going up indicate the increasing heat energy overload. Everyone reading this should check out Dr. E. Chaisson's article titled* 'Long-Term Global Warming from Energy Usage' *in EOS, Trans. Amer. Geophys. Union, V. 89, No. 28, Pgs. 253-4(2008) to learn that nuclear energy, be it fission or fusion, being developed should be dropped with money put into it being put to developing renewable energy supplies using the sun, wind and hydrogen.*
>
> *The hydrogen needs to be generated from splitting water using sunlight with the best one or two of seven catalysts reported in the last two years. Or with excess solar or wind collection generating electricity, that could be used to generate hydrogen by electrolysis of water.*
>
> *There is no way that nuclear power can avoid releasing trapped energy to increase the energy overload, so it should be forgotten.*
>
> *To remove some of the energy as well as some of the carbon overload in the biosphere, we need to turn to pyrolysis of massive ever-expanding organic waste streams to remake charcoal that will be removing some of both overloads. It will require using renewable energy and the pyrolysis process expels about 50% of the carbon as small organic chemicals that can be collected, refined and used foe fuel that is a renewable one. For more about using pyrolysis, search my name on GreenInc blog or google it for other blog comments on pyrolysis.*
>
> *Dr. j. Singmaster*

References

1. DOE/NE-0088, "The History of Nuclear Energy" U.S. Department of Energy Office of Nuclear Energy, Science and Technology
2. Richard Rhodes, "The Making of the Atomic Bomb," Simon and Schuster, 1986.
3. Alvin Weinberg, "The First Nuclear Era," AIP Press, 1994
4. Zohuri B (2015) Combined Cycle Driven Efficiency for Next Generation Nuclear Power Plants: An Innovative Design Approach
5. Zohuri B, McDaniel P, De Oliveira C (2015) Advanced Nuclear Open-Air Brayton Cycles for Highly Efficient Power Conversion, Nuclear Technology Journal
6. B. Zohuri; Hybrid Energy Systems: Driving Reliable Renewable Sources of Energy Storage 1st ed. 2018 Edition, Springer Publishing Company
7. E. Moniz, http://energy.mit.edu/news/why-we-still-need-nuclear-power/
8. B. Zohuri, Neutronic Analysis For Nuclear Reactor Systems, Nov 3, 2016, Springer Publishing Company
9. Flanagan, 2012; ANS 20.1 working group, 2014
10. Furukawa, K., et al., 2007. Molten salt reactor for sustainable nuclear power MSR FUJI. IAEA-TECDOC-1536 (Status of small reactor designs without on-site refueling), pp. 821856
11. GE Energy Flex Efficiency 50 Combined Cycle Power Plant, e-brochure, 2012
12. Horlock, J. H., Cogeneration-Combined Heat and Power(CHP), Krieger Publishing Company, Malabar FL, 1997
13. Mattingly, J. D. Elements of Gas Turbine Propulsion, McGraw-Hill, Inc., New York, 1996
14. Mattingly, J. D., ibid.

15. Zohuri, B. and McDaniel, P. and De Oliveira, C. Advanced Nuclear Open-Air-Brayton Cycles for Highly Efficient Power Conversion, Nuclear Technology Journal, January 23, 2015
16. Zohuri, B. Inertial Confinement Fusion Driven Thermonuclear Energy Jan 29, 2017, Springer Publishing Company
17. Zohuri, B. Plasma Physics and Controlled Thermonuclear Reactions Driven Fusion Energy Nov 17, 2016, Springer Publishing Company
18. Johnson, K. "Is Nuclear Power Renewable Energy," Wall Street Journal, 21 May 09
19. Cohen, B.L. "Breeder Reactors: A Renewable Energy Source," Am. J. Phys. 51, 75 (1983)
20. Zohuri; B. Hybrid Energy Systems: Driving Reliable Renewable Sources of Energy Storage 1st ed. 2018 Edition, Springer Publishing Company
21. E. Moniz, http://energy.mit.edu/news/why-we-still-need-nuclear-power/
22. Fraas, A. P., Heat Exchanger Design, 2nd Ed., John Wiley & Sons, New York, 1989
23. Kanter, J. "Is Nuclear Power Renewable," New York Times, 3 Aug 09
24. Chowdhury, D. "Is Nuclear Energy Renewable Energy", Stanford Physics Department, March 22, 2012
25. Bodansky, David, "Nuclear Energy, Principles, Practices, and Prospects", 2n edition, Springer Publishing Company, 2004

Chapter 3
Nuclear Energy Research and Development Roadmap

Abstract The US Department of Energy (DOE) as part of its activity prepared a report to Congress in April 2010, based on ongoing nuclear energy research and development roadmap among the national laboratories under auspices of DOE in conjunction with industry and companies involved in such activities. In this report, they best characterized the current prospects of nuclear power in a world that is confronted with a burgeoning demand for energy, higher energy prices, energy supply, a new source of renewable energy, and security concerns.

3.1 Introduction

The concept of the atom has existed for many centuries. But we only recently began to understand the enormous power contained in the tiny mass. The history of our discoveries about atoms. We begin with the ideas of the Greek philosophers. Then we follow the path to the early scientists who discovered radioactivity. Finally, we reach modern-day use of atoms as a valuable source of energy.

Worldwide, there are 435 nuclear power reactors in operation, totaling 367 GW(e) of generation capacity, and out of that number of reactors, 103 of them do exist in the United States that are now in operation under the technology of Generations II and III, by employing light-water reactor (LWR) technologies, which use ordinary water as both a moderator and a coolant.

The next wave of nuclear plants has taken over GEN-II concepts to the next level, by improving both safety and efficiency, since these generations of power plant are at the end of their life cycle. Utilities plan to build GEN-III at the end of the decade, and now conceptual studies of new designs are in horizon that are known as GEN-IV and fall under new roadmap of small modular reactors, with better safety and efficiency in mind as part of research and development of this roadmap.

In a Generation II-type reactor of pressurized water reactor (PWR), water circulates through core (see Chap. 1 of this book), where it is heated by fuel's chain reaction. The hot water is then piped to a steam generator, and the steam spins a turbine that produces electricity. However, the Generation III evolutionary PWR improves upon the Generation II of PWR-type design primarily by enhancing safety feature

B. Zohuri, *Small Modular Reactors as Renewable Energy Sources*,
https://doi.org/10.1007/978-3-319-92594-3_3

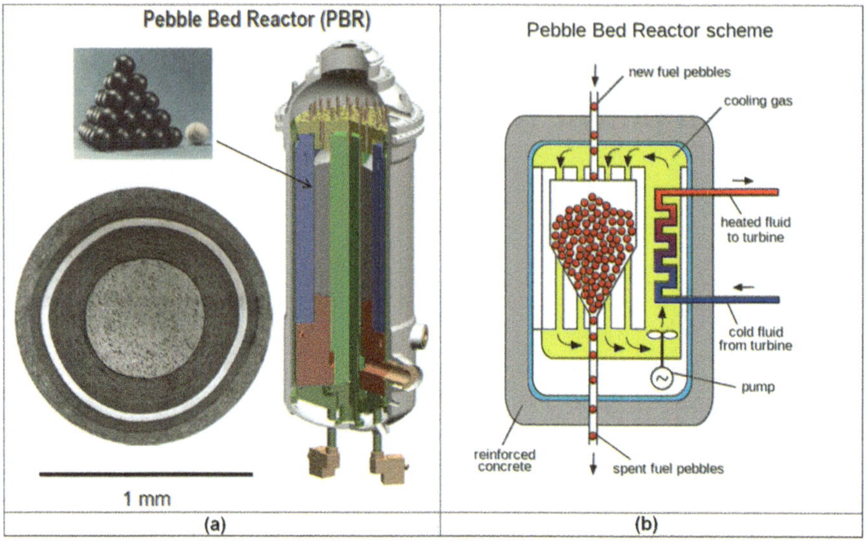

Fig. 3.1 Pebble-powered reactor schematic. (**a**) Pebble Type HTGRs. (**b**) Sketch of a Pebble-Bed Reactor

from viewpoint of probability of risk analysis (PRA) and other mechanical and operational perspectives. Two separate 51-inch-thick concrete walls, in the form of inner one lined up with metal, are each strong enough to withstand the impact of heavy obstacle such as commercial airplane since the aftermath of 9/11 episode and destruction of twin towers in New York.

The reactor vessel sits on a 20 ft slab of concrete with a leak-tight "core catcher," where the molten core would collect and cool in the event of a meltdown. There are also four safeguard buildings with independent pressurizers and steam generators, each capable of providing emergency cooling of the reactor core. One of the new roadmaps for new construction of GEN-III considers the pebble-powered fuel, which is a smooth graphite sphere about the size of tennis ball as it is shown in Fig. 3.1, with assumption that it could take years to assess the pros and cons of all six GEN-IV designs that are mentioned in Chap. 1 of this book. With continuous rise on electricity demand worldwide due to population growth, at least in the United States, the Congress might not wait that long.

In addition to replacing the aging fleet of Generation II reactors that are coming to the end of their life cycles, the government wants to make progress on another front, and that is the production of hydrogen as part of hybrid energy to fuel the dream of exhaust-free cars running independent of gasoline and dependent on foreign oil [1].

As a result, the frontrunner for pebble-bed reactor (PBR) design and production implementation, the initial $1.25 billion demonstration plant, is in progress in Idaho, which is a helium-cooled, graphite-moderated reactor whose extremely high outlet temperature around 1650 °F to 1830 °F, also known as very-high-temperature reactor (VHTR), would be ideal for thermal-efficiency output and also efficiently producing hydrogen [1–5].

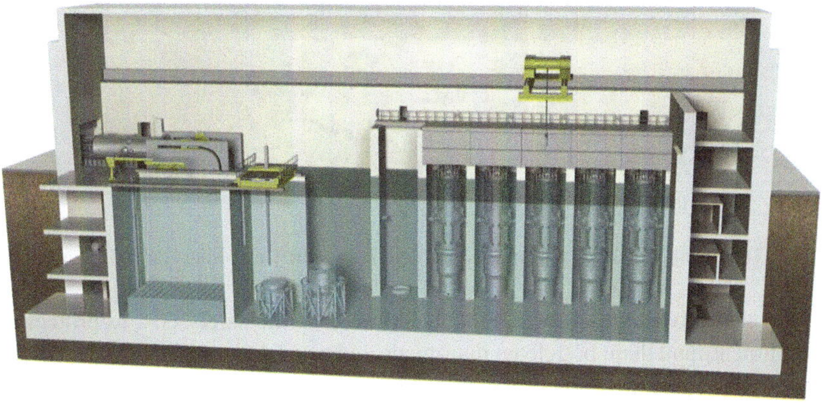

Fig. 3.2 NuScale power reactor building of SMR

However, at the dawn of newer generation of nuclear reactors at smaller footprint and lower cost of the construction and production, small modular reactors (SMRs), as part of Generation IV, have also a roadmap, and that is to achieve energy security and greenhouse gas (GHG) emission reduction objective. The United States must develop and deploy clean, affordable, domestic energy source as quickly as possible, not only to satisfy the issue of GHG but also a new source of renewable energy, in order to meet the demand and supply for electricity going forward in time and changes that new generations are facing.

Nuclear power is continuously a key player of this demand and will continue to be a key component of a portfolio of technologies that meets our energy goals. Thus, the Nuclear Energy (NE) branch of the Department of Energy (DOE) continuously is looking into research, development, and demonstration activities that will ensure that, through small- and medium-sized Generation III or modular reactors of Generation IV, nuclear energy remains a viable energy option for at least the United States, and the International Atomic Energy Agency (IAEA) is in support of worldwide activities within this domain.

Small- and medium-sized or modular reactors are an option to fulfill the need for flexible power generation for a wider range of users and applications. Small modular reactors, deployable either as single or multi-module plant, offer the possibility to combine nuclear with alternative energy sources, including renewables. Small modular reactors are an excellent candidate for meeting demands for need of electricity due to population growth in time, and by virtue of being modular, the demand for supply electricity can be increased by adding more modules to the main infrastructure of the first reactor in site, as well as covering the cost of construction of the entire series of these modules one at a time; thus the cost of ownership does not put the owner of the nuclear energy production facilities into financial debts for a long time. See Fig. 3.2, which is an illustration of small modular reactor infrastructure/building, suggested by NuScale company.

Fig. 3.3 Three major nuclear accident sites. (**a**) Chernobyl. (**b**) Three-Mile Island. (**c**) Fukushima

The other advantage of these modular reactors is the safety cost to protect the population around them both from man-made and natural disasters, so we would not have another Chernobyl, Three Mile Island, and as recent as Fukushima Daiichi in our hand. Fig. 3.3 is an illustration of all three major nuclear disasters that have taken place around the world so far.

The safety and capital cost challenges involved with traditional nuclear power plants may be considerable, but a new class of reactors in the development stage holds promise for addressing them. These reactors, called small modular reactors (SMRs), produce anywhere from 10 to 300 megawatts, rather than the 1000 megawatts produced by a typical reactor. An entire reactor, or at least most of it, can be built in a factory and shipped to a site for assembly, where several reactors can be installed together to compose a larger nuclear power station. SMRs have attractive safety features, too. Their design often incorporates natural cooling features that can continue to function in the absence of external power, and the underground placement of the reactors and the spent-fuel storage pools is more secure.

Since small modular reactors (SMRs) are smaller than conventional nuclear plants, the construction costs for individual projects are more manageable, and thus the financing terms may be more favorable. And because they are factory-assembled, the on-site construction time is shorter. The utility company can build up its nuclear power capacity step by step, adding additional reactors as needed, which means that it can generate revenue from electricity sales sooner. This helps not only the plant owner but also customers, who are increasingly being asked to pay higher rates today to fund tomorrow's plants. Thus, it is fair to say that the small modular reactors are very flexible and affordable power generation and they are an excellent source of renewable energy as well.

Also suggested by Zohuri et al. [6], advanced version of these power plants using the technology of liquid-metal fast reactor (LMFR) is a good way of recycling fuel coming out a uranium-235 fuel type from reactors such as light-water reactor (LWR), with enough residues of plutonium-239 that can be used as fuel in these SMRs, regardless of some arguments about these like ^{239}Pu are weapon grade-type fuel and nobody will use them in any reactor to burn them as fuel. A combined cycle power conversion system for small modular liquid-metal fast-breeder reactor (LMFBR) research by McDaniel et al. [7] was suggesting that the air Brayton combined cycle results obtained earlier for molten salt reactors and lead-cooled reactors have been extended to temperatures typical of liquid-metal cooled fast reactors. A split compressor with intercooler has been added to the Brayton cycle that serves as a "feed water heater" for the steam bottoming cycle. Adding this component to the system provides about a 3% increase in efficiency.

Advanced small modular reactors (SMRs) are a key part of the Nuclear Energy (NE)s goal under the Department of Energy (DOE) which is to develop safe, clean, and affordable nuclear power options. The advanced SMRs currently under development in the United States represent a variety of sizes, technology options, and deployment scenarios. These advanced reactors, envisioned to vary in size from a couple of megawatts up to hundreds of megawatts, can be used for power generation, process heat, desalination, or other industrial uses. SMRs can employ light water as a coolant or other non-light-water coolants such as a gas, liquid metal, or molten salt.

Advanced SMRs offer many advantages, such as relatively small size, reduced capital investment, ability to be sited in locations not possible for larger nuclear plants, and provisions for incremental power additions. SMRs also offer distinct safeguards, security, and nonproliferation advantages.

The Department of NE has long recognized the transformational value that advanced SMRs can provide to the nation's economic, energy security, and environmental outlook. Accordingly, the Department has provided substantial support to the development of light-water-cooled SMRs, which are under licensing review by the Nuclear Regulatory Commission (NRC) and will likely be deployed in the next 10–15 years. The Department is also interested in the development of SMRs that use nontraditional coolants such as liquid metals, salts, and helium because of the safety, operational, and economic benefits they offer.

3.2 US Industry Opportunities for Advanced Nuclear Technology Development

The Department recently issued a multi-year cost-shared funding opportunity to support innovative, domestic nuclear industry-driven concepts that have high potential to improve the overall economic outlook for nuclear power in the United States. This funding opportunity will enable the development of existing, new, and next-generation reactor designs, including SMR technologies.

The scope of the funding opportunity is very broad and solicits activities involved in finalizing the most mature SMR designs; developing manufacturing capabilities and techniques to improve cost and efficiency of nuclear builds; developing plant structures, systems, components, and control systems; addressing regulatory issues; and other technical needs identified by industry. The funding opportunity will provide awards sized and tailored to address a range of technical and regulatory issues impeding the progress of advanced reactor development.

Initiated in FY2012, the SMR Licensing Technical Support (LTS) Program works with industry partners, research institutions, the national laboratories, and academia to accelerate the certification, licensing, and siting of domestic advanced SMR designs and to reduce economic, technical, and regulatory barriers to their deployment. FY2017 was the last year of planned funding for this successful program, but activities will be completed over the next several years as certification and licensing efforts are completed.

As we mentioned in Chap. 2 of this book, there are about 50 SMR designs and concepts globally. Most of them are in various developmental stages, and some are claimed as being near-term deployable, and besides the Nuclear Energy department of DOE, the IAEA is coordinating the efforts of its member states to develop SMRs of various types by taking a systematic approach to the identification and development of key enabling technologies, with the goal to achieve competitiveness and reliable performance of such reactors. The Agency also helps them address common infrastructure issues that could facilitate the SMRs' deployment.

However, the small modular reactors offer a lower initial capital investment, greater scalability, and siting flexibility for locations unable to accommodate more traditional larger reactors. They also have the potential for enhanced safety and security compared to earlier designs. Deployment of advanced SMRs can help drive economic growth.

3.3 Benefits of Small Modular Reactors

Associated with smaller modular reactors (SMRs), there exist certain benefits, which these SMRs offer, and to start with is their cost-effectiveness form investment and total cost of ownership (TCO), thus the return on investment (ROI) point of view.

Small modular reactors offer a lower initial capital investment, greater scalability, and siting flexibility for locations unable to accommodate more traditional larger reactors. They also have the potential for enhanced safety and security compared to earlier designs. Deployment of advanced SMRs can help drive economic growth. These low initial costs are coming with certain associated benefits, which are listed here, and they are:

1. *Modularity*
   The term "modular" in the context of SMRs refers to the ability to fabricate major components of the nuclear steam supply system in a factory environment and ship to the point of use. Even though current large nuclear power plants incorporate factory-fabricated components (or modules) into their designs, a substantial amount of field work is still required to assemble components into an operational power plant. SMRs are envisioned to require limited on-site preparation and substantially reduce the lengthy construction times that are typical of the larger units. SMRs provide simplicity of design, enhanced safety

features, the economics and quality afforded by factory production, and more flexibility (financing, siting, sizing, and end-use applications) compared to larger nuclear power plants. Additional modules can be added incrementally as demand for energy increases.

2. *Lower Capital Investment*
   SMRs can reduce a nuclear plant owner's capital investment due to the lower plant capital cost. Modular components and factory fabrication can reduce construction costs and duration.
3. *Siting Flexibility*
   SMRs can provide power for applications where large plants are not needed or sites lack the infrastructure to support a large unit. This would include smaller electrical markets, isolated areas, smaller grids, sites with limited water and acreage, or unique industrial applications. SMRs are expected to be attractive options for the replacement or repowering of aging/retiring fossil plants or to provide an option for complementing existing industrial processes or power plants with an energy source that does not emit greenhouse gases.
4. *Greater Efficiency*
   SMRs can be coupled with other energy sources, including renewables and fossil energy, to leverage resources and produce higher efficiencies and multiple energy end-products while increasing grid stability and security. Some advanced SMR designs can produce a higher temperature process heat for either electricity generation or industrial applications.
5. *Safeguards and Security/Nonproliferation*
   SMR designs have the distinct advantage of factoring in current safeguards and security requirements. Facility protection systems, including barriers that can withstand design basis aircraft crash scenarios and other specific threats, are part of the engineering process being applied to new SMR design. SMRs also provide safety and potential nonproliferation benefits to the United States and the wider international community. Most SMRs will be built below grade for safety and security enhancements, addressing vulnerabilities to both sabotage and natural phenomena hazard scenarios. Some SMRs will be designed to operate for extended periods without refueling. These SMRs could be fabricated and fueled in a factory, sealed and transported to sites for power generation or process heat, and then returned to the factory for defueling at the end of the life cycle. This approach could help to minimize the transportation and handling of nuclear material. Light-water-based SMRs are expected to be fueled with low enriched uranium, i.e., approximately 5% U-235, which is similar to existing large nuclear power plants. The "security by design" concepts being applied to these technologies are expected to increase SMR resistance to theft and diversion of nuclear material. Also, reactor cores for these light-water SMRs can be designed to burn plutonium as a mixed oxide (MOX) fuel. Further, SMRs based on non-light-water reactor coolants could be more effective at dispositioning plutonium while minimizing the wastes requiring disposal.

6. *US Industry, Manufacturing, and Job Growth*
   The case for SMR economic competitiveness is rooted in the concept that mass manufacture of modular parts and components will reduce the cost per kilowatt of electricity on par with current generating sources. There is both a domestic and international market for SMRs, and US industry is well positioned to compete for these markets. DOE hopes that the development of standardized SMR designs will also result in an increased presence of US companies in the global energy market. If a sufficient number of SMR units were ordered, it would provide the necessary incentive to develop the appropriate factory capacity to further grow domestic and international sales of SMR power plants.
7. *Economic Development*
   SMR deployment to replace retiring electricity generation assets and meet growing generating needs would result in significant growth in domestic manufacturing, tax base, and high-paying factory, construction, and operating jobs. A 2010 [8] study on economic and employment impacts of SMR deployment estimated that a prototypical 100 MWe SMR costing $500 million to manufacture and install would create nearly 7000 jobs and generate $1.3 billion in sales, $404 million in earnings (payroll), and $35 million in indirect business taxes. The report examines these impacts for multiple SMR deployment rates, i.e., low (1–2 units/year), moderate (30 units/year), high (40 units/year), and disruptive (85 units/year). The study indicates significant economic impact would be realized by developing an SMR manufacturing enterprise at even moderate deployment levels.

If we consider these above benefits that are coming from SMR-type nuclear power plants, we do see their cost-effectiveness and initial capital investment, in comparison with today's analysis of cost of electivity from nuclear power plants.

1. *Cost of Electricity from Nuclear Power*
   The newest estimates for the cost of nuclear power in the United States are (costs indicated are per delivered kWh):

   11.2 US-cents per kWh nuclear power (MIT 2003)
   14.1 US-cents per kWh nuclear power (Keystone June 2007)
   18.4 US-cents per kWh nuclear power (Keystone midrange estimate)

   In the United Kingdom, the cost of nuclear electricity was estimated to be 8.2 US-cents per kWh (for an interest rate of 10%) and US-cents 11.5 per kWh (for an interest rate of 15%). Here transmission and distribution of the electricity (usually about 3 US-cents per kWh) have to be added (Source: Thomas, Bradford, Froggatt, Milborrow, 2007) [9]. So, in the United Kingdom, the costs of nuclear energy are estimated to be in the same range as indicated above.
   Comparison with other technologies, e.g., large wind power farms: 7 US-cents per kWh delivered.
   Of course most of the environmentalists are arguing over wind and solar energy as a cheap way of producing electricity and call them as an alternative renewable source of energy, which are getting less expensive almost daily, while nuclear

power is getting more and more expensive. Isn't this alone a very strong indication that nuclear power is a technology from the past?
However, one needs to bear in mind that wind does not blow 24 x 7 nor sun will shine all the time during day and night, providing during cloudy day, we may have this issue, while at night there is no sun out.

2. *Cost of Nuclear Technology Is Too High*
The reactor types you are describing are not available. A lot of research and development would be needed to develop this technology for productive applications. We probably talk about a time frame of 30–50 years, if not even longer. But we need a solution now! It is urgent.
The costs for the development are very difficult to estimate, and therefore the costs indicated above might be (and probably are) wrong. In the history of nuclear power, the cost of development has always been tremendously underestimated.
We should put full effort into developing sustainable energy technologies further and phase-out nuclear technology. This will open many opportunities and have many positive side effects. It is time for change!
In addition, we will inevitably have to change our behavior: We should only use as much energy as we are able to produce with sustainable technologies. Demand of energy has to follow available supply of sustainable energy. This will have many positive side effects, too.

3.4 Cooling Water Requirement for Nuclear Power Reactors

As one can probably imagine, the question of need of freshwater for cooling system of all the power plants, including nuclear, gas, or fossil fuel facility, is always in front of our eyes and is not an easy one to answer in a general way; however estimates that can be found by searching on the Internet indicate at least there is usage of 50% of freshwater for such activities, given that we are in shortage of water in recent years.

Cooling water from nearby lakes or rivers is normally heated up by max. 30 °F (max. 17 °C). For example, the typical 1000 Mwe nuclear power reactor with a 30 °F ΔT needs approximately 476,500 gallons of water per minute (130,000 m^3 water per hour). If the temperature rise is limited to 20 °F, the cooling water need rises to 714,750 gallons per minute (195,000 m^3 per hour).

Nuclear power reactors are about 33% efficient, i.e., for every three units of thermal energy generated by the reactor core, one unit of electrical energy goes out to the grid as electricity, and two units of waste heat go into the environment. Two modes of cooling are used to remove the waste heat from electrical generation: once-through cooling using water from rivers, ocean, or lakes and closed-cycle cooling. In the latter, water from rivers, ocean, or lakes is used to make up the water evaporated in the cooling tower.

Table 3.1 Water consumption (evaporation) [10, 11]

	Water withdrawal [m^3 water per Mwe]	Water consumption [m^3 water per Mwe]
Once-through cooling	830,000	13,000
Pond cooling	17,000	13,000
Cooling tower	27,000	24,000

Nuclear power reactor with closed-cycle cooling system of typical 1000 Mwe reactor with a cooling tower consumes by virtue of evaporation about 2000 m^3 of water per hour (about 10,300 gallons per minute). See the table below for a summary of the cooling water requirements for nuclear power units depending on the type of cooling applied. All figures indicated are in m^3 water per MW electricity (Table 3.1).

With new roadmap of research and development [12] for new generation (GEN-IV) of nuclear power ahead of us, this author and others are suggesting an innovative approach for combined cycle (CC), in the form of open-air cooling system, where air is used as cooling media for these new generations of these modular reactors and is described in the next few sections here in this chapter.

3.5 Open-Air Brayton Gas Power Cycle

Power generation is an important issue today, especially on the West Coast. Demand is outweighing supply because of lack of incentives for the utilities industry to build additional power plants over the past 10–20 years.

The Brayton cycle (or Joule cycle) represents the operation of a gas turbine engine. The cycle consists of four processes, as shown in Fig. 8.1 alongside a sketch of an engine:

- *a–b* Adiabatic, quasi-static (or reversible) compression in the inlet and compressor.
- *b–c* Constant pressure fuel combustion (idealized as constant pressure heat addition).
- *c–d* Adiabatic, quasi-static (or reversible) expansion in the turbine and exhaust nozzle, with which we:
  - Take some work out of the air and use it to drive the compressor.
  - Take the remaining work out and use it to accelerate fluid for jet propulsion or to turn a generator for electrical power generation.
- *d–a* Cool the air at constant pressure back to its initial condition (Fig. 3.4).

The components of a Brayton cycle device for jet propulsion are shown in Fig. 3.5. We will typically represent these components schematically, as in Fig. 3.6. In practice, real Brayton cycles take one of two forms. Figure 3.7a shows an "open"

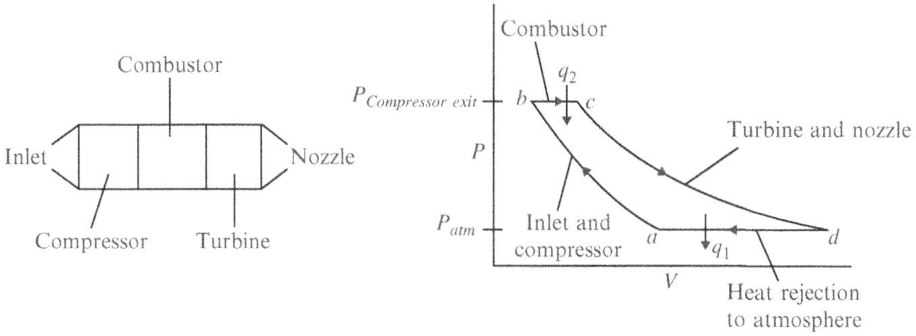

Fig. 3.4 Sketch of the jet engine components and corresponding thermodynamic states

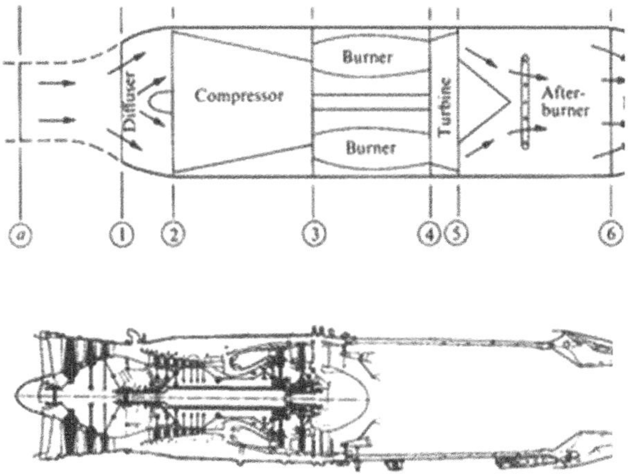

Fig. 3.5 Schematics of typical military gas turbine engines. Top, turbojet with afterburning; bottom, GE F404 low bypass ratio turbofan with afterburning (Hill and Peterson 1992) [13]

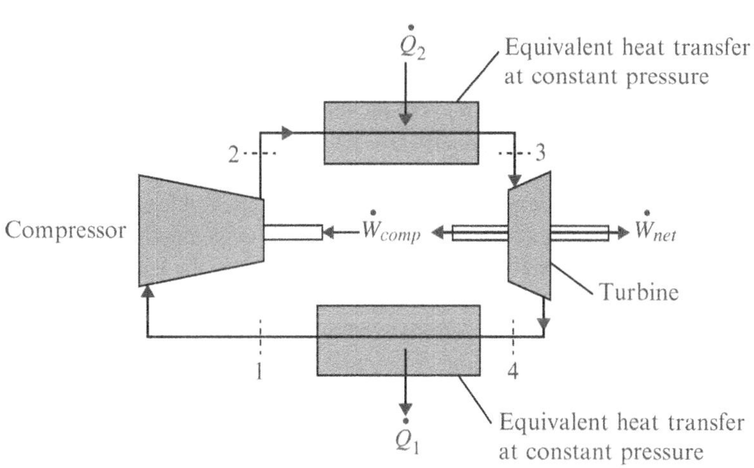

Fig. 3.6 Thermodynamic model of gas turbine engine cycle for power generation

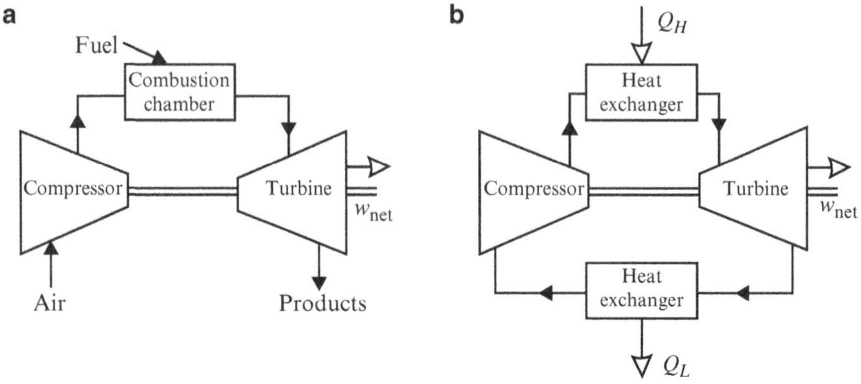

Fig. 3.7 Options for operating Brayton cycle gas turbine engines. (**a**) Open Cycle Operation. (**b**) Closed Cycle Operation

cycle, where the working fluid enters and then exits the device. This is the way a jet propulsion cycle works. Figure 3.7b shows the alternative, a closed cycle, which recirculates the working fluid. Closed cycles are used, for example, in space power generation.

The major growth in the electricity production industry in the last 30 years has centered on the expansion of natural gas power plants based on gas turbine cycles. The most popular extension of the simple Brayton gas turbine has been the combined cycle power plant with the air Brayton cycle serving as the topping cycle and the steam Rankine cycle serving as the bottoming cycle. The air Brayton cycle is an open-air cycle and the steam Rankine cycle is a closed cycle. The air Brayton cycle for a natural gas-driven power plant must be an open cycle, where the air is drawn in from the environment and exhausted with the products of combustion to the environment. The hot exhaust from the air Brayton cycle passes through a heat recovery steam generator (*HRSG*) prior to exhausting to the environment in a combined cycle. The HRSG serves the same purpose as a boiler for the conventional steam Rankine cycle.

In 2007 gas turbine combined cycle plants had a total capacity of 800 GW and represented 20% of the installed capacity worldwide. They have far exceeded the installed capacity of nuclear plants, though in the late 1990s, they had less than 5% of the installed capacity worldwide of 22%. There are several reasons for this. First natural gas is abundant and cheap. Second combined cycle plants achieve the greatest efficiency of any thermal plant. And third, they require the least amount of waste heat cooling water of any thermal plant.

A typical gas turbine plant consists of a compressor, combustion chamber, turbine, and an electrical generator. A combined cycle plant takes the exhaust from the turbine and runs it through a heat recovery steam generator (*HRSG*) before exhausting to the local environment. The HRSG serves the function of the boiler for a typical closed-cycle steam plant. The steam plant consists of a steam turbine, a condenser, a water pump, an evaporator (boiler), and an electrical generator. In a

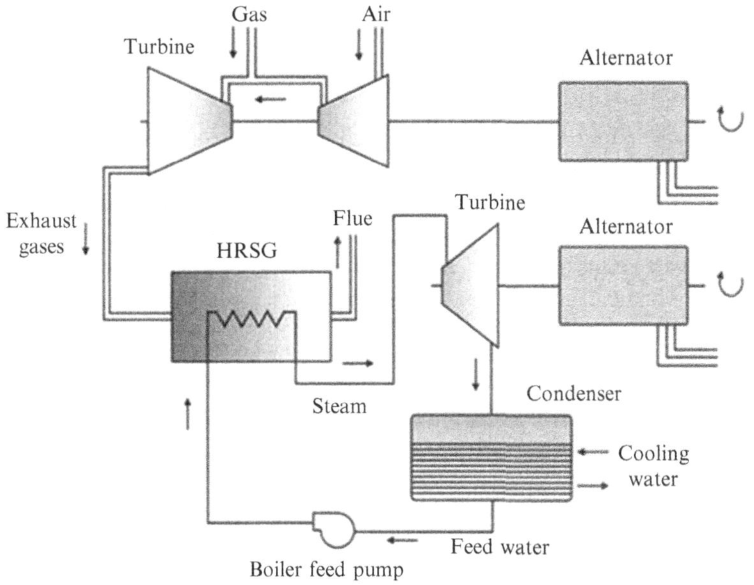

Fig. 3.8 Working principle of combined cycle gas turbine (CCGT) plant

combined cycle plant, the gas turbine and steam turbine can be on the same shaft to eliminate the need for one of the electrical generators. However, the two shaft generator systems provide a great deal of more flexibility at a slightly higher cost. In addition to the closed loop for the steam, an open loop circulating water system is required to extract the waste heat from the condenser. The waste heat extracted by this "circulating" water system is significantly less per megawatt for a combined cycle system as the open Brayton cycle exhausts its waste heat directly to the air.

The layout for the components of a typical combined cycle power plant is given in Fig. 3.8.

3.5.1 Computer Code Development

This effort was undertaken to investigate the possibility of using a nuclear reactor-driven heat exchanger, or group of heat exchangers, to drive a Brayton-like cycle gas turbine as an open-cycle power conversion system. Since in a nuclear reactor-driven system, the core fuel elements of the reactor must exist at a higher temperature than any other component in the system, such a system is usually severely limited in the peak temperatures that can be produced in the gas turbine working fluid. For this study, a peak temperature of 660 °C (933 K) was chosen as a reasonable upper limit on the temperature that the working fluid could attain prior to being expanded through a turbine. This turbine inlet temperature is roughly half of the state-of-the-art aircraft jet engine turbine inlet temperatures. Given this limitation, it is necessary to use several turbines with a reheat heat exchanger between each. The baseline for this study was four turbines operating with an inlet temperature of 660 °C and an exit temperature of 537 °C. To demonstrate that as many as four turbines may be required, the analysis started with a system using only one turbine. The effects of a bottoming steam cycle and a recuperator were investigated.

GE currently markets a system that will produce 61% efficiency at design power and better than 60% efficiency down to 87% of design power [14] for gas turbine combined cycle plants.

An approximate efficiency can be calculated for a combined cycle power plant by the following simple argument [15].

$$\text{Brayton cycle efficiency} = \frac{W_B}{Q_{\text{in}}} = \eta_B$$

$$\text{Heat to Rankine cycle} = Q_R = (1-\eta_B)Q_{\text{in}}$$

$$\text{Rankine cycle efficiency} = \frac{W_R}{Q_R} = \eta_R$$

$$\text{Overall efficiency} = \frac{W_B + W_R}{Q_{\text{in}}} = \eta_T = \frac{\eta_B Q_{\text{in}} + \eta_R Q_R}{Q_{\text{in}}}$$
$$= \frac{\eta_B Q_{\text{in}} + \eta_R (1-\eta_B) Q_{\text{in}}}{Q_{\text{in}}} = \eta_B + \eta_R - \eta_B \eta_R$$

$$\eta_T = \eta_B + \eta_R - \eta_B \eta_R \tag{3.1}$$

This efficiency must be corrected for pressure losses and assumes that all of the heat in the Brayton exhaust is used in the *HRSG*. For a combustion gas turbine, this is not usually possible if condensation of the water in the exhaust products is to be avoided. The detailed models developed in this effort give a more accurate answer.

For a nuclear system to take advantage of combined cycle technology, there are a number of changes to the plant components that have to be made. The most significant of course is that the combustion chamber must be replaced by a heat exchanger in which the working fluid from the nuclear reactor secondary loop is used to heat the air. The normal Brayton cycle is an internal combustion one where the working fluid is heated by the combustion of the fuel with the air in the combustion chamber. The walls of the combustion chamber can be cooled, and peak temperatures in the working fluid can be significantly above the temperature that the walls of the chamber can tolerate for any length of time.

For the nuclear reactor system, the heat transfer is in the opposite direction. All reactor components and fluids in the primary and secondary loops must be at a higher temperature than the peak temperature of the gas exiting the heat exchanger.

This severely restricts the peak temperature that can be achieved for the air entering the turbine. However, all is not lost.

In a typical combustion system, there are pressure losses approaching 5% of the total pressure to complete the combustion process [16]. Heat exchangers can be built with significantly lower pressure drops than 5% approaching 1%. So, the most straightforward method to overcome this severe temperature limitation is to borrow a technique from steam power plants and implement multiple reheat cycles. That is, the first heat exchanger heats the air to its peak temperature. Then the air is expanded through the first turbine. The air is then reheated to the same peak temperature and expanded through the second turbine. Based on the relative pressure losses that appear possible, up to five turbines might be considered. All five turbines will be driving the same compressor. Multiple compressors on concentric shafts [3] driven by different sets of turbines might be possible, but that has not been considered here.

Multiple reheat cycles allow more heat to be put into the working fluid at a higher temperature. This improves the efficiency of the overall cycle. Interestingly enough, it does not improve the efficiency of the Brayton cycle, but because the exit temperature from the last turbine is higher, it does improve the efficiency of the Rankine part of the cycle with a net gain for the overall cycle. For this to work, the reactor coolant temperature must reach temperatures significantly higher than current light-water reactor temperatures. Even sodium-cooled reactor exit temperatures in the 550 °C range are not quite high enough to get a nuclear air Brayton combined cycle to compete with the efficiency of a pure steam Rankine cycle. But when the coolant exit temperatures reach the 650–700 °C ranges, the combined cycle systems with multiple turbines surpass the performance of steam Rankine systems. So, the analysis that follows is targeted at a molten salt reactor or a lead coolant reactor. A pressurized sodium reactor that could reach these temperatures is another possibility, but no one is proposing such a system at this time. It could also apply to a high-temperature gas reactor, but the heat exchangers would be quite different. Gas to gas heat exchangers have not been considered as the primary heat exchangers at this point but will be addressed in the recuperated systems.

Liquid-metal and molten salt heat exchangers were developed and tested successfully during the Aircraft Nuclear Propulsion program in the late 1950s [17–20]. They were conventional tube and plate exchangers and were tested for over 1000 h at temperatures up to 1100 K. The largest size tested transferred 55 MW of heat in a package of approximately 1.2 m^3. The heat-transfer area on the air side had a surface area per unit volume of 1180 m^2/m^3. Certainly, some development will be needed to bring this technology up to modern standards and get NRC approval for a power-producing reactor. But the tasks involved do not appear insurmountable. Several additional heat exchangers were designed in this work in an attempt to estimate sizes of components and validate that pressure drop criteria could be met. The heat from the exhaust of the Brayton cycle transfers heat to vaporize the steam in the Rankine cycle in an HRSG of fairly, conventional design. This includes air to steam superheaters as well as an economizer and evaporator section. A condenser of conventional design is included. For this work all heat exchangers were considered to be counterflow designs.

The heat exchanger design procedures and experimental data were taken from the text by Kays and London [21]. All of the data presented in this text was developed from steam to air heat exchangers and should be particularly applicable to the types of heat exchangers developed here. The only ones not using these two fluids are the molten salt or liquid metal to air heat exchangers referenced above; thus there is a reality basis for all of the design calculations performed here to estimate power conversion system performance and sizing.

One of the significant advantages of the combined cycle power system over current LWR power systems is its reduced requirement for circulating water in the waste heat rejection loop for the Rankine cycle. The typical combined cycle plant considered here produces approximately 50% of its power from the Rankine cycle and 50% of its power from the Brayton cycle. This automatically reduces the cooling water requirement by half. In addition, the combined cycle plant achieves 45% efficiency so that only 55% of the heat generated must be released as waste. So, a typical 25 MW system will only need to get rid of 6.9 MW of heat via a circulating water system. A current LWR plant generating 25 MW at an efficiency of 33% would need to dump 16.8 MW of heat. This represents major savings in circulating water requirements.

Since the combined cycle reduced the circulating water requirements, so significantly the natural question arises as to whether they can be eliminated completely. With a recuperated air Brayton cycle, they can be. All of the waste heat can be rejected directly to the atmosphere. At first it was thought that the efficiency of a multi-turbine recuperated cycle could not compete with a combined cycle plant. However, after performing the detailed analysis, the efficiencies of a recuperated cycle come within 1% or 2% of predicted combined cycle efficiencies. This would seem to be a minor penalty to pay for being free of a circulating water requirement. However, achieving these high efficiencies requires a very effective recuperator which can become quite large.

If the readers are interested in more details, they should refer to the book by Zohuri and McDaniel, Chap. 8 [4].

3.6 Modeling the Nuclear Air Brayton Combined Cycle

Given that the combined cycle (CC) code does a good job of modeling current generation gas turbine combined cycle (GTCC) plants, it is useful to extrapolate its capabilities to nuclear air Brayton combined cycle (NACC) power plants and nuclear air Brayton recuperated cycle (NARC) power plants. The combined cycle plants will be dealt with in this chapter and the recuperated plants in the next chapter. In the nuclear air Brayton power plants, the combustion chamber of the gas turbine system is replaced by the nuclear reactor and a heat exchanger. The nuclear reactor will heat a working fluid, and that working fluid will in turn pass through a heat exchanger to heat the air for the turbine. Because the heat-transfer process for a nuclear system is in the opposite direction (solid to gas) from that in the gas turbine (gas to solid), the peak temperatures achievable in a nuclear air Brayton system will never be as high as those in a gas turbine system. However, the nuclear system can reheat the air multiple times and expand it across multiple turbines to increase the available power.

In comparing nuclear systems, there are multiple advantages for the nuclear air Brayton cycle over conventional light-water cycles. The ability to go to higher temperatures and not be trapped under the water vapor dome adds a major flexibility. The higher gas temperatures allow for the implementation of a combined cycle approach to improve system efficiency. There is also a much larger base of gas turbine manufacturers than large steam turbine manufacturers for system designers to draw from. All downstream components after the turbines themselves are currently being used in GTCC plants. A NACC plant will require significantly less water for waste heat removal than a current light-water reactor (LWR) system because the efficiency is greater, and a significant amount of the waste heat is directly dumped into the atmosphere without going through a cooling tower. Finally, the NARC system dumps all of its heat into the atmosphere and as a result is not tied to a location near a water source. Readers should refer to Chap. 9 of the book by Zohuri and McDaniel [4].

3.7 A Combined Cycle Power Conversion System for Small Modular LMFBR

In previous publications, this author and others have addressed applying an open-air Brayton power conversion cycle to next-generation nuclear power plants [6, 22, 23]. Both a combined cycle power conversion system and a recuperated Brayton cycle look very promising at the temperatures anticipated for molten salt reactors and liquid lead-cooled reactors. However, at coolant temperatures more typical of a liquid-metal fast-breeder reactors (LMFBR), the combined cycle did not seem to hold an advantage over current conversion systems giving cycle efficiencies in the 39–40% range. Recently taking a clue from standard steam cycle efficiency enhancement techniques, we added a "feed water heater" to the bottoming Rankine cycle [24]. This is not the classic feed water heater because it draws its heat by cooling the compressed air from the first stage of a split compressor in the Brayton topping cycle. This frees up most of the hot air that would be used to raise the water to the saturation temperature for the Rankine cycle so that it can be passed through a recuperator to preheat air prior to entering the main heat exchangers. The net effect of the process is to raise the thermal efficiency of the cycle about 3% making a combined cycle more competitive at LMFBR outlet temperatures.

3.7.1 The Air Brayton Cycle Pros and Cons

The fastest growing power conversion systems in the electric utility market today are internal combustion gas turbines. Any external combustion or heat engine system is always at a disadvantage to an internal combustion system. The internal combustion systems used in current jet engine and gas turbine power systems can operate at very high temperatures in the fluid and cool the structures containing the fluid to achieve high thermodynamic efficiencies. In an external energy generation system, like a reactor, all of the components from the core to the heat exchangers heating the working fluid must operate at a higher temperature than the fluid. This severely limits the peak cycle temperature compared to an internal combustion system. One way this liability can be overcome is by using multiple expansion turbines and designing highly efficient heat exchangers to heat the working fluid between expansion processes.

Typically, the combustion chamber in a gas turbine involves a pressure drop of 3% to 5% of the total pressure. Efficient liquid metal to air heat exchangers can theoretically be designed with pressure drops of less than 1%. This allows three to five expansion cycles to achieve a pressure drop comparable to a combustion system. Multiple turbines operating at different pressures have been common in steam power plants for a number of years. In this study three to five gas turbines operating on a common shaft were considered. Multiple expansion turbines allow a larger fraction of the heat input to be provided near the peak temperature of the cycle. The exhaust from the last gas turbine still contains significant amounts of thermal energy. This gas is provided to the heat recovery steam generator (HRSG) to produce the steam used in the Rankine bottoming cycle. In a traditional gas turbine combined cycle, this gas gives up its thermal energy by heating steam in a superheater, then vaporizing the high-pressure water to make steam, and finally heating the high-pressure water to the liquid saturation point prior to evaporating it. At the liquid saturation point or "pinch point," the hot air must be at a temperature greater than the boiling temperature of the high-pressure water. The heat remaining in the gas is then used to heat the liquid water or exhausted to the atmosphere.

3.7.2 The Feed Water Heater

Traditional Rankine steam cycles bleed steam from the high-pressure and high-temperature steam in the turbine(s) to heat the high-pressure water from low temperature to its boiling point. Because the temperature drop between the heating fluid and heated fluid in this case is less than in the case where the liquid water is heated by the boiler directly, the overall efficiency of the cycle increases. For the combined cycle, there is another source of heat that can be used to heat this low-temperature, high-pressure water. If the air compressor in the Brayton cycle is split into two parts such that the first part does about 56% of the work, its air can be used to heat the

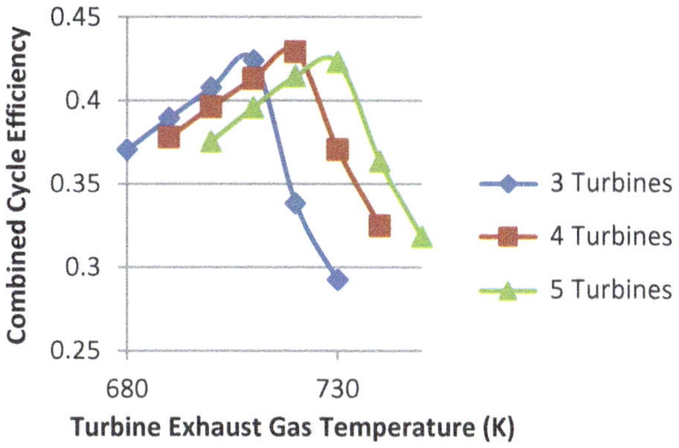

Fig. 3.9 Combined cycle efficiency vs. turbine exhaust gas temperature

low-temperature water. The heat exchanger that cools the air exiting the first compressor is normally called an intercooler. Traditionally an outside source of cold water is used to perform this function. However, the cold water in the Rankine bottoming cycle can perform this function almost as well.

This frees up the hot gas exiting the "pinch point" to preheat the air out of the second compressor in a traditional recuperator. Intercooling without recuperating does not normally improve the efficiency of a cycle [24, 25]. The major limitation on the size of the steam system is the enthalpy available from high-temperature air above the pinch point where the high-pressure water working fluid starts to vaporize. Below this point, there is still a significant enthalpy in the air which is readily available to heat the high-pressure water or to heat the compressed air in the Brayton cycle. There does not appear to be an advantage to including traditional feed water heaters in the cycle to bring the high-pressure water up to the saturation point. The possibility that an intercooler could be inserted between the two stages of a split compressor was considered. The cooling fluid for the intercooler was the high-pressure water coming out of the water pump. This process would combine the function of the traditional intercooler with the preheating of a typical feed water heater.

3.7.3 Results of Modeling

The results of modeling this process for a temperature typical of an LMFBR (510 °C or 783 K) are presented in Fig. 3.9.

Basically, the addition of the intercooler-feed water heater increased the efficiency of the cycle by about 3%. The peak efficiency achieved by the four-turbine case is 42.95%. The four-turbine case is once again optimal as it was for the simple combined cycle, though the margin over three or five turbines is less than a percent. There were significant differences from the simple combined cycle (CC) however.

Table 3.2 Comparison of air and steam pressures

	Compression ratio	Steam pressure
Three turbines	3.846	21.0 Mpa
Four turbines	4.642	15 Mpa
Five turbines	5.000	13.5 Mpa

Table 3.3 Comparison of intercooled with simple CC

	Exhaust	Compress	Steam	Overall
	Temp	Ratio	Pressure	Efficiency
Simple CC	820 K	14.125	2 MPa	46.1
IC-CC	820 K	10.488	15 Mpa	49.5

Table 3.4 Performance parameters for 50 MW system

Total power	50 MW(e)
Brayton system	31.2 MW(e)
Rankine system	18.8 MW(e)
Total heat input	116.0 MW(t)
Mass flow rate (air)	291.4 kg/sec
Mass flow rate (water)	12.65 kg/sec
Heat rejection	29.4 MW(t)

Because the Brayton cycle is now recuperated, the overall compression ratios are much smaller. Also, the operating pressures in the Rankine cycle were much higher fed by the need to provide hot gas from the "pinch point" to the recuperator.

Consider the following Table 3.2.

A more meaningful comparison can be made with the four-turbine case presented previously for the molten salt reactor. Simple CC is the previous case; intercooled combined cycle (IC-CC) is the split compressor intercooled combined cycle in Table 3.3.

The cycle optimizes with the same turbine exit temperature, but a significantly different compression ratio. It is worth pointing out that with the addition of an intercooler-feed water heater, the molten salt efficiency is now very close to 50%.

Returning to the LMFBR case, for the optimum four-turbine configuration, the major performance parameters for a 50 MW(e) system would be as presented in Table 3.4.

Note once again that the heat rejection to a circulating water system is significantly less than the heat rejected by a current coal fired plant at 40% efficiency. It is approximately 39.2% of that required by this kind of steam plant.

Perhaps even more interesting are the state points for the combined cycle system. The following table gives the typical state points for both the air Brayton cycle and the steam Rankine cycle (Table. 3.5).

Table 3.5 Result of the computer model run output

State points	Pressure	Temperature
Brayton	(Mpa)	(K)
Atmosphere	0.1013	288.0
1st compressor exit	0.2489	382.9
Intercooler exit	0.2464	319.9
1st heater inlet	0.4610	604.6
1st heater exit	0.4563	783.0
1st turbine exit	0.3204	720.0
4th turbine exit	0.1076	720.0
Superheater exit	0.1066	664.1
Evaporator exit	0.1053	625.3
Recuperator inlet	0.1044	625.3
Recuperator exit	0.1034	402.4
Exhaust to atmosphere	0.1023	402.4
Rankine		
Pump entrance	0.0073	313.0
Pump exit	15.0	316.6
Intercooler exit	15.0	615.3
Evaporator exit	15.0	615.3
HP turbine inlet	15.0	705.0
HP turbine exit	3.75	519.7
MP turbine inlet	3.75	705.0
MP turbine exit	0.9375	519.7
LP turbine inlet	0.9375	705.0
LP turbine exit quality	0.914	

It is probably worth mentioning that the intercooler-feed water heater could probably be adapted to current generation gas turbine systems if the low-temperature gas exiting the recuperator does not condense the water added to the air as the result of combustion. That question is beyond the scope of our effort.

3.8 Summary

In order for nuclear power to continue to be a viable energy option in any country, including the United States, nuclear safety, security, and safeguards must be maintained at the highest levels on a global scale. DOE will help to achieve consensus criteria for safe reactor operation through international organizations, such as the World Association of Nuclear Operators, and seek to enhance safety standards for nuclear power, promote appropriate infrastructure at the national and international levels, and minimize proliferation risks from the expansion of nuclear power through its participation with the IAEA and related organizations.

References

1. Zohuri, B. "Hybrid Energy Systems: Driving Reliable Renewable Sources of Energy Storage" Nov 26, 2017, Published by Springer Publishing Company
2. Zohuri, B. "Nuclear Energy for Hydrogen Generation through Intermediate Heat Exchangers: A Renewable Source of Energy" Springer Publishing Company, Jul 15, 2016
3. Zohuri, B. "Combined Cycle Driven Efficiency for Next Generation Nuclear Power Plants, An Innovative Design Approach", 1st Edition, Springer Publishing Company 2014
4. Zohuri, B. McDaniel, P. "Combined Cycle Driven Efficiency for Next Generation Nuclear Power Plants: An Innovative Design Approach" 2nd Edition, Springer Publishing Company, 2018
5. Zohuri, B. "Application of Compact Heat Exchangers For Combined Cycle Driven Efficiency In Next Generation Nuclear Power Plants: A Novel Approach" 1st Edition, Springer Publishing Company, Dec. 2016
6. Zohuri, B., McDaniel, P. and De Oliveira, C. "Advanced Nuclear Open Air-Brayton Cycles for Higley Efficient Power Conversion", Nuclear Technology, Vol. 192, October 2015
7. Patrick McDaniel, Bahman Zohuri, Cassiano de Oliveira, "A Combined Cycle Power Conversion System for small modular LMFBRs" American Nuclear Transition, Summer 2014, Reno Nevada.
8. Economic and Employment Impacts of Small Modular Reactors, June 2010, Energy Policy Institute of the Center for Advanced Energy Studies
9. Thomas, S. Bradford, P., Froggatt, A. and Milborrow, D. "The economics of nuclear power report 2007", Published by greenpeace.org
10. The table about has been extracted from an Australian Study with the title "Water requirements of nuclear power stations"
11. Another source of information is from the Union of concerned scientist with the title "Got water?"
12. Nuclear Energy, Research and Development Roadmap, Report to Congress April 2010, U.S. Department of Energy, Nuclear Energy Department
13. Hill and Peterson, GE F404 low bypass ratio turbofan with afterburning, 1992
14. GE Energy Flex Efficiency 50 Combined Cycle Power Plant, e-brochure, 2012
15. Horlock, J. H., Cogeneration-Combined Heat and Power (CHP), Krieger Publishing Company, Malabar FL, 1997
16. Mattingly, J. D. Elements of Gas Turbine Propulsion, McGraw-Hill, Inc., New York, 1996
17. MacPherson, R. E., J. C. Amos, and H.W. Savage, "Development Testing of Liquid Metal and Molten Salt Heat Exchangers", Nuclear Science and Engineering, 8, pp 14–20, 1960
18. Fraas, A. P., "Design Precepts for High Temperature Heat Exchangers", Nuclear Science and Engineering, 8, pp 21–31, 1960
19. Yarosh, M. M., "Evaluation of the Performance of Liquid Metal and Molten Salt Heat Exchangers", Nuclear Science and Engineering, 8, pp 32–43, 1960
20. B. Zohuri, P. J. McDaniel, and C. de Olivera, "Air Brayton Cycles for Nuclear Power Plants", Nuclear Technology, Submitted for Review, April 2014
21. Kays, W. M. and A. L. London, Compact Heat Exchangers, 2nd Ed. McGraw-Hill Book Company, New York, 1964
22. McDaniel, P. and De Oliveira, C, Zohuri, B., and Cole, J., "A Combined Cycle Power Conversion System for the Next Generation Nuclear Plant", *ANS Transactions*, December 2012
23. Zohuri, B., McDaniel, P. and De Oliveira, C, "A Comparison of a Recuperated Open Cycle (Air) Brayton Power Conversion System with the Traditional Steam Rankine Cycle for the Next Generation Nuclear Power Plant", ANS Transactions, June 2014
24. Wilson, D. G. and T., Korakianitis, *The Design of High-Efficiency Turbomachinery and Gas Turbines,* 2nd Ed., Prentice-Hall, Upper Saddle River NJ, 1998
25. Haywood, R. W., *Analysis of Engineering Cycles,* 4th Ed., Pergamon Press, Oxford, 1991

Chapter 4
A Combined Cycle Power Conversion System for Small Modular Reactor

Abstract This chapter presents the innovative approach to combined cycle system for power conversion for small modular reactor designed in the form of a liquid-metal fast-breeder reactor infrastructure and the need for nuclear power plant for production of electricity. The foundation has been structured and consequently some technology of ongoing research that makes production of electricity from the nuclear power plant more cost-effective. For the nuclear reactors to be more comparative with fossil- and gas-fueled power plants, they need to be as efficient as the traditional power plants are when it comes to output thermal efficiency. As this chapter suggests, utilizing the combined cycles to drive and produce electricity via nuclear fuel makes more sense to own them from both return on investment and total cost of ownership and efficient for their owners, namely, electricity companies. Results of modeling a combined cycle Brayton-Rankine power conversion system are presented in this chapter, and base model reactor for this purpose was chosen to be the molten salt reactor type. The Rankine bottoming cycle appears to offer significant advantages over the recuperated Brayton cycle. The overall cycle in the modeling, for purpose of writing this chapter, was optimized as a unit, and lower-pressure Rankine systems did seem to be more efficient. The combined cycle requires a lot less circulating water for a heat dump than current power plants.

4.1 Introduction

The discovery of nuclear fission in 1939, following the Manhattan Project, was an event that opened up the prospect of entirely new source of power utilizing the internal energy of the atom. The basic materials that can be used for the release of nuclear energy driven by a fission product are uranium, plutonium, and thorium, where larger quantities of these materials can be found in the earth's crust, and sometimes they are very expensive to be extracted due to their low concentration at the source. Such cost can be balanced by electrical energy produced by them that makes extraction of such materials very cost-effective. In the present century, the world's consumption of energy has grown rapidly due to the increase in the earth's population; thus the per capita increase in the use of energy for day-to-day operation of

B. Zohuri, *Small Modular Reactors as Renewable Energy Sources*,
https://doi.org/10.1007/978-3-319-92594-3_4

industries, agriculture, and transportation is one of the few examples that we can mention. For example, in the United States, the population growth rate from 1940 to 1970 was about 1.6% per annum, but the total consumption of energy increased at an average annual rate of approximately 3.5%.

Naturally, our primary interest is in commercializing the nuclear power energy, which in return produces larger and larger proportion of nuclear energy in the form of electrical power. In the United Sates, the total consumption of electricity in industries, commerce, and the home had been growing by some 7% each year from 1940 to about 1973. An annual growth rate of only 3% would require the construction of a new electrical power plant, over the 25-year period that was ending in the year 2000, having a capacity equal to that of all the plants operating in 1975. This estimate did not consider the need for replacing obsolete facilities or possible shifts to electricity as an energy source to replace oil, e.g., in electric vehicles [1].

In 2002, nuclear power supplied 20% of the United States and 17% of the world's electricity consumption. Experts project worldwide electricity consumption will increase substantially in the coming decades, especially in the developing world, accompanying economic growth and social progress. However, official forecast calls for a mere 5% increase in nuclear electricity generating capacity worldwide by 2020, and even this is questionable, while electricity use could grow by as much as 75% [2]. These projections demand for construction of new nuclear power plants and reflect both economic considerations and growing antinuclear sentiment within the public in key countries having this technology in their disposal.

The limited prospects for nuclear power and producing energy from it are attributable, ultimately, to four unresolved problems [2].

1. *Costs:* nuclear power has higher overall lifetime costs compared to natural gas with combined cycle gas turbine (CCGT) technology and coal, at least in the absence of a carbon tax or an equivalent "cap-and-trade" mechanism for reducing carbon emissions.
2. *Safety:* nuclear power has perceived adverse safety, environmental, and health effects, heightened by the 1979 Three Mile Island and 1986 Chernobyl reactor accidents but also by accidents at fuel cycle facilities in the United States, Russia, and Japan. There is also a growing concern about the safe and secure transportation of nuclear materials and the security of nuclear facilities from terrorist attacks.
3. *Proliferation:* nuclear power entails potential security risks, notably the possible misuse of commercial or associated nuclear facilities and operations to acquire technology or materials as a precursor to the acquisition of nuclear weapons capability. Fuel cycles that involve the chemical reprocessing of spent fuel to separate weapons-usable plutonium and uranium enrichment technologies are of special concern, especially as nuclear power spreads around the world.
4. *Waste:* nuclear power has unresolved challenges in long-term management of radioactive wastes. The United States and other countries have yet to implement final disposition of spent fuel or high-level radioactive waste streams created at various stages of the nuclear fuel cycle. Since these radioactive wastes present some danger to present and future generations, the public and its elected

representatives, as well as prospective investors in nuclear power plants, properly expect continuing and substantial progress toward solution to the waste disposal problem. Successful operation of the planned disposal facility at Yucca Mountain would ease, but not solve, the waste issue for the United States and other countries if nuclear power expands substantially.

Aside from global warming and its impact reduction using nuclear power plant that the MIT [1] report argues with, today, nuclear power generation is not an economically cost-effective and competitive choice. Taking into consideration the above four factors, we need the government's involvement at least in three areas, safety, proliferation, and waste, respectively. However, if we push the issue of free carbon dioxide emission as source of producing energy for electricity usage, then the cost of ownership of nuclear power plant per kilo watt will be possibly justified and indeed important and vital future of source of energy.

The generation of electricity from fossil fuels, notably natural gas and coal, is a major and growing contributor to the emission of carbon dioxide – a greenhouse gas that contributes significantly to global warming. We share the scientific consensus that these emissions must be reduced and believe that the United States will eventually join with other nations in the effort to do so.

At least for the next few decades, there are only a few realistic options for reducing carbon dioxide emissions from electricity generation:

- Increase efficiency in electricity generation and use.
- Expand use of renewable energy sources such as wind, solar, biomass, and geothermal.
- Capture carbon dioxide emissions at fossil-fueled (especially coal) electric-generating plants and permanently sequester the carbon.
- Increase use of nuclear power.

Increasing energy demand, global climate change, and constrained energy supplies are likely to impact how energy affects your business in the future. Is your company prepared for the energy challenges that lie ahead? The groundbreaking report, *Energy Strategy for the Road Ahead*, reveals what 20 leading US companies recommend businesses should do now to prepare for the risks and opportunities of our energy future [2].

Market trends suggest that the demand for energy resources will rise dramatically over the next 25 years:

- Global demand for all energy sources is forecast to grow by 57% over the next 25 years.
- US demand for all types of energy is expected to increase by 31% within 25 years.
- By 2030, 56% of the world's energy use will be in Asia.
- Electricity demand in the United States will grow by at least 40% by 2032.
- New power generation equal to nearly 300 (1000 MW) power plants will be needed to meet electricity demand by 2030.
- Currently, 50% of US electrical generation relies on coal, a fossil fuel, while 85% of US greenhouse gas emissions result from energy-consuming activities supported by fossil fuels.

Sources: Annual Energy Outlook (DOE/EIA-0383(2007)), International Energy Outlook 2007 (DOE/EIA-0484(2007), Inventory of U.S. Greenhouse Gas Emissions and Sinks: 1990–2005 (April 2007) (EPA 430-R-07-002).

If energy prices also rise dramatically due to increased demand and constrained supply, business impacts could include:

- Reduced profits due to high operating costs
- Decline of sales of energy-using products
- Loss of competitiveness in energy-intensive businesses
- Disruptions in supply chains as suppliers are unable to meet cost obligations or go bankrupt

Recent history also demonstrates that catastrophic weather events, terrorism, and shifting economic centers are not just events of our imagination but realities of our lifetime.

Given this challenging landscape, what steps do US businesses need to take today to survive a potentially disruptive energy future? How also can we make production of energy more cost-effective so the total cost of ownership and return on investment for these businesses make total sense?

The Global Business Network (GBN), a member of the Monitor Group, in cooperation with the US Environmental Protection Agency (EPA), gathered senior executives from 20 major US companies to consider the potential energy impacts that US businesses may face over the next decade. Based on four plausible scenarios of the world in 2020, the Energy Strategy for the Road Ahead report identifies a set of strategies that will help businesses to act now to prepare for future energy-related risks.

Considering changes in global economic patterns and shifts in US policy and regulation toward climate change as key factors that would affect the shape of the future ahead, the following four scenarios were created by the corporate executives who participated in the GBN workshops:

- *The same road* — where the world continues much in the same direction it appears to be going now in regard to energy and environmental concerns around climate change
- *The long road* — where the world undergoes a significant shift in the economic, geopolitical, and energy centers of gravity
- *The broken road* — where the world continues much in the direction of today but is then hit by a severe event that overturns established systems and rules
- *The fast road* — where reasoned decisions and investments about energy and climate risk are made early enough to make a difference

"The world can expect energy prices to continue their generally upward spiral in the years ahead if global energy policies remain the same," the International Energy Agency (IEA) reported [3].

Rapid economic development in China and India, coupled with relatively consistent energy use in industrialized nations, will likely strain the world's ability to meet a projected rise in energy demand of some 1.6% a year until 2030, the Agency predicted in its annual *World Energy Outlook* report [PDF] [4].

The IEA significantly increased its projections of future oil costs in this year's report due to the changing outlook for demand and production costs. It now expects crude oil to average \$100 per barrel over the next two decades and more than \$200 per barrel in 2030, in nominal terms. Last year's forecast estimated that a 2030 barrel would amount to only \$108.

4.2 Nuclear Reactors for Power Production

In the United States, most reactors' design and development for the generation of electrical power were branched from early nuclear navy research, when it was realized that a compact nuclear power plant would have a great advantage for submarine-driven nuclear propulsion system. To have such power plant on board would make possible long voyages cross the oceans at high speeds without the necessity for resurfacing at frequent intervals, and Argonne National Laboratory was assigned with such task of designing such reactor. So, the first generation of pressurized water reactor (PWR) was born, where the use of highly enriched uranium as the fuel and water under pressure as the moderator as well as coolant allowed small version of this type of high-power reactor. Consequently, the first prototype of such reactor, namely, STR Mark 1, started operation at Arco, Idaho, in March 1953, and production version of it was installed on USS Nautilus, the first nuclear power submarine after May 31, 1953.

As result of experience gained in successful operation of the submarine reactors, the first commercial version of PWR was designed in Pennsylvania shipping port and went into operation in December 2, 1957 with the water pressure of 13.8 MPa, i.e., 2000 psi, and steam produced in the heat exchanger with a temperature of about 254 °C (490 °F) and pressure of close to 4.14 MPa (600 psi). In order to make such reactor cost-effective and reduce the cost of the power produced, only a small number of the fuel elements are highly enriched in Uranium-235 (U^{235}) as an alloy with zirconium, the remainder being of normal Uranium as the dioxide. The change in core design required bigger real estate for footprint of commercialized version PWR that was not an issue for a land-based facility. The output power of this reactor was about 60 MW electrical equivalents to 230 MW thermal equivalents, and further enhancement in core design increased the power out to 150 MW electrical and 505 thermal equivalents. Pressurized water reactors, using slightly 2–4% enriched uranium dioxide as fuel, are now commonly used in the United States and other countries around the globe for commercial power generation. The most recent plants have electrical capacities in the neighborhood of 1000 MW (3000 MW thermal) which is due to a modification of the pressurized water concept where steam is produced directly by utilizing fission heat to boil water within the reactor core, rather than in an external heat exchanger.

Later on, other reactor designs based on different fuel materials, moderators, and coolants with various electrical and thermal power output were born, and examples are thermal reactors such as:

- Boiling water reactor (BWR) initiated in 1953
- Water-cooled graphite moderated in 1954
- High-temperature gas-cooled reactor (HTGR)
- Liquid-metal fast-breeder reactors (LMFBR)

Basically, all commercial reactor power plants of present interest are systems for generating steam utilizing the heat of nuclear fission to boil water and produce steam for turbine, and they are often referred to as **n**uclear **s**team **s**upply **s**ystems or **NSSS**. The steam is expanded in a turbine which drives a generator to produce electricity in the conventional manner. The exhaust steam from the turbine passes on to a condenser where it is converted into liquid water, and this is returned as feed water to the steam generator of **NSSS**.

The proportion of the heat supplied in a power plant that is actually converted into electrical energy is called the *thermal efficiency* of the system; thus, in a nuclear installation:

$$\text{Thermal Efficiency} = \frac{\text{Electrical Energy Generated}}{\text{Heat Produced in the Reactor}} \tag{4.1}$$

The maximum possible value of the thermal efficiency is the *ideal thermodynamic efficiency*, which is given by following relationship:

$$\text{Ideal Thermodynamic Efficiency} = \frac{T_2 - T_1}{T_2} \tag{4.2}$$

where:

T_1 = the absolute temperature of the steam entering the turbine (°K Kelvin).
T_2 = the temperature at which heat is rejected to the condenser (°K Kelvin).

The ideal thermodynamic efficiency can be increased by having T_2 as high as possible and T_1 as low as possible. In practice, T_1 is more or less fixed by the ambient temperature; the thermal efficiency of a steam electric plant is then largely determined by the steam temperature, which should be as high as feasible.

Conditions in PWRs and BWRs are such that the steam temperature is lower than in modern fossil-fueled power plants, in which the heat is produced by burning coal, oil, or gas. Here, the thermal efficiencies of these reactor plants are only about 33%, compared with 40% for the best fossil-fueled facilities. With the HTGRs and fast breeder reactors, however, the thermal efficiencies should equal to those of the best fossil-fueled plants, i.e., about 40%.

4.3 Future of Nuclear Power Plant System

In response to the difficulties in achieving suitability, a sufficiently high degree of safety, and a competitive economical basis for nuclear power, the US Department of Energy initiated the Generation IV program in 1999. Generation IV refers to the broad division of nuclear designs into four categories as follows:

1. Early prototype reactor (Generation I)
2. The large central station nuclear power plants of today (Generation II)
3. The advanced light-water reactors and other systems with inherent safety features that have been designed in recent years (Generation III)
4. The next-generation system to be designed and built two decades from now (Generation IV)

By 2000 international interest in the Generation IV project had resulted in a nine-country coalition that includes:

(i) Argentina
(ii) Brazil
(iii) Canada
(iv) France
(v) Japan
(vi) South Africa
(vii) South Korea
(viii) The United Kingdom
(ix) The United States

Participating countries are mapping out and collaborating on the research and development (R&D) of future nuclear energy systems.

Although the Generation IV program is exploring a wide variety of new systems, a few examples serve to illustrate the broad approaches reactor designs are developing to meet their objectives.

The next-generation systems are based on three general classes of reactors:

1. Gas cooled
2. Water cooled
3. Fast spectrum

All these categories and their brief designs, extensively, were explained in Chap. 1 of this book and briefly are covered in the next sections.

4.4 Next Generation of Nuclear Power Reactions for Power Production

Experts are projecting worldwide electricity consumption will increase substantially in the coming decades, especially in the development world, accompanying economic growth and social progress that has direct impact on rising electricity prices have focused fresh attention on nuclear power plants. New, safer, and more economical nuclear reactors could not only satisfy many of our future energy needs but could combat global warming as well. Today's existing nuclear power plants on line in the United States provide fifth of the nation's total electrical output.

Taking into account the expected increase in energy demand worldwide and the growing awareness about global warming, climate change issues, and sustainable development, nuclear energy will be needed to meet future global energy demand.

Nuclear power plant technology has evolved as distinct design generations as we mentioned in previous section and briefly summarized here again as follows:

First generation (GEN-I): prototypes and first realizations (~1950–1970)
Second generation (GEN-II): current operating plants (~1970–2030)
Third generation (GEN-III): deployable improvements to current reactors (~2000 and on)
Fourth generation (GEN-IV): advanced and new reactor systems (2030 and beyond)

The **G**eneration IV **I**nternational **F**orum, or **GIF**, was chartered in July 2001 to lead the collaborative efforts of the world's leading nuclear technology nations to develop next-generation nuclear energy systems to meet the world's future energy needs.

Eight technology goals have been defined for Generation IV systems in four broad areas:

1. Sustainability
2. Economics
3. Safety and Reliability
4. Proliferation resistance and physical protection

These ambitious goals are shared by a large number of countries as they aim at responding to economic, environmental, and social requirements of the twenty-first century. They establish a framework and identify concrete targets for focusing GIF R&D efforts.

Figure 4.1 shows the evolution of the historical generation of all nuclear powers, from past to present and future.

Table 4.1 further describes these ambitious goals in more detail for the Generation IV nuclear energy systems as well.

The next generation ("Generation IV") of nuclear energy systems is intended to meet the above goals (while being at least as effective as the "third" generation in terms of economic competitiveness, safety, and reliability) in order to provide a sustainable development of nuclear energy.

In principle, the Generation IV systems should be marketable or deployable from 2030 onward. The systems should also offer a true potential for new applications compatible with an expanded use of nuclear energy, particularly in the fields of hydrogen or synthetic hydrocarbon production, sea water desalination, and process heat production.

It has been recognized that these objectives, widely and officially shared by a large number of countries, should be at the basis of an internationally shared R&D program, which allows keeping open and consolidating the technical options and avoiding any early or premature down selection.

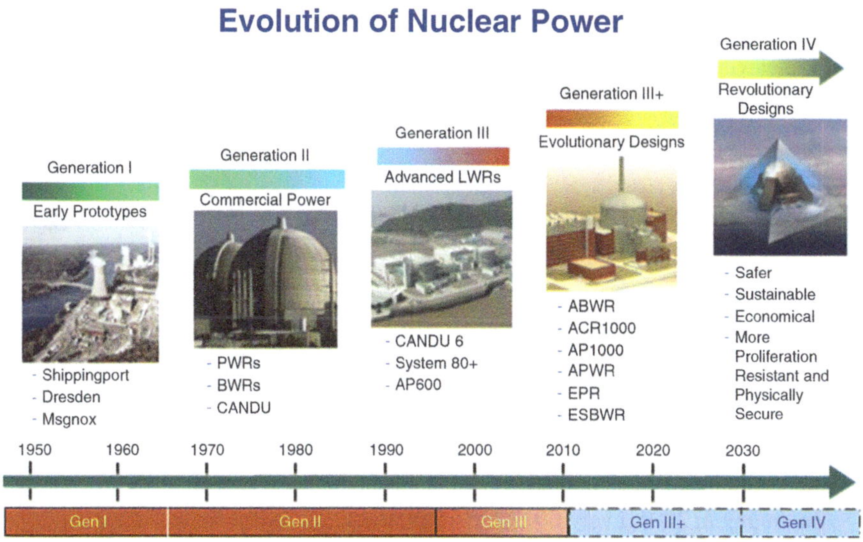

Fig. 4.1 Evolution of nuclear power

Table 4.1 Generation IV nuclear energy systems

Sustainability 1 – Generation IV nuclear energy systems will provide sustainable energy generation that meets clean air objectives and provides long-term availability of systems and effective fuel utilization for worldwide energy production
Sustainability 2 – Generation IV nuclear energy systems will minimize and manage their nuclear waste and notably reduce the long-term stewardship burden, thereby improving protection for the public health and the environment
Economics 1 – Generation IV nuclear energy systems will have a clear life-cycle cost advantage over other energy sources
Economics 2 – Generation IV nuclear energy systems will have a level of financial risk comparable to other energy projects
Safety and reliability 1 – Generation IV nuclear energy systems operations will excel in safety and reliability
Safety and reliability 2 – Generation IV nuclear systems will have a very low likelihood and degree of reactor core damage
Safety and reliability 3 – Generation IV nuclear energy systems will eliminate the need for offsite emergency response
Proliferation resistance and physical protection – Generation IV nuclear energy systems will increase the assurance that they are very unattractive and the least desirable route for diversion or theft of weapons-usable materials and provide increased physical protection against acts of terrorism

In fact, because the next-generation nuclear energy systems will address needed areas of improvement and offer great potential, many countries share a common interest in advanced R&D that will support their development. Such development benefits from the identification of promising research areas and collaborative efforts that should be explored by the international research community. The collaboration on R&D by many nations on the development of advanced next-generation nuclear energy systems will in principle aid the progress toward the realization of such systems, by leveraging resources, providing synergistic opportunities, avoiding unnecessary duplication, and enhancing collaboration.

4.5 Technology Roadmap for Generation IV Nuclear Energy Systems

The technology roadmap defines and plans the necessary research and development (R&D) to support the next generation of innovative nuclear energy systems known as Generation IV. The roadmap has been an international effort of 10 countries, including Argentina, Brazil, Canada, France, Japan, Republic of Korea, South Africa, Switzerland, the United Kingdom, and the United States, the International Atomic Energy Agency, and the OECD Nuclear Energy Agency.

Beginning in 2001, over 100 experts from these countries and international organizations began work on defining the goals for new systems, identifying many promising concepts and evaluating them, and defining the R&D needed for the most promising systems. By the end of 2002, the work resulted in a description of the six most promising systems and their associated R&D needs, and they are listed below.

1. Gas-cooled fast reactor (GFR): Features a fast-neutron-spectrum, helium-cooled reactor and closed fuel cycle
2. Very-high-temperature reactor (VHTR): A graphite-moderated, helium-cooled reactor with a once-through uranium fuel cycle
3. Supercritical water-cooled reactor (SCWR): A high-temperature, high-pressure, water-cooled reactor that operates above the thermodynamic critical point of water
4. Sodium-cooled fast reactor (SFR): Features a fast-spectrum, sodium-cooled reactor and closed fuel cycle for efficient management of actinides and conversion of fertile uranium
5. Lead-cooled fast reactor (LFR): Features a fast-spectrum, lead/bismuth eutectic liquid-metal-cooled reactor and a closed fuel cycle for efficient conversion of fertile uranium and management of actinides
6. Molten salt reactor (MSR): Produces fission power in a circulating molten salt fuel mixture with an epithermal-spectrum reactor and a full actinide recycling fuel cycle

These systems offer significant advances in sustainability, safety and reliability, economics, proliferation resistance, and physical protection. These six systems

Table 4.2 Summary of six Generation IV systems

System	Neutron spectrum	Coolant	Temp. °C	Fuel cycle	Size (MWe)
VHTR (very-high-temperature gas reactor)	Thermal	Helium	900–1000	Open	250–300
SFR (sodium-cooled fast reactor)	Fast	Sodium	550	Closed	30–150 300–1500 1000–2000
SCWR (supercritical water-cooled reactor)	Thermal/ fast	Water	510–625	Open/ closed	300–700 1000–2000
GFR (gas-cooled fast reactor)	Fast	Helium	850	Closed	1200
LFR (lead-cooled fast reactor)	Fast	Lead	480–800	Closed	20–180 300–1200 600–1000
MSR (molten salt reactor)	Epithermal	Fluoride salt	700–800	Closed	1000

feature increased safety, improved economics for electricity production and new products such as hydrogen for transportation applications, reduced nuclear wastes for disposal, and increased proliferation resistance.

In 2009, the Experts Group published an outlook on Generation IV R&D to provide a view of what GIF members hope to achieve collectively in the period 2010–2014. All Generation IV systems have features aiming at performance improvement, new applications of nuclear energy, and/or more sustainable approaches to the management of nuclear materials. High-temperature systems offer the possibility of efficient process heat applications and eventually hydrogen production. Enhanced sustainability is achieved primarily through adoption of a closed fuel cycle with reprocessing and recycling of plutonium, uranium, and minor actinides using fast reactors; this approach provides significant reduction in waste generation and uranium resource requirements. Table 4.2 summarizes the main characteristics of the six Generation IV systems.

An extensive description of the six most promising nuclear power systems of Generation IV (GEN-IV) is given in Chap. 1 with their artistic infrastructures.

4.6 Power Conversion Study and Technology Options Assessment

A study done in September 2004 by a team of experts at the University of California, Nuclear Engineering Department showed in their executive summary that the electrical power conversion system (PCS) for the Next Generation Nuclear Plant (NGNP) will take advantage of significantly higher reactor outlet temperature to provide greater efficiency than can be achieved by the current generation of light-water reactors. In anticipation of the design, development, and procurement of an advanced power conversion system for NGNP, this study was initiated to identify the major design and technology options and the trade-offs that must be considered in the evaluation of PCS options to support future research and procurement decisions. These PCS technology options affect cycle efficiency, capital cost, system reliability and maintainability and technical risk, and therefore the cost of electricity from Generation IV systems. A reliable evaluation and estimate of actual costs require an optimized, integrated PCS design. At this early stage of the NGNP project, it is useful to identify the technology options that will be considered in the design of proposed PCS systems, identify the system performance and cost implications of these design options, and provide a general framework for evaluating the design choices and technology trade-offs.

The ultimate measure of the value of power conversion options is the cost of electricity produced, which is a function of capital and operating cost recovery and the system efficiency and reliability. Evaluating cost is difficult to do without detailed integrated designs, but it is possible to identify the factors that influence component and system performance, cost, and technical risk. In this study, several existing Brayton conversion system designs were studied to illustrate and evaluate the implications of the major design choices, to assess performance against the Generation IV economics and sustainability goals, and to identify areas of technical incompleteness or weakness. Several reference system designs were considered to provide a semiquantitative basis for performing comparisons. The reference systems include the GT-MHR, PBMR, GTHTR-300, Framatome indirect cycle design, and AHTR high-temperature Brayton cycle designs. Where appropriate, Generation II, III, and III + light-water reactors (two 1970s designs, the EPR and the ESBWR) were also considered.

The design choices and technology options considered relevant for the assessment of NGNP power conversion include the cycle types and operational conditions, such as working fluid choices, direct vs. indirect, system pressure, and interstage cooling and heating options. The cost and maintainability of the PCS are also influenced by the PCS layout and configuration including distributed vs. integrated PCS designs, single vs. multiple shafts, shaft orientation, and the implications for the pressure boundary.

From the summary table below, it is apparent that high-temperature gas reactor power conversion design efforts to date have resulted in very different design choices based on project-specific requirements and performance or technical risk requirements.

In the review of existing designs and the evaluation of the major technology options, it immediately becomes apparent that optimized design involves a complex trade-off of diverse factors, such as cost, efficiency, development time, maintainability, and technology growth path, which must be considered in an integrated PCS system context before final evaluation (see Table 4.3 as well). General observations

Table 4.3 Summary of PCS design features for representative gas reactor systems

Feature	PBMR (horizontal)	GT-MHR	GTHTR300	Framatome indirect	AHTR-IT
Thermal power (MWt)	400	600	600	600	2400
Direct vs. indirect cycle	Direct	Direct	Direct	Indirect	Indirect
Recuperated vs. combined cycle	Recuperated	Recuperated	Recuperated	Combined	Recuperated
Intercooled vs.non-intercooled	Intercooled	Intercooled	Non-intercooled	Intercooled	Intercooled/reheat
Integrated vs. distributed PCS	Distributed	Integrated	Distributed	Distributed	Distributed (modular)
Single vs. multipleTM shafts	Single (previously multiple)	Single	Single	Single	Multiple (modular)
Synchronous vs.asynchronous	Reduction to synchronous	Asynchronous	Synchronous	Synchronous	Synchronous
Vertical vs. horizontal TM	Horizontal	Vertical	Horizontal	Horizontal	Vertical
Submerged vs. external generator	External	Submerged	Submerged	External	Submerged

derived from the review of the reference systems, including comparisons with light-water reactors systems where applicable, are as follows:

- There are key PCS design choices that can have large effects on PCS power density and nuclear island size, making careful and detailed analysis of design trade-offs important in the comparison of PCS options.
- Considering the major construction inputs for nuclear plants – steel and concrete – high-temperature reactors appear to be able to break the economy-of-scale rules for LWRs and achieve similar materials-inputs performance at much smaller unit sizes.
- For HTRs, a much larger fraction of total construction inputs go into the nuclear island. To compete economically with LWRs, HTRs must find approaches to reduce the relative costs for nuclear-grade components and structures.

Power conversion system technology options also include variations on the cycle operating conditions and the cycle type that can have an important impact on performance and cost. These options include:

- Working fluid choice – He, N^2, CO^2, or combinations have been considered. Working fluid physical characteristics influence cycle efficiency and component design.
- System pressure – higher pressures lead to moderate efficiency increases and smaller PCS components but increase the pressure boundary cost, particularly for the reactor vessel, which introduces a component design and a system cost and performance trade-off.

- Direct vs. indirect – indirect cycles involve an intermediate heat exchanger (IHX) and resulting efficiency reduction and more complex control requirements but facilitate maintenance.
- Interstage cooling (or heating) results in higher efficiency but greater complexity.

Some of the observations from this assessment of these factors include the following:

- Differences between He versus N^2 working fluids were not considered critical for turbomachinery design, because both involve similar differences from current combustion turbines, with the primary difference being in the heat exchanger size to compensate for the lower N^2 thermal conductivity.
- N^2 allows 3600 rpm compressor operation at thermal powers at and below 600 MW(t), while He compressors must operate at higher speeds requiring reduction gears, asynchronous generators, or multiple-shaft configurations. However, power-up rating to approximately 800 MW(t) would permit 3600 rpm He compressor operation, providing a potentially attractive commercialization approach. Turbomachinery tolerances for He systems do not appear to be a key issue.
- Direct/indirect – efficiency loss can be 2–4%, depending on design, and the IHX becomes a critical component at high temperatures. Direct cycles have an extended nuclear-grade pressure boundary. Maintainability is considered a key design issue for direct cycles.
- Interstage cooling, as well as bottoming cycles (Rankine), can result in significant efficiency improvements but at a cost of complexity and lower-temperature differences for heat rejection, affecting the potential for dry cooling and reduced environmental impact from heat rejection.

The PCS configuration and physical arrangement of the system components have important effects on the volume and material inputs into structures, on the pressure boundary volume and mass, on gas inventories and storage volume, on the uniformity of flow to heat exchangers, on pressure losses, and on maintainability. The major factors considered in this study include:

- Distributed vs. integrated PCS design approach – PCS components can be located inside a single pressure vessel (e.g., GT-MHR) or can be divided between multiple pressure vessels (e.g., PBMR, HTR-300). This is a major design choice, with important impacts in several areas of design and performance.
- Shaft orientation (vertical/horizontal) – orientation affects the compactness of the system and the optimal design of ducting between turbomachinery and heat exchangers. Vertical turbomachinery provides a reduced PCS footprint area and building volume and can simplify the ducting arrangement to modular recuperator and intercooler heat exchangers.
- Single vs. multiple shafts – single shafts may include flexible couplings or reduction gears. In multiple-shaft systems, turbo-compressors are separated from synchronous turbo-generators, allowing the compressors to operate at higher speed and reducing the number of compressor stages required. Multiple shafts

and flexible couplings reduce the weight of the individual turbomachines that bearings must support.
- Pressure boundary design – the pressure vessels that contain the PCS typically have the largest mass of any PCS components and provide a significant (~33%) contribution to the total PCS cost.

4.6.1 Heat Exchanger Components

Heat exchanger components are defined and required designs are summarized as follows:

- Heat exchanger designs have significant impacts on both the efficiency and cost of the PCS. For a given heat exchanger type, higher effectiveness must be balanced against the increased size or pressure drop implications. Using small passages increases heat exchanger surface area per unit volume, but those same small passages tend to reduce the heat-transfer coefficient due to laminar flow. Higher pressures may be utilized to force those flows back into the turbulent region, but those higher pressures force construction of a more robust pressure boundary and increase pumping power.
- The recuperator effectiveness and total heat exchanger (HX) pressure drop is a significant impact on the cycle efficiency, and there is significant leverage in optimizing the recuperator design for both high heat-transfer effectiveness and minimum pressure drop. For modular recuperators, careful attention must be paid to the module configuration and duct design to obtain equal flow rates to each module.
- Material limitations may limit the operating temperatures for many components, including the reactor vessel and heat exchangers. But because of the large flexibility of the Brayton cycle, high-efficiency systems can still be designed within these limitations. Fabrication techniques will probably differ between intermediate, pre- and intercooler, and recuperator heat exchangers, because of their operating temperature ranges. It would appear that transients could be tolerated by most of these heat exchanger designs.

4.6.2 Turbomachinery

Turbomachinery, used in new-generation nuclear power system (GEN-IV), plays a significant role in building them for the commercial application in order to cater to the rising need for electricity in the future.

- First-order estimates of key turbine and compressor design and performance characteristics can be made with low-level analysis. For the reference systems, key turbomachinery design parameters (speed, stages, stage diameters, blade

heights, blade clearances) will be similar to current commercial gas turbine engines.
- At lower reactor thermal powers, He compressors will require greater than 3600 rpm operation to achieve efficiency goals (800 MW(t) allows 3600 rpm operation).
- Maximum system temperatures in the reference designs are near the limit for uncooled turbines.
- For both direct and indirect designs, the seals and housing and bearing components will be fundamentally different than current gas turbines, requiring extensive development with the associated cost and risk.

These observations illustrate the complex interactions of the many design choices that will be considered in the NGNP power conversion system. It is clear that detailed and integrated design efforts must be performed on candidate designs before quantitative evaluations are possible. The assessment described in this study helps illuminate those critical design choices and the resulting implications for the cost and performance of the future Next Generation Nuclear Plant (NGNP) PCS design.

4.6.3 Advanced Computational Materials Science Proposed for GEN-IV Systems

A renewed interest in nuclear reactor technology has developed in recent years, in part as a result of international interest in sources of energy that do not produce CO^2 as a by-product. One result of this interest was the establishment of the Generation IV International Forum, which is a group of international governmental entities whose goal is facilitating bilateral and multilateral cooperation related to the development of new nuclear energy systems.

Historically, both the fusion and fission reactor programs have taken advantage of and built on research carried out by the other program. This leveraging can be expected to continue over the next 10 years as both experimental and modeling activities in support of the GEN-IV program grow substantially. The GEN-IV research will augment the fusion studies (and vice versa) in areas where similar materials and exposure conditions are of interest. However, in addition to the concerns that are common to both fusion and advanced fission reactor programs, designers of a future DT fusion reactor [5–7] have a unique problem of anticipating the effects of the 14 MeV neutron source term. For example, advances in computing hardware and software should permit improved (and in some cases the first) descriptions of relevant properties in alloys based on ab initio calculations. Such calculations could provide the basis for realistic interatomic potentials for alloys, including alloy-He potentials that can be applied in classical molecular dynamic simulations. These potentials must have a more detailed description of many-body interactions than accounted for in the current generations which are generally based on a simple embedding function. In addition, the potentials used under fusion reactor conditions (very high PKA energies) should account for the effects of local electronic excitation and electronic energy loss. The computational cost of using more complex potentials also requires the next generation of massively parallel computers. New results of ab initio and atomistic calculations can be coupled with ongoing advances in kinetic and phase-field models to dramatically improve predictions of the nonequilibrium, radiation-induced evolution in alloys with unstable microstructures. This includes phase stability and the effects of helium on each microstructural component.

However, for all its promise, computational materials science is still a house under construction. As such, the current reach of the science is limited. Theory and modeling can be used to develop understanding of known critical physical phenomena, and computer experiments can, and have been used to, identify new phenomena and mechanisms and aid in alloy design. However, it is questionable whether the science will be sufficiently matured in the foreseeable future to provide a rigorous scientific basis for predicting critical material's properties or for extrapolating well beyond the available validation database.

Two other issues remain even if the scientific questions appear to have been adequately answered. These are licensing and capital investment. Even a high degree of scientific confidence that a given alloy will perform as needed in a particular Gen-IV or fusion environment is not necessarily transferable to the reactor licensing or capital market regimes. The philosophy, codes, and standards employed for reactor licensing are properly conservative with respect to design data requirements.

Experience with the US Nuclear Regulatory Commission suggests that only modeling results that are strongly supported by relevant, prototypical data will have an impact on the licensing process. In a similar way, it is expected that investment on the scale required to build a fusion power plant (several billion dollars) could only be obtained if a very high level of confidence existed that the plant would operate long and safely enough to return the investment.

These latter two concerns appear to dictate that an experimental facility capable of generating a sufficient, if limited, body of design data under essentially prototypic conditions (i.e., with ~14 MeV neutrons) will ultimately be required for the commercialization of fusion power. An aggressive theory and modeling effort will reduce the time and experimental investment required to develop the advanced materials that can perform in a DT fusion reactor [5–7] environment. For example, the quantity of design data may be reduced to that required to confirm model predictions for key materials at critical exposure conditions. This will include some data at a substantial fraction of the anticipated end-of-life dose, which raises the issue of when such an experimental facility is required. Long lead times for construction of complex facilities, coupled with several years of irradiation to reach the highest doses, imply that the decision to build any fusion-relevant irradiation facility must be made on the order of 10 years before the design data is needed.

Two related areas of research can be used as reference points for the expressed need to obtain experimental validation of model predictions. Among the lessons learned from ASCI, the importance of code validation and verification was emphasized at the workshops among the courtiers involved with such research.

Because of the significant challenges associated with structural materials applications in these advanced nuclear energy systems, the *Workshop on Advanced Computational Materials Science: Application to Fusion and Generation IV Fission Reactors* was convened by the US Department of Energy's Office of Science and Office of Nuclear Energy, Science, and Technology to ensure that research funded by these programs takes full advantage of ongoing advancements in computational science and the Department's investment in computational facilities. In particular, participants in the workshop were asked to:

1. Examine the role of high-end computing in the prediction of materials behavior under the full spectrum of radiation, temperature, and mechanical loading conditions anticipated for advanced structural materials that are required for future Generation IV fission and fusion reactor environments.
2. Evaluate the potential for experimentally validated computational modeling and simulation to bridge the gap between data that is needed to support the design of these advanced nuclear technologies and both the available database and data that can be reasonably obtained in currently available irradiation facilities.

The need to develop materials capable of performing in the severe operating environments expected in Generation IV reactors represents a significant challenge in materials science. There is a range of potential GEN-IV fission reactor design concepts, and each concept has its own unique demands. Improved economic performance is a major goal of the GEN-IV designs. As a result, most designs call for significantly higher operating temperatures than the current generation of LWRs to obtain higher thermal efficiency. In many cases, the desired operating temperatures rule out the use of the structural alloys employed today. The very high operating temperature (up to 1000 °C) associated with the Next Generation Nuclear Plant (NGNP) is a prime example of an attractive new system that will require the development of new structural materials.

However, the Department of Energy (DOE) and Idaho National Laboratory (INL) established the NGNP Project as required by Congress in Subtitle C of Title VI of the Energy Policy Act of 2005. The mission of the NGNP Project is to develop, license, build, and operate a prototype modular high-temperature gas-cooled reactor (HTGR) plant that would generate high-temperature process heat for use in hydrogen production and other energy-intensive industries while generating electric power at the same time. As stipulated by the Energy Policy Act of 2005, pre-licensing activities for the NGNP prototype began with the development of the NGNP Licensing Strategy Report to Congress that was jointly issued by the Nuclear Regulatory Commission (NRC) and DOE in August 2008. Subsequent NRC interactions with DOE and INL centered primarily on the NRC's review and assessment of a series of NGNP white paper submittals

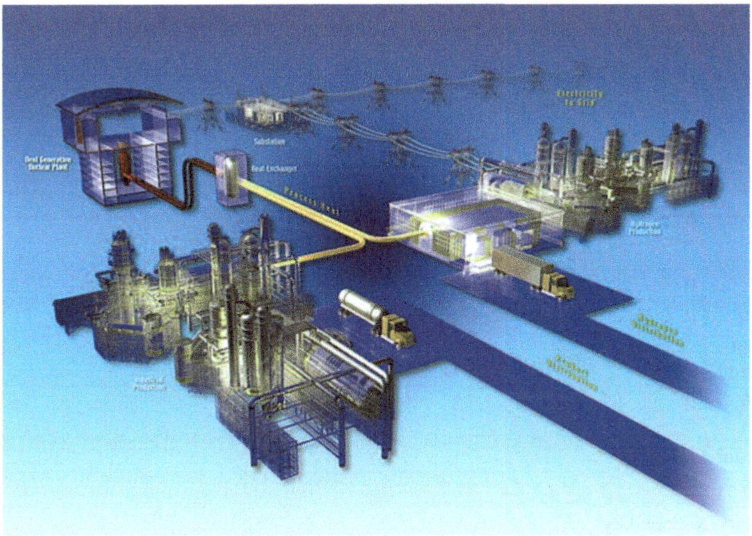

Fig. 4.2 Artistic layout of GEN-IV power plant station

that describe the approaches that DOE and INL propose to pursue in establishing the technical safety bases and criteria for licensing the NGNP prototype. NGNP pre-licensing interactions began in 2006 and were suspended in 2013 after DOE decided in 2011 not to proceed into the detailed design and license application phases of the NGNP Project. DOE's decision cited impasses between DOE and the NGNP Industry Alliance in cost-sharing arrangements for the public-private partnership required by Congress. See Fig. 4.2 for a conceptual and artistic layout of GEN-IV power plant station.

The operating temperatures, neutron exposure levels, and thermomechanical stresses for proposed GEN-IV fission reactors are huge technological challenges among material scientist and engineers. In addition, the transmutation products created in the structural materials by the high-energy neutrons produced in this generation of nuclear power reactors can profoundly influence the microstructural evolution and mechanical behavior of these materials.

4.6.4 Material Classes Proposed for GEN-IV Systems

Table 4.4 shows the suggested materials for new generation of nuclear reactors by the Department of Energy. The types of materials that are proposed in the Department of Energy (DOE) workshop on March of 2004 are tabulated as follows:

Table 4.4 Materials suggested for GEN-IV by the Department of Energy

Structural materials							
System	Ferritic-martensitic stainless-steel alloys	Austenitic stainless-steel alloys	Oxide dispersion strengthened steels	NI-based alloys	Graphite	Refractory alloys	Ceramics
GFR	P	P	P	P		P	P
LFR	P	P	S			S	S
MSR				P	P	S	S
SFR	P	P	P				
SCWR – thermal spectrum	P	P	S	S			
SCWR – fast spectrum	P	P	S	S			
VHTR	S			P	P	S	P

P primary, *S* secondary

4.6.5 *Generation IV Materials: Challenges*

The summary of these challenges for the generation of nuclear power plant is presented here.

- **Higher-temperature/larger-temperature ranges**
  - Examples
    - VHTR coolant outlet temperature near 1000 °C
    - GFR transient temps to 1600–1800 °C, gradient across core of ~400 °C
    - LFR to 800 °C steady-state outlet
  - Issues
    - Creep
    - Fatigue
    - Toughness
    - Corrosion/SCC
- *Must drive modeling toward a predictive capability of materials properties in complex alloys across a wide temperature range*
- **High fluence/dose**
  - Examples
    - LFR, SFR cladding
    - SCWR core barrel
    - GFR matrix

  - Issues
    - Swelling
    - Creep and stress relaxation
- *Must drive modeling toward a predictive capability of materials properties in complex alloys to large radiation dose*
- **Unique chemical environments**
  - Examples
    - Pb and Pb-Bi Eutectic
    - Supercritical water
    - High-temperature oxidation in gas-cooled systems
    - Molten salts
  - Issues
    - Corrosion
    - SCC/IASCC
    - Liquid-metal embrittlement
- *Must drive modeling toward a predictive capability of chemical interactions in complex alloys to large radiation dose*

4.7 Generation IV Materials: Fundamental Issues

The coevolution of all components of the microstructure and their roles in the macroscopic response in terms of swelling, anisotropic growth, irradiation creep, and radiation-induced phase transformations should be studied within the science of complex systems.

Figure 4.3 is an illustration of top level material's study from a microstructure and macroscopic viewpoint.

In summary, we can conclude that:

- Six concepts have been identified with the potential to meet the Generation IV goals
- Concepts operate in more challenging environments than current LWRs, and significant material development challenges must be met for any of the Generation IV systems to be viable.
- Experimental programs cannot cover the breadth of materials and irradiation conditions for the proposed GEN-IV reactor designs.
- Modeling and microstructural analysis can provide the basis for a material selection that is performed based on an incomplete experimental database and that requires considerable judgment to carry out the necessary interpolation and extrapolation.

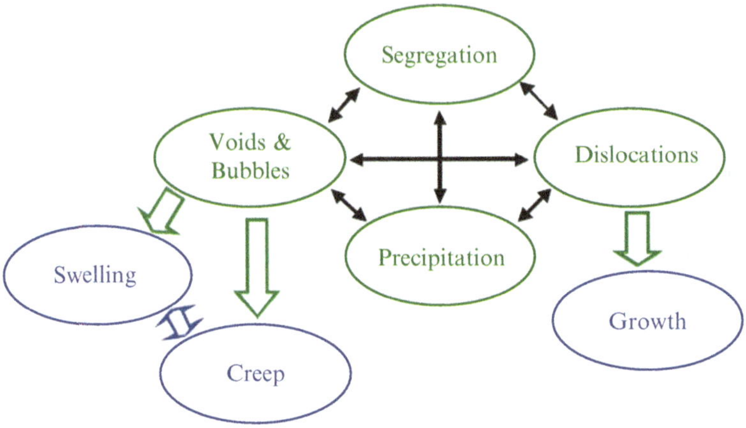

Fig. 4.3 Top level material's study from a microstructure and macroscopic viewpoint

4.8 A Combined Cycle Power Conversion System Driven Next Generation IV

Today's global energy market places many demands on power generation technology including high thermal efficiency, low cost, rapid installation, reliability, environmental compliance, and operation flexibility.

Several technologies are being investigated for the Next Generation Nuclear Plant or Generation IV (GEN-IV) that will produce heated fluids at significantly higher temperatures than current generation power plants (i.e., third generation that is known as GEN-III).

Higher temperatures offer the opportunity to significantly improve the thermodynamic efficiency of the energy conversion cycle. One of the concepts currently under study is the molten salt reactor. The coolant from the molten salt reactor may be available at temperatures as high as 800–1000 °C. At these temperatures, an open Brayton cycle combined with and Rankine bottoming cycle appears to have some strong advantages. Thermodynamic efficiencies approaching 50% appear possible. Requirements for circulating cooling water will be significantly reduced.

However, to realistically estimate the efficiencies achievable, it is essential to have good models for the heat exchangers involved as well as the appropriate turbomachinery. This chapter will present studies that have concentrated on modeling all power conversion equipment from the fluid exiting the reactor to the energy releases to the environment.

Gas turbine engines, both aircraft and industrial power generation, represent one of the most aggressive applications for structural materials. With ever-growing demands for increasing performance and efficiencies, all classes of materials are being pushed to higher-temperature capabilities. These materials must also satisfy stringent durability and reliability criteria. As materials are developed to meet these demanding requirements, the processing of these materials often becomes very complicated and expensive. As a result, the cost of materials and processes has become a much larger consideration in the design and application of high-performance materials. Both the aircraft engine and power generation industries are highly cost competitive, and market advantage today relies on reducing cost as well as increasing performance and efficiency.

Development of high-temperature/high-strength materials, corrosion-resistant coatings, and improved cooling technology has led to increases in gas turbine firing temperatures. This increase in firing temperature is the primary development that has led to increases in combined cycle gas turbine (CCGT) thermal efficiencies. The improvements in combined cycle thermal efficiencies and the commercial development of combined cycle power plants have proceeded in parallel with advances in gas turbine technologies.

Heat exchangers, filters, turbines, and other components in integrated next-generation nuclear power plant (GEN-IV) combined cycle system must withstand demanding conditions of high temperatures and pressure differentials. Under the highly sulfating conditions of the high temperature such as inlet hot steam or other related environmental effects, the performance of components degrades significantly with time unless expensive high-alloy materials are used. Deposition of a suitable coating on a low-cost alloy may improve its resistance to such sulfidation attack and decrease capital and operating costs. A review of the literature indicates that the corrosion reaction is the competition between oxidation and sulfidation reactions. The Fe- and Ni-based high-temperature alloys are susceptible to sulfidation attack unless they are fortified with high levels of Cr, Al, and Si. To impart corrosion resistance, these elements need not be in the bulk of the alloy and need only be present at the surface layers.

A number of technologies are being investigated for the Next Generation Nuclear Plant that will produce heated fluids at significantly higher temperatures than current generation power plants. Higher temperatures offer the opportunity to significantly improve the thermodynamic efficiency of the energy conversion cycle. One of the concepts currently under study is the molten salt reactor. The coolant from the molten salt reactor may be available at temperatures as high as 800–1000 °C. At these temperatures, an open Brayton cycle combined with and Rankine bottoming cycle appears to have some strong advantages. Thermodynamic efficiencies approaching 50% appear possible. Requirements for circulating cooling water will be significantly reduced. However, to realistically estimate the efficiencies achievable it is essential to have good models for the heat exchangers involved as well as the appropriate turbomachinery. This study has concentrated on modeling all power conversion equipment from the fluid exiting the reactor to the energy releases to the environment.

The Rankine bottoming cycle appears to offer significant advantages over the recuperated Brayton cycle as part of computer simulation written by this author. The overall cycle was optimized as a unit, and lower-pressure Rankine systems seem to be more efficient. The combined cycle requires a lot less circulating water for a heat dump than current power plants. Molten salt reactors and lead-cooled reactors have been extended to temperatures typical of liquid-metal cooled fast reactors. A split compressor with intercooler has been added to the Brayton cycle that serves as a "feed water heater" for the steam bottoming cycle as part of computer simulation. Adding this component to the system provides about a 3% increase in efficiency. For purpose of simplicity, an in-house computer code was developed, where the consideration was on the steady-state situation; however a more sophisticated computer code is needed, which will take the transient approach under consideration for better accuracy and optimization of these analysis.

Results of modeling a combined cycle Brayton-Rankine power conversion system are presented. The Rankine bottoming cycle appears to offer significant advantages over the recuperated Brayton cycle. The overall cycle was optimized as a unit, and the lower-pressure Rankine systems seem to be more efficient. The combined cycle requires a lot less circulating water for a heat dump than current power plants [8].

4.8.1 Modeling the Brayton Cycle

Any external combustion or heat engine system is always at a disadvantage to an internal combustion system. The internal combustion systems used in current jet engine and gas turbine power systems can operate at very high temperatures in the fluid and cool the structures containing the fluid to achieve high thermodynamic efficiencies. In an external energy generation system, like a reactor powered one, all of the components from the core to the heat exchangers heating the working fluid must operate at a higher temperature than the fluid. This severely limits the peak cycle temperature compared to an internal combustion system. This liability can be overcome to a certain extent by using multiple expansion turbines and designing highly efficient heat exchangers to heat the working fluid between expansion processes.

Typically, the combustion chamber in a gas turbine involves a pressure drop of 3–5% of the total pressure. Efficient liquid salt to air heat exchangers can theoretically be designed with a pressure drop of less than 1%. This allows three to five expansion cycles to achieve a pressure drop comparable to a combustion system. Multiple turbines operating at different pressures have been common in steam power plants for number of years. In this study three to five gas turbines operating on a common shaft were considered.

Multiple expansion turbines allow a larger fraction of the heat input to be provided near the peak temperature of the cycle significantly improving the thermodynamic efficiency. The exhaust from the last turbine is provided to the Heat Recovery Steam Generator (HRSG) to produce the steam used in the Rankine bottoming cycle. The hot air after it passes through the HRSG is exhausted to the atmosphere. A detailed comparison of this system was made with a recuperated stand-alone Brayton cycle, and the dual cycle appears to be more efficient for open systems.

4.8.2 *Modeling the Rankine Cycle*

The Rankine cycle was modeled with the standard set of components including the HSRG, a steam turbine, condenser, and high-pressure pump. Multiple reheat processes were considered. There is a slight efficiency advantage to include two reheat processes as per fairly standard design practices in today's power plants.

The major limitation on the size of the steam system is the enthalpy available from high-temperature air above the pinch point where the high-pressure water working fluid starts to vaporize. Below this point, there is still a significant enthalpy in the air which is readily available to heat the high-pressure water. There does not appear to be an advantage to including feed water heaters in the cycle to bring the high-pressure water up to the saturation point. The possibility that an intercooler could be inserted between the two stages of a split compressor was considered. The cooling fluid for the intercooler was the high-pressure water coming out of the water pump.

This process would combine the function of the traditional intercooler with the preheating of a typical feed water heater. The effect of this addition to the 2 cycles had a marginal effect on the overall system efficiency and likely is not worth the cost, or effort, to implement.

4.8.3 *The Combined Brayton-Rankine Cycle*

The combined cycle unit combines the Rankine (steam turbine) and Brayton (gas turbine) thermodynamic cycles by using heat recovery boilers to capture the energy in the gas turbine exhaust gases for steam production to supply a steam turbine as shown in figure "Combined Cycle Cogeneration Unit" (see Fig. 4.4). Process steam can be also provided for industrial purposes [8].

Fossil fuel-fired (central) power plants use either steam or combustion turbines to provide the mechanical power to electrical generators. Pressurized high-temperature steam or gas expands through various stages of a turbine, transferring energy to the rotating turbine blades. The turbine is mechanically coupled to a generator, which produces electricity.

The Brayton cycle efficiency is quite low primarily because a substantial amount of the energy input is exhausted to surroundings. This exhausted energy is usually at a relatively high temperature, and thus it can be used effectively to produce power. One possible application is the combined Brayton Rankine cycle in which the high-temperature exhaust gases exiting the gas turbine are used to supply energy to the boiler of the Rankine cycle, as illustrated in Fig. 4.5. Note that the temperature T_9 of the Brayton cycle gases exiting the boiler is less than the temperature T_3 of the Rankine cycle steam exiting the boiler; this is possible in the counterflow heat exchanger, the boiler.

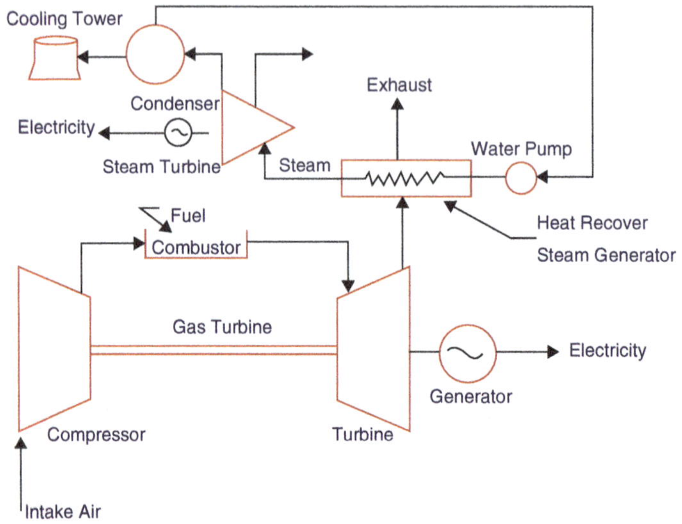

Fig. 4.4 Combined cycle illustration

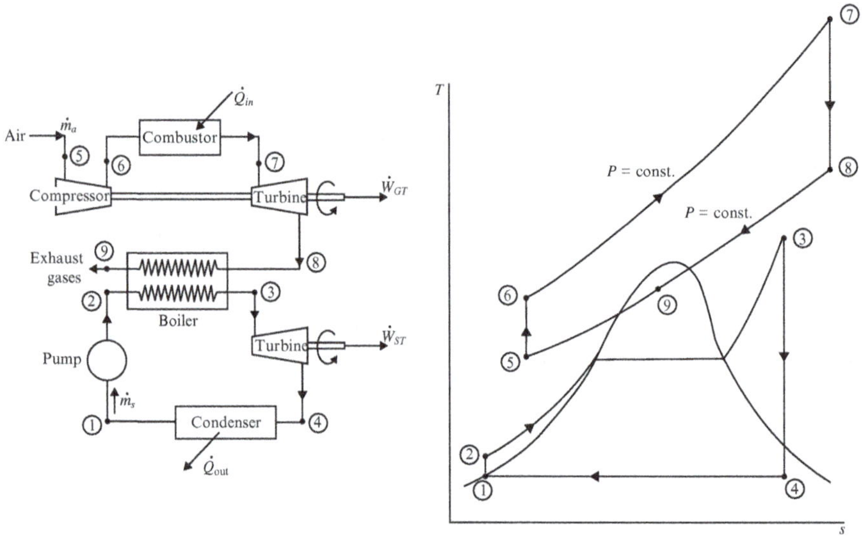

Fig. 4.5 The combined Brayton-Rankine cycle

To relate the air mass flux $\dot{m}_a$ of the Brayton cycle to the steam mass flux $\dot{m}_s$ of the Rankine cycle, we use an energy balance in the boiler; it gives (see Fig. 4.5):

$$\dot{m}_a\left(h_8 - h_9\right) = \dot{m}_s\left(h_3 - h_2\right) \tag{4.3}$$

This equation is valid, assuming no additional energy addition in the boiler, which would be possible with an oil burner, for example.

The cycle efficiency would be found by considering the purchased energy as $\dot{Q}_{\text{in}}$, the energy input in the combustor. The output is the sum of the net output $\dot{W}_{\text{GT}}$ from the gas turbine and the output $\dot{W}_{\text{ST}}$ from the steam turbine. The combined cycle efficiency is thus given by

$$\eta = \frac{\dot{W}_{\text{GT}} - \dot{W}_{\text{ST}}}{\dot{Q}_{\text{in}}} \tag{4.4}$$

An example will illustrate the increase in efficiency of such a combined cycle.

Example A simple steam power plant operates between pressures of 10 kPa and 4 MPa with a maximum temperature of 400 °C. The power output from the steam turbine is 100 MW. A gas turbine provides the energy to the boiler; it accepts air at 100 kPa and 25 °C and has a pressure ratio of 5 and a maximum temperature of 850 °C. The exhaust gases exit the boiler at 350 K. Determine the thermal efficiency of the combined Brayton-Rankine cycle.

Solution If we neglect the work of the pump, the enthalpy remains unchanged across the pump. Hence, $h_2 = h_1 = 192$ kJ/kg. At 400 °C and 4 MPa, we have $h_3 = 3214$ kJ/kg and $s_3 = 6.7698$ kJ/kg.K. State 4 is located by noting that $s_4 = s_3$ so that the quality is

$$x_4 = \frac{s_4 - s_f}{s_{\text{fg}}} = \frac{6.798 - 0.6491}{7.5019} = 0.8159$$

Thus, $h_4 = h_f + x_4 h_{\text{fg}} = 192 + (0.8159)(2393) = 2144$ kJ/kg. The steam mass flux is found using the turbine output as follows:

$$\dot{W}_{\text{ST}} = \dot{m}_s (h_3 - h_4) 100{,}000.00 = \dot{m}_s (3214 - 2144) \dot{m}_s = 93.46 \text{ kg/s}$$

Considering the gas turbine cycle,

$$T_6 = T_5 \left(\frac{P_6}{P_5} \right)^{(k-1)/k} = (298)(5)^{0.2857} = 472.0\,\text{K}$$

and also

$$T_8 = T_7 \left(\frac{P_8}{P_7} \right)^{(k-1)/k} = (1123) \left(\frac{1}{5} \right)^{0.2857} = 709.1\,\text{K}$$

thus, we have, for the boiler:

$$\dot{m}_s\left(h_3 - h_2\right) = \dot{m}_a C_p\left(T_8 - T_9\right)(93.46)(3214 - 192) = \left(\dot{m}_a\right)(1.00)(709.1 - 350)$$

$$\dot{m}_a = 786.5\ \text{kg/s}$$

The output of the gas turbine is (note that this is not $\dot{W}_{\text{GT}}$):

$$\dot{W}_{\text{turb}} = \dot{m}_a C_p\left(T_7 - T_8\right) = (786.5)(1.00)(1123 - 709.1) = 325.5\,\text{MW}$$

The energy needed by the compressor is:

$$\dot{W}_{\text{comp}} = \dot{m}_a C_p\left(T_6 - T_5\right) = (786.5)(1.00)(472 - 298) = 136.9\,\text{MW}$$

Hence, the net gas turbine output is $\dot{W}_{\text{GT}} = \dot{W}_{\text{turb}} - \dot{W}_{\text{comp}} = 325.5 - 136.9 = 188.6$ MW. The energy input by the combustor is:

$$\dot{Q}_m = \dot{m}_a C_p\left(T_7 - T_6\right) = (786.5)(1.00)(1123 - 472) = 512\,\text{MW}$$

The above calculations allow us to determine the combined cycle efficiency as:

$$\eta = \frac{\dot{W}_{\text{ST}} + \dot{W}_{\text{GT}}}{\dot{Q}_m} = \frac{100 + 188.6}{512} = 0.564\,\text{or}\,56.4\%$$

4.8.4 *Study Results*

Multiple turbines and highly efficient heat exchangers can produce highly efficient systems. A typical example for a turbine inlet temperature of 800 °C (1073 K) and a 100 MWe system using four gas turbine expansion processes and three steam turbine expansion processes produces a system with the following characteristics [9, 10]:

Air turbine inlet temp	1073.0 K
Air turbine exit temp	903.0 K
Compressor ratio	21.8
Brayton $\dot{m}_{\text{Air}}$	205.6 kg/s
Brayton efficiency	31.3%
Brayton power	60.2 MW
Rankine $\dot{m}_{\text{Water}}$	21.2 kg/s
Rankine efficiency	42.5%
Rankine power	39.8 MW
Overall efficiency	51.7%
Heat rejection to water	53 MW

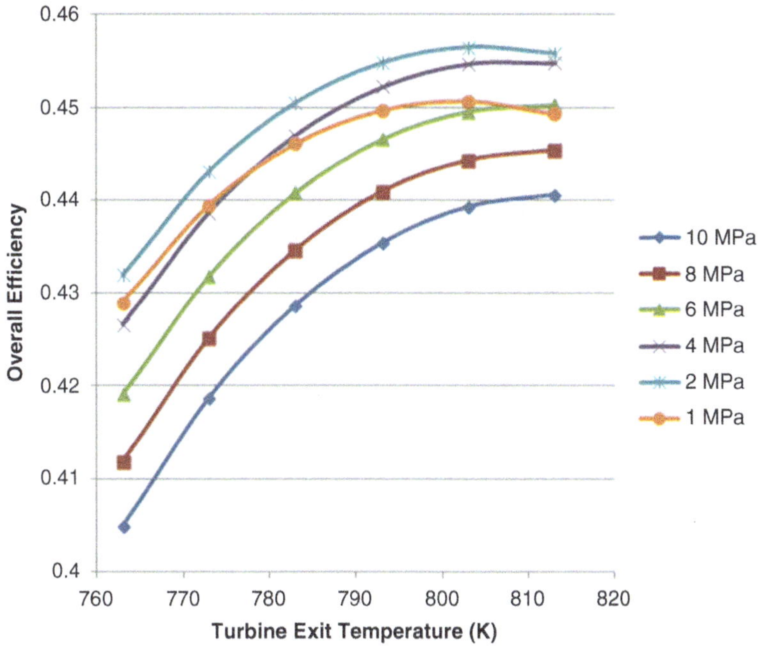

Fig. 4.6 System efficiency versus turbine exit temperature [9]

Note that the bottoming cycle produces 40% of the power and has a mass flow rate that is approximately 10% of the air mass flow rate. Also, the heat rejected to cooling water is 53 MW for a 100 MW power plant. This is roughly one fourth of the heat rejection requirement for a current 33% efficient power plant operating at the same electrical power level. It may be possible to increase the condenser temperature for the bottoming cycle, at a decrease in overall efficiency, so that the condenser could be cooled by air. This would mean that this type of power plant would be free of the requirement to have a body of water nearby.

At a more modest turbine inlet temperature of 660 °C, a comparison of possible turbine exit temperatures and bottoming cycle pressures was made to determine the optimum overall efficiency as a function of these two parameters. The results are displayed in Fig. 4.6.

The optimum overall efficiency occurs at a turbine exit temperature of 803 K and a steam pressure of 2 megapascal. Higher Rankine cycle pressures might produce higher cycle efficiencies, but the mass flow rate must be lower due to the narrowing of the range between the turbine exit temperature and the water saturation temperature at the higher pressures. Thus, the combined cycle optimizes at a lower steam pressure than might be expected.

Additional work will be aimed at varying the expansion ratio per turbine and the penalties associated with air cooling the condenser.

The detailed above analysis provided in the report by McDaniel et al. [9] and a FORTRAN computer code, using steady-state approach, were written to provide the inputs for the Excel sheet for further analysis of combined bottoming Brayton and Rankine cycles. The results were used to calculate overall system efficiencies for this report [9].

Note that the bottoming cycle produces 40% of the power and has a mass flow rate that is approximately 10% of the air mass flow rate. Also, the heat rejected to cooling water is 53 MW for a 100 MW power plant. This is roughly one fourth of the heat rejection requirement for a current 33% efficient power plant operating at the same electrical power level. It may be possible to increase the condenser temperature for the bottoming cycle, at a decrease in overall efficiency, so that the condenser could be cooled by air. This would mean that this type of power plant would be free of the requirement to have a body of water nearby.

At a more modest turbine inlet temperature of 660 °C, a comparison of possible turbine exit temperatures and bottoming cycle pressures was made to determine the optimum overall efficiency as a function of these two parameters. The results are displayed in Fig. 4.6. The optimum overall efficiency occurs at a turbine exit temperature of 803 K and a steam pressure of 2 megapascal. Higher Rankine cycle pressures might produce higher cycle efficiencies, but the mass flow rate must be lower due to the narrowing of the range between the turbine exit temperature and the water saturation temperature at the higher pressures. Thus, the combined cycle optimizes at a lower steam pressure than might be expected.

Additional work will be aimed at varying the expansion ratio per turbine and the penalties associated with air cooling the condenser. This research will be continued by this author and his collaborators using a more sophisticated transient analysis, which makes the fluid mechanics more complex to deal with mathematically [11–12].

4.9 Open Air Brayton Gas Power Cycle

Power generation is an important issue today, especially on the West Coast. Demand is outweighing supply because of lack of incentives for the utilities industry to build additional power plants over the past 10–20 years.

4.9.1 Introduction to Open Air Brayton Gas Power Cycle

This effort looks at extending an initial spreadsheet set of notes describing a multi-turbine Brayton-like system driven by four turbines with a bottoming Rankine cycle. The effort was undertaken to investigate the possibility of using a nuclear reactor driven heat exchanger, or group of heat exchangers, to drive a Brayton-like cycle gas turbine as an open cycle power conversion system. Since in a nuclear reactor driven system the core fuel elements of the reactor must exist at a higher temperature than any other component in the system, such a system is usually severely limited in the peak temperatures that can be produced in the gas turbine working fluid. For this study, a peak temperature of 660 °C was chosen as a reasonable upper limit on the temperature that the working fluid could attain prior to being expanded through a turbine. This turbine inlet temperature is roughly half of the state-of-the-art aircraft jet engine turbine inlet temperature. Given this limitation, it is necessary to use several turbines with a reheat heat exchanger between each. The baseline for this study was four turbines operating with an inlet temperature of 660 °C, and an exit temperature of 500 °C. To demonstrate that as many as four turbines may be required, the analysis started with a system using only one turbine. The effects of a recuperator and a bottoming steam cycle were investigated.

4.9.2 System Description

A simple system layout of open air Brayton gas power cycle is illustrated in Fig. 4.7. This simple model was used as base of the analysis presented above for combined cycle.

The analysis of this simple mode was presented by this author to the faculties and students of the Department of Nuclear Engineering, University of California, Berkeley, in fall of 2011, as part of continuous research funded by DOE and NRC. Interested readers should get in touch with this author for more detailed information or refer to book by Zohuri [8].

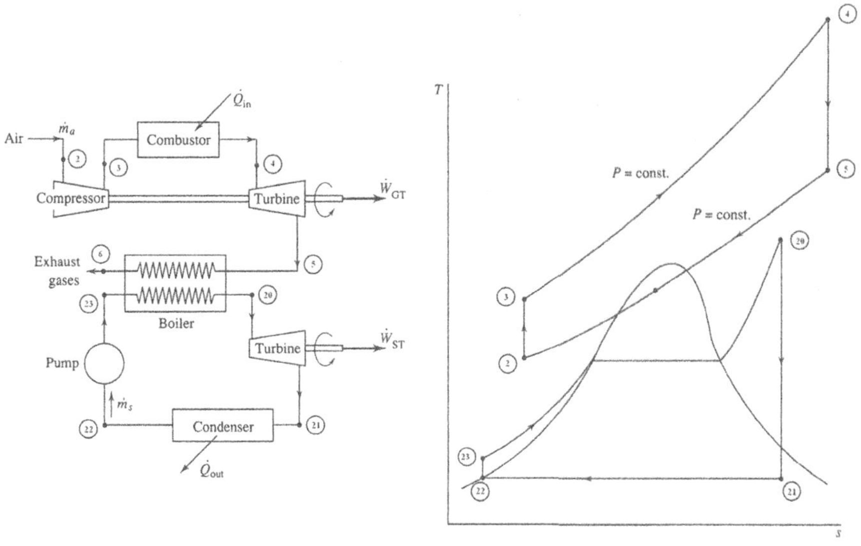

Fig. 4.7 Simplified combined system layout with thermodynamic cycle plot

4.9.3 *Nature of Problem Solved*

The single-turbine system is modeled in Spreadsheets using excel software, analyzes the system with a recuperator and the recuperator exit temperature is so low that a bottoming steam system does not make sense. The recuperator is assumed to have a thermal efficiency of 95% and a pressure ratio of 0.96. That is, about 4% of the pressure coming out of the compressor is lost in pumping the fluid through the recuperator to acquire the heat transfer. This same pressure ratio is used for all of the heat exchangers in this analysis. This is a conservative value and further analysis will relieve some of this conservatism. The temperature increase of the high-pressure fluid is 95% of the possible temperature difference between the turbine exit temperature and the compressor exit temperature. The recuperator boosts the system efficiency from about 13–16%, both of which are probably unacceptable.

The bottoming steam cycle is presented in spreadsheet and shows a very significant increase in efficiency over the recuperated cycle. A 2 MPa peak pressure in the steam cycle appears optimum, and it reaches approximately 32.5% at 480 °C. Lower turbine exit temperatures may show even further increases. This more than doubles the system efficiencies achieved with the recuperator system.

Plots of the single-turbine system performance are presented in Figs. F1, F5, F9, and F13, generated by Excel using output of FORTRAN code that was developed for this purpose and presented in section titled supporting figures down below.

4.9.3.1 Two-Turbine Systems

The two-turbine system is analyzed in Spreadsheet S3. The transition to two turbines ups the system efficiency to nearly 40%. This is a major increase and the recuperator seems to have a beneficial effect in this case. Given the results obtained in the single-turbine case, this was a bit of a surprise. The Brayton cycle efficiency increases dramatically from around 15% to greater than 30%. The minimum compressor pressure ratio required is a little above 6.0. There is a slight increase in the Rankine cycle efficiency also. The overall efficiency appears to optimize at a fairly low Rankine cycle pressure between 0.3 and 0.6 MPa.

Plots of the two-turbine system performance are presented in Figs. F2, F6, F10, and F14.

4.9.3.2 Three-Turbine Systems

The three-turbine system is analyzed in Spreadsheet S4. The transition to three turbines produces only a modest increase in efficiency of 1–2%. The recuperator heat transfer continues to play a significant role in keeping the efficiency up. The minimum compressor ratio required to operate is now about 13–14. The Rankine cycle pressure appears to optimize about 1 MPa.

Plots of the three-turbine system performance are presented in Figs. F3, F7, F11, and F15.

4.9.3.3 Four-Turbine Systems

The four-turbine system is analyzed, and the addition of the fourth turbine appears to increase the overall system efficiency approximately 1%.

However, now the compressor pressure ratio (~30) is high enough that a recuperator does not help. The compressor exit temperature is above the last turbine exit temperature. The Rankine cycle contribution appears to be the most significant at about 2 MPa though the efficiency falls off only a little over 1% in upping the steam cycle pressure to 10 MPa.

Based on the data in the note by Boss [24], two reheat cycles were considered in the bottoming cycle. The full system pressure was chosen as 10 MPa, and the first reheat was attempted at 2.5 MPa and the second reheat at 0.3 MPa following the data presented by Boss. The overall system efficiency dropped by ~1.5% with one reheat and as much as 7% with two reheats. It appears that the 10 MPa system pressure pushed the pinch point too high in temperature, and the reduction in steam flow to get all of the added heating done above the pinch point reduced the overall system efficiency. It would appear that a lower steam cycle pressure might be more beneficial to reduce the steam temperature at the pinch point where the high-pressure water begins to vaporize.

Plots of the four-turbine system performance are presented in Figs. F4, F8, F12, F16, and F17.

4.9.3.4 Additional Useful Analysis

Heat exchangers (1) Each of the heat exchangers was assumed to have an outlet-to-inlet pressure ratio of 0.96. This is probably conservative and should be investigated with a detailed design at what appears to be optimum conditions for the system efficiency.

Heat exchangers (2) Using four separate heat exchangers to heat the fluid before each turbine could be a plumbing nightmare. A detailed layout should be developed to demonstrate this is a very feasible design.

Power turbines The systems analyzed assume all of the gas turbines are fixed on the same shaft as the compressor. This may not be the optimal system for a number of reasons. Typical turbo-shaft engines and helicopter engines operate with a free turbine that is not connected to the main shaft. In some cases, the power turbine shaft passes down the center of the compressor-turbine shaft, and some designs have it emerging from the opposite end of the engine assembly.

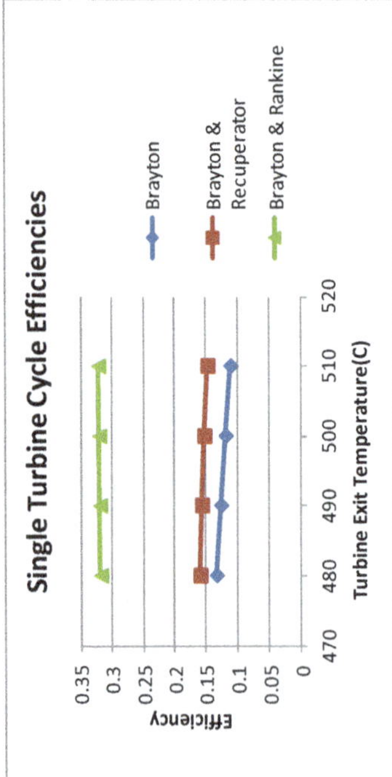

F1. Cycle efficiencies for a single turbine

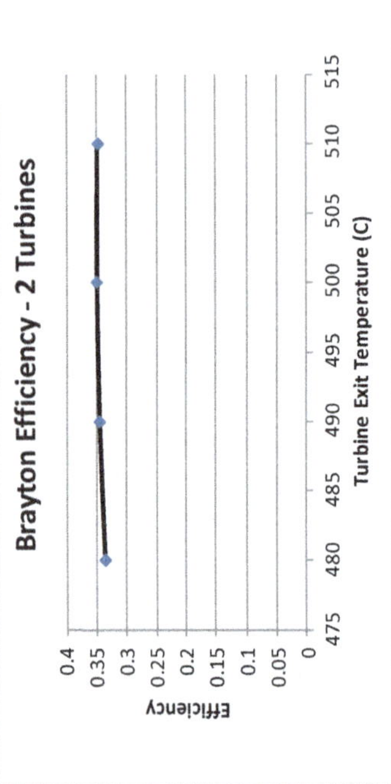

F2. Brayton cycle efficiencies for two turbines

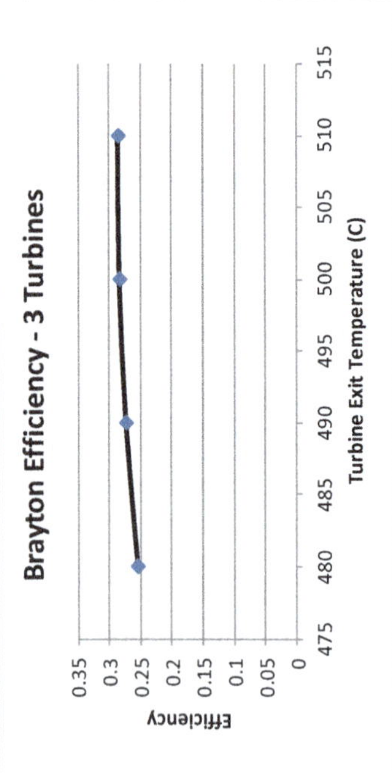

F3. Brayton cycle efficiencies for three turbines

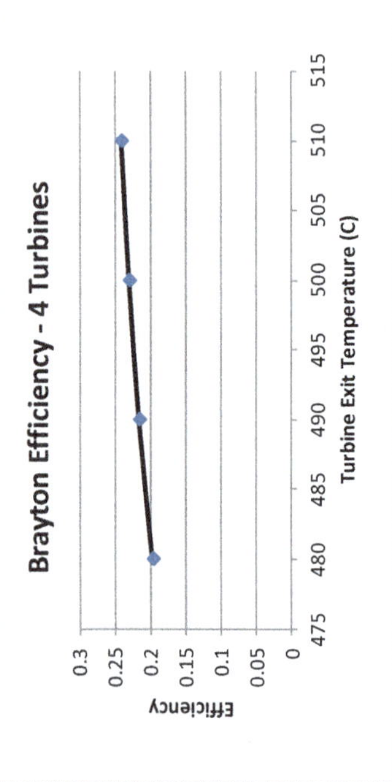

F4: Brayton cycle efficiencies for four turbines

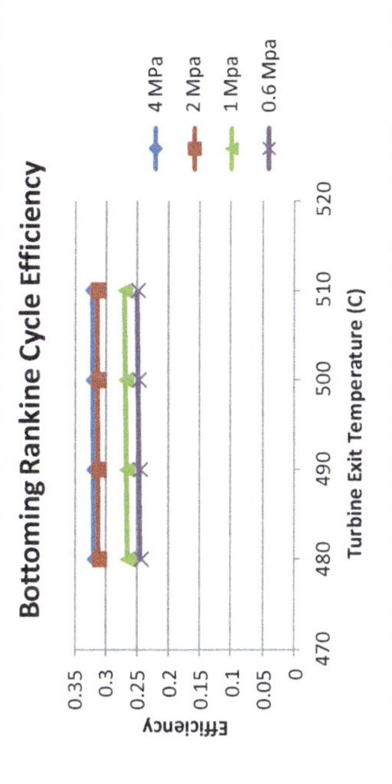

F5. Bottoming Rankine cycle efficiencies for one turbine

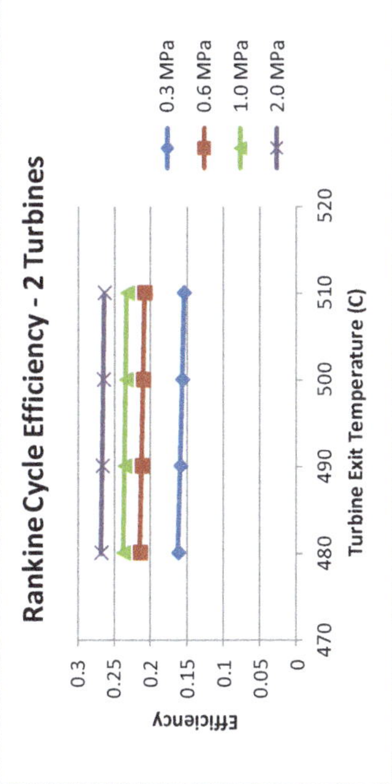

F6. Bottoming Rankine cycle efficiencies for two turbines

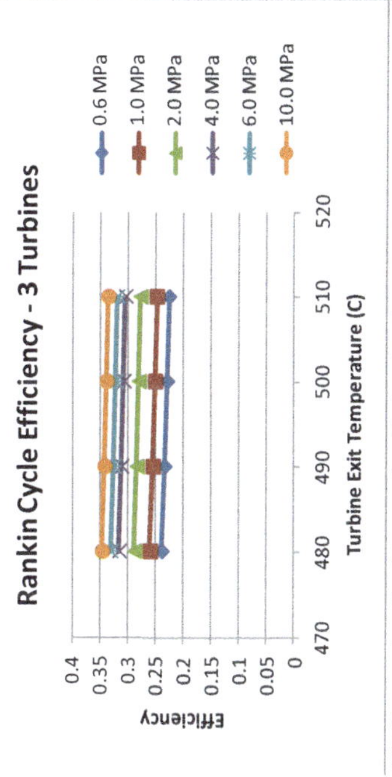

F7. Bottoming Rankine cycle efficiencies for three turbines

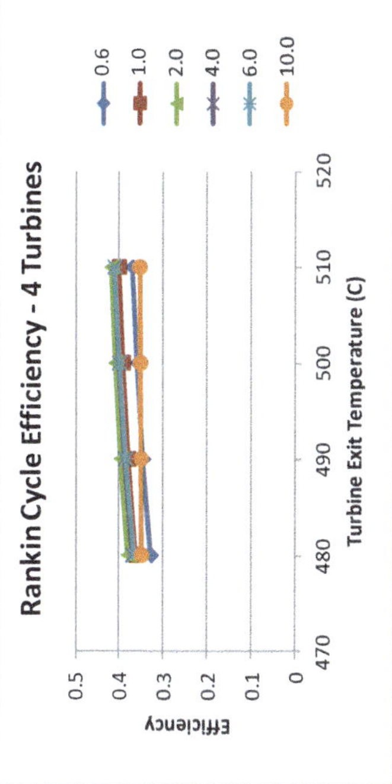

F8. Bottoming Rankine cycle efficiencies for four turbines

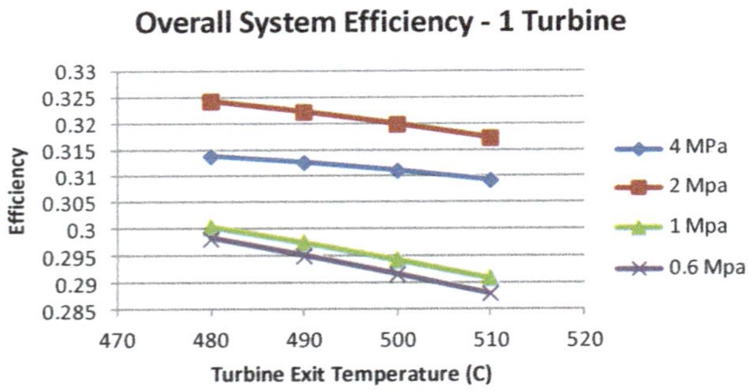

F9. Overall system efficiency for one turbine

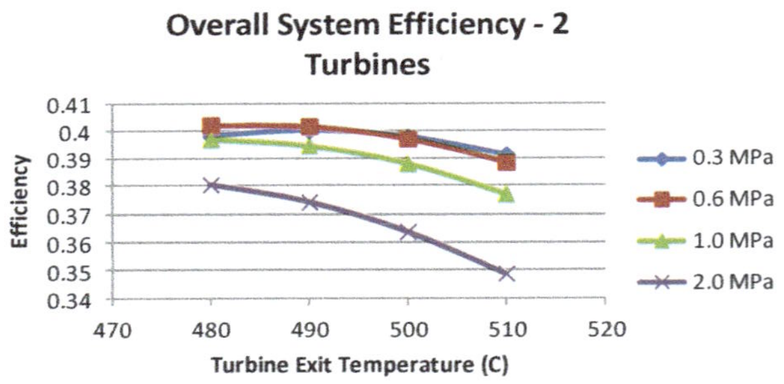

F10. Overall system efficiency for two turbines

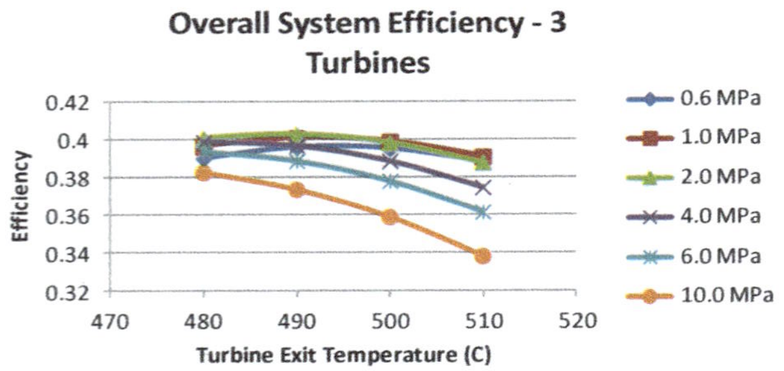

F11. Overall system efficiency for three turbines

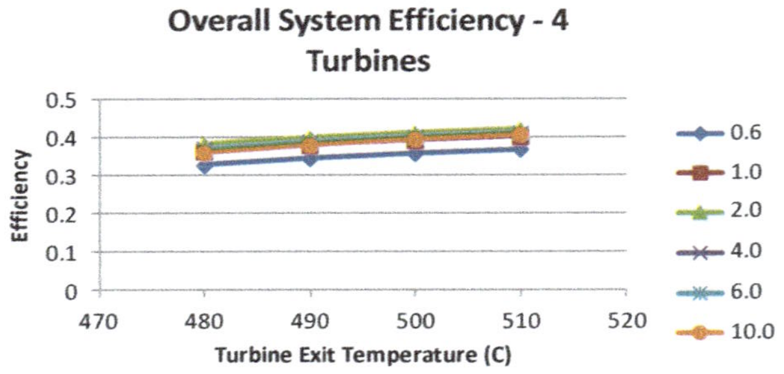

F12. Overall system efficiency for four turbines

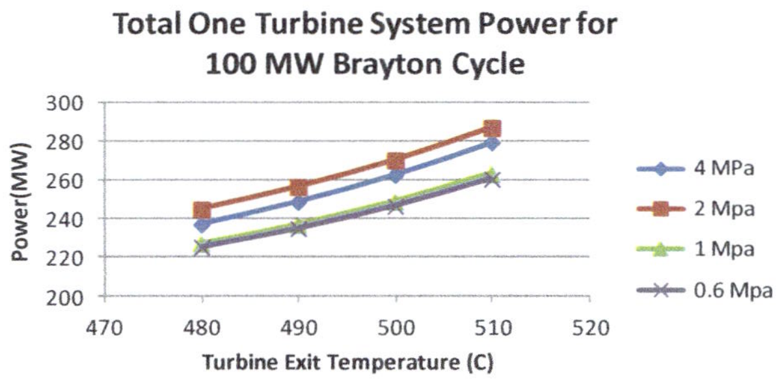

F13. System total power for one turbine

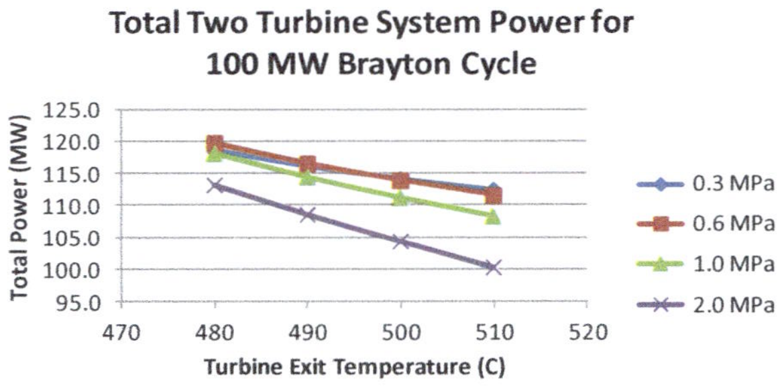

F14. System total power for two turbines

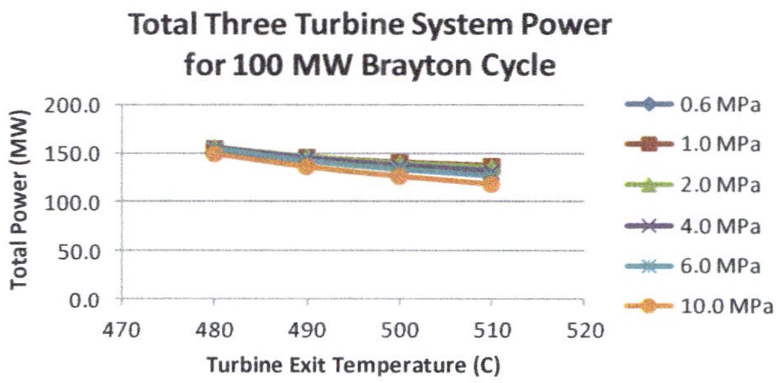

F15. System total power for three turbines

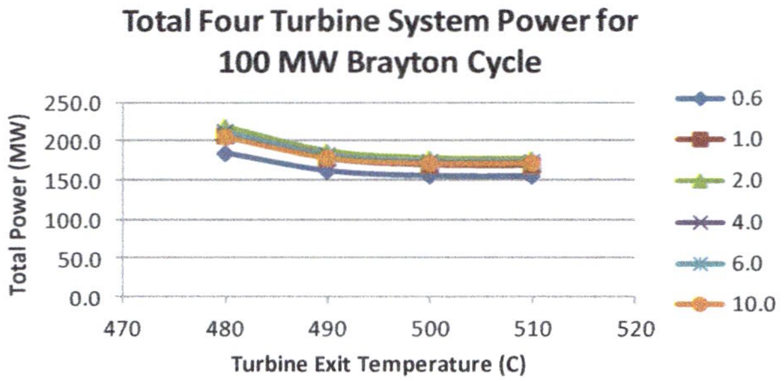

F16. System total power for four turbines

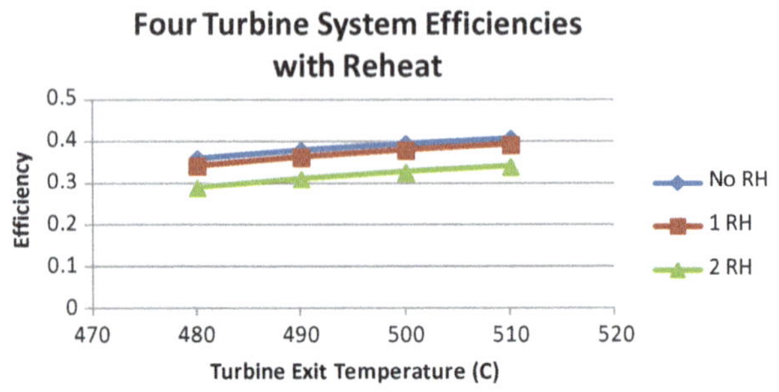

F17. Overall four-turbine system efficiency with Rankine reheat cycles

Reheat cycles at lower pressures For the 10 MPa design considered, a single or dual reheat cycle did not make sense. Since adding more heat to the steam above the pinch point with a reheat cycle decreases the steam-to-air mass ratio, the overall system efficiency is lowered. Lowering the pinch point temperature by operating the Rankine cycle at a lower pressure may improve system efficiency. This should be investigated.

Variable operating conditions All of the design conditions presented here are for standard sea-level conditions of 20 °C and 101.325 kPa. These should be varied to determine performance at other elevations above sea level and for summer and winter conditions.

Ultimate heat rejection The assumption made in this study is that a circulating water system is available to provide ultimate heat dump to the environment for the Rankine bottoming cycle. A condenser temperature of 50 °C was assumed in all cases. In some regions it may be useful to try to provide the Rankine cycle heat dump with an air radiator system. In this case the condenser temperature will likely be significantly higher than 50 °C. The effect on system performance should be investigated.

Component sizing and layout thermodynamic analysis is only one aspect of system design. Component sizing and layout are essential to understanding the economics of any system design. Some effort should be expended to optimize component sizing and layout from an economics, and, possibly, safety standpoint.

All the plots produced by the above Additional Useful Analysis are shown in the following figures:
For these analyses, this author has considered that for a nuclear system to take advantage of combined cycle technology, there are several changes to the plant components that have to be made. The most significant of course is that the combustion chamber must be replaced by a heat exchanger in which the working fluid from the nuclear reactor secondary loop is used to heat the air. The normal

Brayton cycle is an internal combustion one where the working fluid is heated by the combustion of the fuel with the air in the combustion chamber. The walls of the combustion chamber can be cooled, and peak temperatures in the working fluid can be significantly above the temperature that the walls of the chamber can tolerate for any length of time.

For the nuclear reactor system, the heat transfer is in the opposite direction. All reactor components and fluids in the primary and secondary loops must be at a higher temperature than the peak temperature of the gas exiting the heat exchanger. This severely restricts the peak temperature that can be achieved for the air entering the turbine. However, all is not lost.

In a typical combustion system, there are pressure losses approaching 5% of the total pressure to complete the combustion process [14]. Heat exchangers can be built with significantly lower-pressure drops than 5% approaching 1% [15]. Therefore, the most straightforward method to overcome this severe temperature limitation is to borrow a technique from steam power plants and implement multiple reheat cycles. That is, the first heat exchanger heats the air to its peak temperature. Then the air is expanded through the first turbine. The air is then reheated to the same peak temperature and expanded through the second turbine. Based on the relative pressure losses that appear possible, up to five turbines might be considered. All five turbines will be driving the same compressor. Multiple compressors on concentric shafts driven [14] by different sets of turbines might be possible, but that has not been considered here.

Multiple reheat cycles allow more heat to be put into the working fluid at a higher temperature. This improves the efficiency of the overall cycle. Interestingly enough, it does not improve the efficiency of the Brayton cycle, but because the exit temperature from the last turbine is higher, it does improve the efficiency of the Rankine part of the cycle with a net gain for the overall cycle. For this to work, the reactor-coolant temperature must reach temperatures significantly higher than current light-water reactor temperatures. Even sodium-cooled reactor exit temperatures in the 550 °C range are not quite high enough to get a nuclear Air Brayton combined cycle to compete with the efficiency of a pure steam Rankine cycle. However, when the coolant exit temperatures reach the 650–700 °C ranges, the combined cycle systems with multiple turbines surpass the performance of steam Rankine systems. Therefore, the analysis that follows is targeted at a molten salt reactor or a lead coolant reactor. A pressurized sodium reactor that could reach these temperatures is another possibility, but no one is proposing such a system at this time. It could also apply to a high-temperature gas reactor, but the heat exchangers would be quite different. Gas-to-gas heat exchangers have not been considered as the primary heat exchangers at this point but will be addressed in the recuperated systems.

Liquid-metal and molten salt heat exchangers were developed and tested successfully during the Aircraft Nuclear Propulsion program in the late 1950s [16–18]. They were conventional tube and plate exchangers and were tested for over 1000 h at temperatures up to 1100 K. The largest size tested transferred 55 MW of heat in a package of approximately 1.2 m^3. The heat-transfer area on the air side had a surface area per unit volume of 1180 m^2/m^3. Certainly, some development will be needed to bring this technology up to modern standards and get Nuclear Regulatory Commission (NRC)s approval for a power-producing reactor. In addition, the tasks involved do not appear insurmountable. Several additional heat exchangers were designed in this work in an attempt to estimate sizes of components and validate that pressure drop criteria could be met. The heat from the exhaust of the Brayton cycle transfers heat to vaporize the steam in the Rankine cycle in a heat recovery steam generator (HRSG) of fairly, conventional design. This includes air to steam superheaters as well as an economizer and evaporator section. A condenser of conventional design is included. For this work all heat exchangers were considered to be counterflow designs.

The heat exchanger design procedures and experimental data were taken from the text by Kays and London [19]. All of the data presented in this text was developed from steam-to-air heat exchangers and should be particularly applicable to the types of heat exchangers developed here. The only ones not using these two fluids are the molten salt or liquid metal to air heat exchangers referenced above. Thus, there is a reality basis for all the design calculations performed here to estimate power conversion system performance and sizing.

4.10 Preliminary Conclusions

System efficiencies around 40% appear very achievable with conservative performance on the various heat exchangers required for each of these systems. Two-turbine systems and three-turbine systems appear to achieve performances comparable to a four-turbine system.

A recuperator does not appear to be a useful component on either a single-turbine system or a four-turbine system. On the single-turbine system, it reduces the Brayton system exhaust temperature to a level that will not allow a bottoming steam cycle. Removing the recuperator and driving the bottoming cycle steam system with the turbine exhaust give the best system efficiency, easily, over 30%. The four-turbine system cannot use a recuperator because the compressor exit temperature is too high. It exceeds the exhaust temperature from the fourth turbine. However, for two- and three-turbine systems, the recuperator adds significant benefit. It is the component, which allows the two- and three-turbine systems to approach the efficiency of the four-turbine system.

Reheat cycles in the bottoming steam cycle do not appear to be advantageous at the pressures (10 MPa) currently being used in combined cycles. Lower pressures in the order of 1–2 MPa may reverse this result and let efficiencies see an improvement with steam reheat cycles. It would appear that the analysis accomplished here only scratched the surface of the possibilities for this type of system and thermal efficiencies easily above 40% should be readily achievable.

4.11 Combined Cycle Power Conversion System for Small Modular LMFBR

In previous section we have addressed applying an open-air Brayton power conversion cycle to next-generation nuclear power plants [9, 13, 20]. Both a combined cycle power conversion system and a recuperated Brayton cycle look very promising at the temperatures anticipated for molten salt reactors and liquid lead-cooled reactors. However, at coolant temperatures more typical of liquid-metal fast-breeder reactors (LMFBR), the combined cycle did not seem to hold an advantage over current conversion systems giving cycle efficiencies in the 39–40% range. Recently taking a clue from standard steam cycle efficiency enhancement techniques, we added a "feed water heater" to the bottoming Rankine cycle [21]. This is not the classic feed water heater because it draws its heat by cooling the compressed air from the first stage of a split compressor in the Brayton topping cycle. This frees up most of the hot air that would be used to raise the water to the saturation temperature for the Rankine cycle so that it can be passed through a recuperator to preheat air prior to entering the main heat exchangers. The net effect of the process is to raise the thermal efficiency of the cycle about 3% making a combined cycle more competitive at LMFBR outlet temperatures.

4.11.1 The Air Brayton Cycle Pros and Cons

The fastest-growing power conversion systems in the electric utility market today are internal combustion gas turbines. Any external combustion or heat engine system is always at a disadvantage to an internal combustion system. The internal combustion systems used in current jet engine and gas turbine power systems can operate at very high temperatures in the fluid and cool the structures containing the fluid to achieve high thermodynamic efficiencies. In an external energy generation system, like a reactor, all of the components from the core to the heat exchangers heating the working fluid must operate at a higher temperature than the fluid.

This severely limits the peak cycle temperature compared to an internal combustion system. One way this liability can be overcome is by using multiple expansion turbines and designing highly efficient heat exchangers to heat the working fluid between expansion processes. Typically, the combustion chamber in a gas turbine involves a pressure drop of 3–5% of the total pressure. Efficient liquid metal to air heat exchangers can theoretically be designed with pressure drops of less than 1%. This allows three to five expansion cycles to achieve a pressure drop comparable to a combustion system. Multiple turbines operating at different pressures have been common in steam power plants for several years. In this study three to five gas turbines operating on a common shaft were considered. Multiple expansion turbines allow a larger fraction of the heat input to be provided near the peak temperature of the cycle.

The exhaust from the last gas turbine still contains significant amounts of thermal energy. This gas is provided to the Heat Recovery Steam Generator (HRSG) to produce the steam used in the Rankine bottoming cycle. In a traditional gas turbine combined cycle, this gas gives up its thermal energy by heating steam in a superheater and then vaporizing the high-pressure water to make steam and finally heats the high-pressure water to the liquid saturation point prior to evaporating it. At the liquid saturation point or "pinch point," the hot air must be at a temperature greater than the boiling temperature of the high-pressure water. The heat remaining in the gas is then used to heat the liquid water or exhausted to the atmosphere.

4.11.2 The Feed Water Heater

Traditional Rankine steam cycles bleed steam from the high-pressure and high-temperature steam in the turbine(s) to heat the high-pressure water from low temperature to its boiling point. Because the temperature drop between the heating fluid and heated fluid in this case is less than in the case where the liquid water is heated by the boiler directly, the overall efficiency of the cycle increases. For the combined cycle, there is another source of heat that can be used to heat this low-temperature, high-pressure water. If the air compressor in the Brayton cycle is split into two parts such that the first part does about 56% of the work, its air can be used to heat the low-temperature water. The heat exchanger that cools the air exiting the first compressor is normally called an intercooler.

Traditionally an outside source of cold water is used to perform this function. However, the cold water in the Rankine bottoming cycle can perform this function almost as well. This frees up the hot gas exiting the "pinch point" to preheat the air out of the second compressor in a traditional recuperator. Intercooling without recuperating does not normally improve the efficiency of a cycle [22–23]. The major limitation on the size of the steam system is the enthalpy available from high-temperature air above the pinch point where the high-pressure water working fluid starts to vaporize. Below this point, there is still a significant enthalpy in the air which is readily available to heat the high-pressure water or to heat the compressed air in the Brayton cycle. There does not appear to be an advantage to including traditional feed water heaters in the cycle to bring the high-pressure water up to the saturation point. The possibility that an intercooler could be inserted between the two stages of a split compressor was considered. The cooling fluid for the intercooler was the high-pressure water coming out of the water pump. This process would combine the function of the traditional intercooler with the preheating of a typical feed water heater.

4.11.3 Results

The results of modeling this process for a temperature typical of an LMFBR (510 °C or 783 K) are presented in Fig. 4.8.

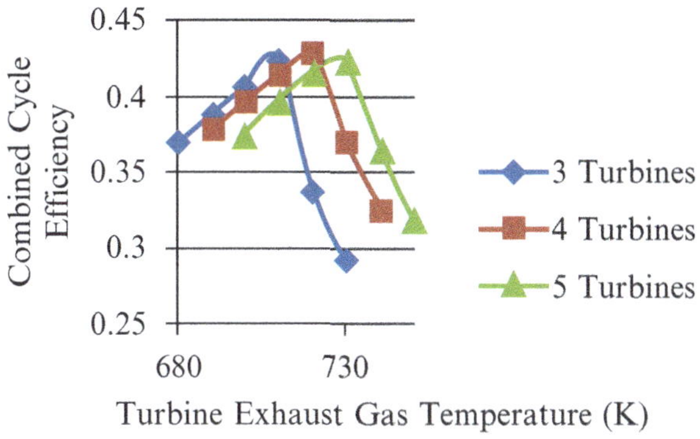

Fig. 4.8 Combined cycle efficiency versus turbine exhaust gas temperature

Table 4.5 Comparison of air and steam pressures

	Compression ratio	Steam pressure
Three turbines	3.846	21.0 Mpa
Four turbines	4.642	15 Mpa
Five turbines	5.000	13.5 Mpa

Table 4.6 Comparison of intercooled with simple CC

	Exhaust temp	Compress ratio	Steam pressure	Overall efficiency
Simple CC	820 K	14.125	2 MPa	46.1
IC CC	820 K	10.488	15 Mpa	49.5

Basically, the addition of the intercooler-feed water heater increased the efficiency of the cycle by about 3%; the peak efficiency achieved by the four-turbine case is 42.95%. The four-turbine case is once again optimal as it was for the simple combined cycle (CC), though the margin over three or five turbines is less than a percent. There were significant differences from the simple combined cycle however. Because the Brayton cycle is now recuperated, the overall compression ratios are much smaller. Also, the operating pressures in the Rankine cycle were much higher fed by the need to provide hot gas from the "pinch point" to the recuperator. Consider Table 4.5.

A more meaningful comparison can be made with the four-turbine case presented previously for the molten salt reactor. Simple CC is the previous case; IC-CC is the split compressor intercooled combined cycle in Table 4.6.

The cycle optimizes with the same turbine exit temperature but a significantly different compression ratio. It is worth pointing out that with the addition of an intercooler-feed water heater, the molten salt efficiency is now very close to 50%.

Returning to the LMFBR case, for the optimum four-turbine configuration, Table 4.7 shows the major performance parameters for a 50 MW(e) system.

Note once again that the heat rejection to a circulating water system is significantly less than the heat rejected by a current coal-fired plant at 40% efficiency. It is approximately 39.2% of that required by this kind of steam plant.

Table 4.7 Performance parameters for 50 MW systems

Total power	50 MW(e)
Brayton system	31.2 MW(e)
Rankine system	18.8 MW(e)
Total heat input	116.0 MW(t)
Mass flow rate (air)	291.4 kg/sec
Mass flow rate (water)	12.65 kg/sec
Heat rejection	29.4 MW(t)

Table 4.8 Summary of parameters

State points Brayton	Pressure (Mpa)	Temperature (K)
Atmosphere	0.1013	288.0
First compressor exit	0.2489	382.9
Intercooler exit	0.2464	319.9
First heater inlet	0.4610	604.6
First heater exit	0.4563	783.0
First turbine exit	0.3204	720.0
Fourth turbine exit	0.1076	720.0
Superheater exit	0.1066	664.1
Evaporator exit	0.1053	625.3
Recuperator inlet	0.1044	625.3
Recuperator exit	0.1034	402.4
Exhaust to atmosphere	0.1023	402.4
Rankine		
Pump entrance	0.0073	313.0
Pump exit	15.0	316.6
Intercooler exit	15.0	615.3
Evaporator exit	15.0	615.3
HP turbine inlet	15.0	705.0
HP turbine exit	3.75	519.7
MP turbine inlet	3.75	705.0
MP turbine exit	0.9375	519.7
LP turbine inlet	0.9375	705.0
LP turbine exit quality	0.914	

Perhaps even more interesting are the state points for the combined cycle system. The following table gives the typical state points for both the air Brayton cycle and the steam Rankine cycle.

It is probably worth mentioning that the intercooler-feed water heater could probably be adapted to current-generation gas turbine systems if the low-temperature gas exiting the recuperator does not condense the water added to the air as the result of combustion. That question is beyond the scope of our effort (Table 4.8).

In summary, nuclear heated open-air Brayton cycles have been investigated both as topping cycles for combined Brayton-Rankine cycles and as stand-alone recuperated Brayton cycles. The peak turbine inlet temperature chosen for the analysis was 933 K for a range of Generation IV molten salt reactors or lead-cooled reactors. A baseline power level of 25 MW (electric) was chosen to be representative of some of the small modular reactor concepts being considered. *Extensions* to higher temperatures and power levels were evaluated. Thermal efficiencies in the 45–46% range can be achieved by both the combined cycle systems and the recuperated systems, though the combined cycle systems achieve about a 1–1.5% improvement over the recuperated systems.

The nuclear heated open-air Brayton systems have several advantages over current light-water reactor and other Generation IV systems. The analysis demonstrates that the cycle thermal efficiencies are higher than other proposed systems. The gas turbine hardware is readily available over a broad range of power levels. And both the combined cycle and recuperated systems require significantly less circulating water for waste heat rejection than any other proposed systems [20].

4.12 Why Combined Cycle Power Conversion System Approach

One of the significant advantages of the combined cycle power system over current LWR power systems is its reduced requirement for circulating water in the waste heat rejection loop for the Rankine cycle. The typical combined cycle plant considered here produces approximately 50% of its power from the Rankine cycle and 50% of its power from the Brayton cycle. This automatically reduces the cooling water requirement by half. In addition, the combined cycle plant achieves a 45% efficiency so that only 55% of the heat generated has to be released as waste. Therefore, a typical 25 MW system will only need to get rid of 6.9 MW of heat via a circulating water system. A current LWR plant generating 25 MW at an efficiency of 33% would need to dump 16.8 MW of heat. This represents a major savings in circulating water requirements.

Since the combined cycle reduced the circulating water requirements so significantly, the natural question arises as to whether they can be eliminated completely. With a recuperated Air Brayton cycle, they can be. All the waste heat can be rejected directly to the atmosphere. At first it was thought that the efficiency of a multi-turbine recuperated cycle could not compete with a combined cycle plant. However, after performing the detailed analysis, the efficiencies of a recuperated cycle come within 1 or 2% of predicted combined cycle efficiencies. This would seem to be a minor penalty to pay for being free of a circulating water requirement. However, achieving these high efficiencies requires a very effective recuperator, which can become quite large.

4.12.1 Typical Cycles

Figures 4.9 and 4.10 provide a schematic of a four-turbine combined cycle system and its thermodynamic cycles on a temperature-entropy plot. This turns out to be the near optimum combined cycle system [11].

Figures 4.11 and 4.12 provide a schematic layout and a temperature-entropy diagram for a three-turbine recuperated configuration system. This turns out to be the optimum recuperated system [11].

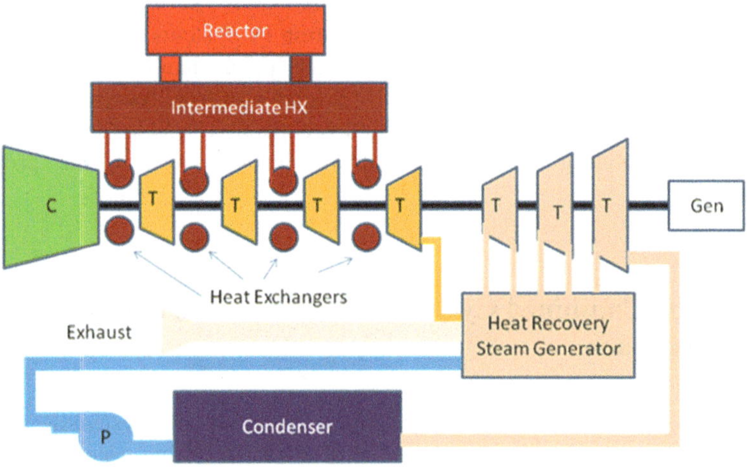

Fig. 4.9 Layout for four-turbine combined cycle

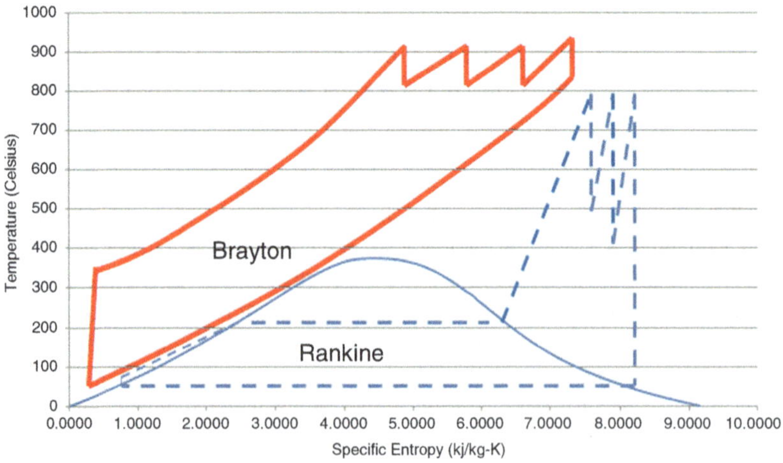

Fig. 4.10 Temperature-entropy diagram for combined cycle

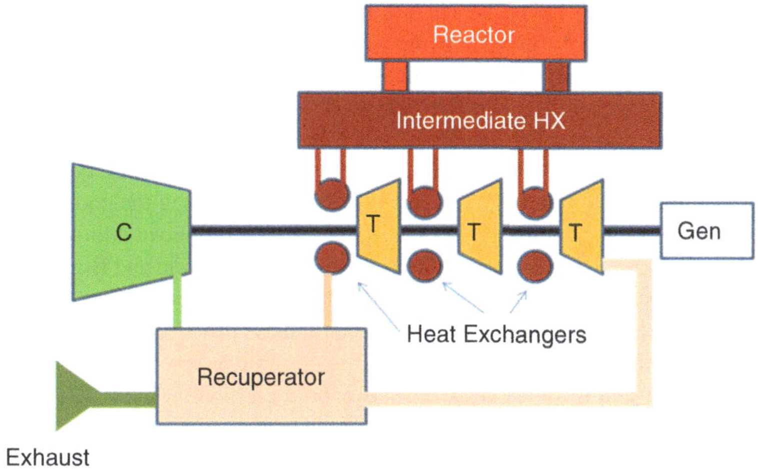

Fig. 4.11 Recuperator system schematic layout

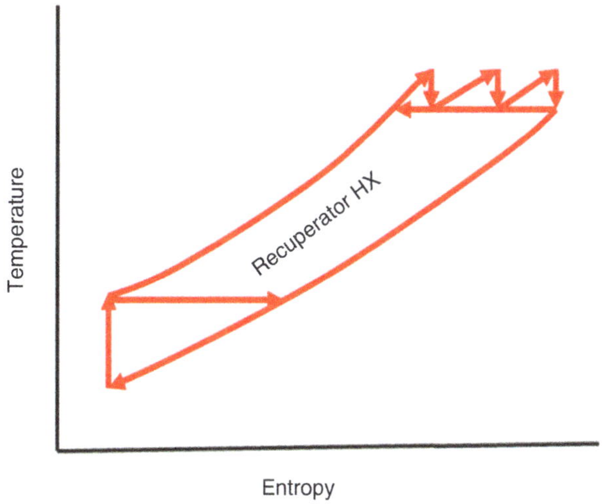

Fig. 4.12 Recuperated system temperature-entropy diagram

4.12.2 Typical Methodology

The approach taken in the combined cycle (CC) code developed for this effort is to model the thermodynamics of the components making up the power conversion systems as real components with non-ideal efficiencies. Pressure drops are included for every component except the connected piping. The compressor design is modeled with a small stage polytropic efficiency to take into account state-of-the-art designs [14]. The gas turbines are likewise modeled with polytropic efficiency. The steam turbines are modeled with a simple overall thermal efficiency. Pressure drops in each of the heat exchangers are included. The input files specify the pressure drops, and the heat exchangers are designed to meet these specifications if possible [8].

The code begins with an estimated compressor pressure ratio and then calculates the state points for both the Brayton and Rankine cycles. Then the code iterates on the compressor ratio to deliver an exit air pressure slightly above atmospheric. In a sub-iteration, it calculates the ratio of mass flows in the Rankine and Brayton cycles. Once the cycle state points have been identified, the output from the cycle is normalized to the desired power level. This sets the total mass flows of air and water. Once the mass flow rates have been calculated, it is possible to size all the components. The compressor and turbines are sized based on a correlation with state-of-the-art components and simple scaling rules. Each of the heat exchangers is designed based on the configuration chosen from the Kays and London text. Finally, the volumes of all the components are summed to get an estimate of system size [19].

In order to optimize the efficiency of the combined cycle, there are two main parameters that must be varied. These are the gas turbines' exit temperatures and the steam cycle peak pressure. The peak gas turbine inlet temperature is set as an input parameter. It can be varied, but it is obvious that the higher the gas turbine inlet temperature, the better the efficiency will be; the same can be said for most of the components. If their efficiency is higher, the efficiency of the cycle will be higher. The same is true of atmospheric conditions. The colder the input air and the circulating water, the better the efficiency will be. Therefore, most of the input parameters are chosen based on nominal values. It is not obvious though what the values of gas turbine exit temperatures and steam pressure should be to get an optimum efficiency. These must be varied to identify the peak efficiency achievable.

The coding for the recuperated system is much simpler than the combined cycle coding because the steam cycle doesn't have to be modeled. The compressor pressure ratio iteration is much simpler but still sets the pressure ratio so as to meet an exit pressure slightly above atmospheric. The new calculation in the recuperated system is the air to air recuperator itself. Since the recuperator will be the largest component in the system, it does not make sense to prescribe anything other than a counterflow heat exchanger. In this case the pressure drops for the hot and cold fluids cannot be set independently for simple heat-transfer correlations. So, the pressure drop on the hot air side was chosen as the flow path setting parameter. In this case, also, most of the parameters in the code models are set to nominal values. The only parameter that must be varied to optimize the efficiency of the system is the gas turbines' exit temperatures. The choice of the exit temperature can be made to achieve the peak efficiency.

For more details, reader should refer to a book by Zohuri, under the title of *Combined Cycle Driven Efficiency for Next Generation Nuclear Power Plants: An Innovative Design Approach* [8].

4.13 Motivation for This Innovative Approach

Energy is broadly defined as the ability to produce a change from the existing conditions. Thus, the term energy implies that a capacity of action present. The evaluation of energy is done by measuring certain effects that are classified by descriptive names, and these effects can be produced under controlled conditions. For example, mass that is located at certain position may have a potential energy of if same mass is in motion then it may possess the kinetic energy or if its characteristics of composition such as temperature or pressure changes, that is going through an energy process which, is called internal energy. The internal energy can be measured by release of an amount by change in potential energy experienced by external load.

For the past half century, fossil fuels, namely, coal, oil, and natural gas, have supplied the major portion of the world's energy requirements. It has long been realized, however, that in the not too distant future, these sources of energy will be largely exhausted. At the present time, the total energy consumption, for all countries, is about 1×10^{17} Btu per year. Since the world's population is steadily growing and the power use per capita is increasing as well, the rate of energy utilization by the year 2020 could well be five to ten times the current value. According to one estimate, the known coal, oil, gas, and oil shale which can be extracted at no more than twice the present cost would be equivalent to roughly 4×10^{19} Btu. This means about 100 years the world's economically useful reserves of fossil fuels may approach exhaustion.

The total amount of basic raw materials as source of fuel for fission power planet, such as uranium and thorium, in the earth's crust, to a depth of 3 miles, is very large, possibly something like 10^{12} tons. However, much of this is present in minerals containing such a small proportion of the desired element that extraction would be very expensive and not very cost-effective in particular for high-grade ore reserves that are believed to be in the order of 2×10^{6} tons; therefore we need to reduce the cost of recovery from moderately low-grade ores to at least \$100 or less per pound of metal with advancing technology in this matter.

Development of plant layout and modularization concepts requires an understanding of both primary and secondary systems.

The present and current design of light-water reactors (LWRs) cannot compete with high-efficiency gas turbine systems economically enough that total cost of ownership (TCO) and return on investment (ROI) make sense to the owners of electricity companies, even at the verge of end of cheap oil and increase of demand for electricity, which is increasing at the rate of 17% per year. With growth of economy and increase of population around the globe and consequently demand for electricity, we need to find a cheap and reasonable way of producing electricity to satisfy the demand increases. The present production of electricity per kWatt hours using traditional fossil-fueled or gas-fueled power plants is about \$3.5 versus similar production driven by nuclear power plant that is about \$7.5 per kWatt hours. Thus, to make the nuclear power plant more cost-effective, one solution is increasing the efficiency of these plants to roughly 60% thermal output, as it is defined by Eq. 4.1.

The following points are motivation for such innovative approach:

- Advanced power conversion systems are required to take advantage of high-temperature capabilities of several GEN-IV designs.
- Molten salt, liquid lead bismuth, and high-temperature gas systems may reach reactor outlet temperatures in the range of 1175–1275 K.
- Brayton power conversion systems using air as a working fluid are a well-developed technology. Traditional studies of improved efficiency have centered on recuperated systems.
- Combined cycle systems using a Rankine bottoming cycle have shown excellent performance with natural gas and coal gasification systems, giving above 50% overall thermodynamic efficiency.

To have some basic ideas, the following applies to a nuclear power operation process, generating energy for electricity production.

- A nuclear reactor produces and controls the release of energy from fission.
- The energy released as heat is used to make steam.
- The steam is used to drive the turbines which produce electricity.

Several components of a nuclear reactor include:

- Fuel
- Moderator
- Control rods
- Coolant
- Steam generator
- Containment structure

A typical top-level light-water reactor (LWR) is illustrated in Fig. 4.13.

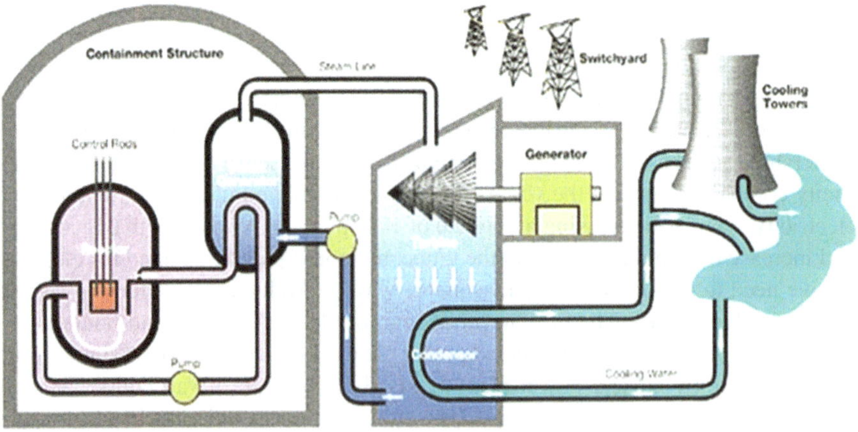

Fig. 4.13 A typical light-water reactor

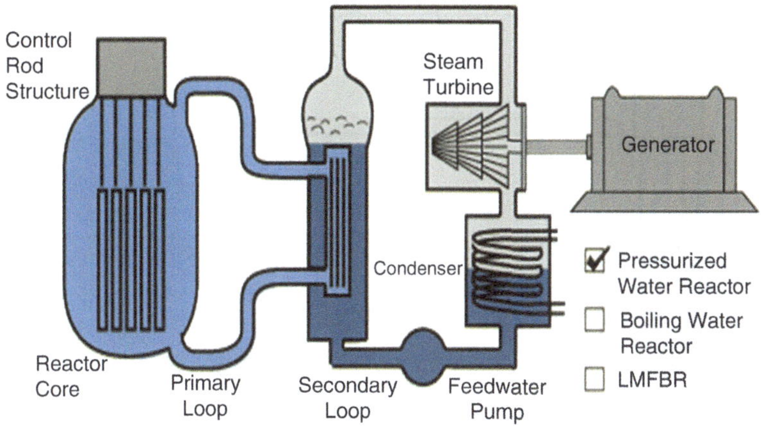

Fig. 4.14 A typical pressurized water reactor

Similarly as part of GEN-III nuclear reactor class, we can describe pressurized water reactor from top level as shown in Fig. 4.14.

In the pressurized water reactor (PWR), the water which passes over the reactor core to act as moderator and coolant does not flow to the turbine but is contained in a pressurized primary loop. The primary loop water produces steam in the secondary loop which drives the turbine. The obvious advantage to this is that a fuel leak in the core would not pass any radioactive contaminants to the turbine and condenser.

Another advantage is that the PWR can operate at higher pressure and temperature, about 160 atmospheres and about 315 °C. This provides a higher Carnot efficiency than the BWR, but the reactor is more complicated and costlier to construct. Most of the US reactors are pressurized water reactors [11].

Another class of GEN-III nuclear power plant (NPP) reactor is called boiling water reactor (BWR), and it is illustrated in Fig. 4.15 with the following operational sequence, in which water passes through the reactor core acting as moderator and coolant is also the steam source for the turbine. The disadvantage of this is that any fuel leak might make the water radioactive and that radioactivity would reach the turbine and the rest of the loop.

A typical operating pressure for such reactors is about 70 atmospheres at which pressure the water boils at about 285 °C. This operating temperature gives a Carnot efficiency of only 42% with a practical operating efficiency of around 32%, somewhat less than the PWR.

In both, a BWR and PWR, the diagram of steam generation is presented in Fig. 4.16 as follows:

The turbine power cycle in terms of Rankine obeys the following steps:

- Rankine (steam) power cycle.
- It directly employs steam to drive the turbines.

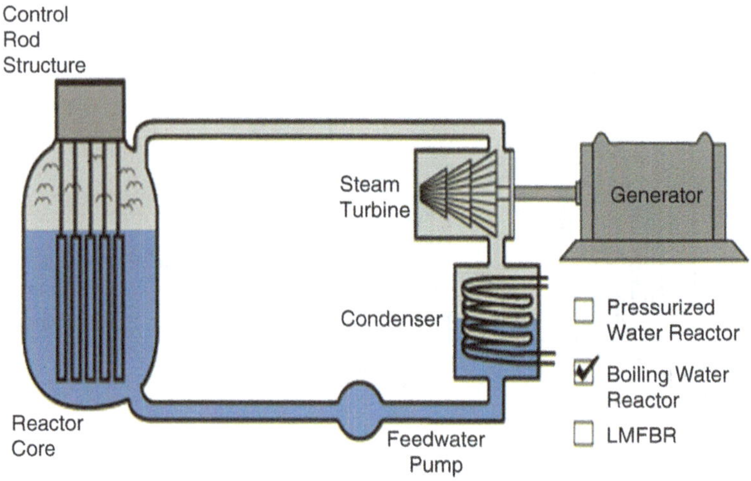

Fig. 4.15 A typical boiling water reactor

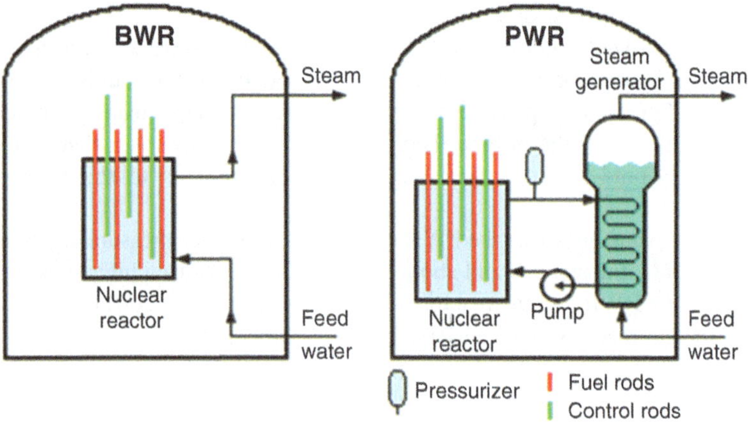

Fig. 4.16 Nuclear power plant steam generation

- Associated problems include lower operating temperatures (lower efficiency), turbine blade fouling, larger equipment, and wet cooling.

These steps are depicted in Fig. 4.17.

The drawback of current NPPs designs can be explained from thermodynamics viewpoint and steam generation, which is the driving factor for combined cycle innovation approach by this author and co-authors [8, 9, 13, 20].

A saturation dome is a graphical representation of the combination of vapor and gas that is used in thermodynamics. It can be used to find either the pressure or the specific volume as long as one already has at least one of these properties.

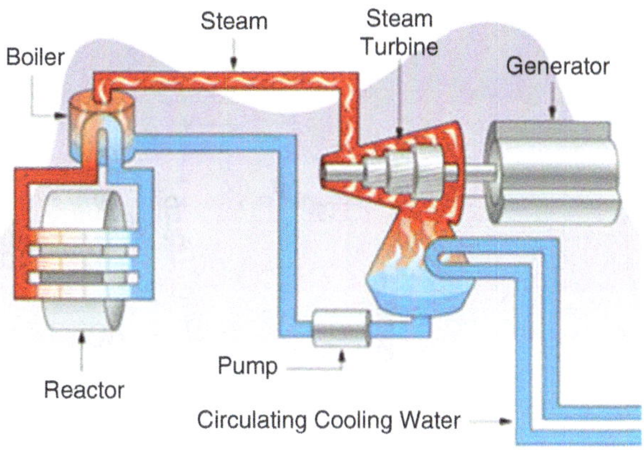

Fig. 4.17 Rankine cycle process

A saturation dome uses the projection of a *P*-*v*-*T* diagram (pressure, specific volume, and temperature) onto the *P*-*v* plane. This gives a *P*-*v* diagram at a constant temperature. The points that create the left-hand side of the dome represent the saturated liquid states, while the points on the right-hand side represent the saturated vapor states (commonly referred to as the "dry" region). On the left-hand side of the dome, there is compressed liquid, and on the right-hand side, there is superheated gas [8].

This matter is graphically depicted as Fig. 4.18. Within the dome itself, there is a liquid-vapor mixture. This two-phase region is commonly referred to as the "wet" region. The percentage of liquid and vapor can be calculated using vapor quality [11]. The triple-state line is where the three phases (pressure, specific volume, and temperature) exist in equilibrium.

The point at the very top of the dome is called the critical point. This point is where the saturated liquid and saturated vapor lines meet. Past this point, it is impossible for a liquid-vapor transformation to occur. It is also where the critical temperature and critical pressure meet. Beyond this point, it is also impossible to distinguish between the liquid and vapor phases.

A saturation state is the point where a phase change begins or ends. For example, the saturated liquid line represents the point where any further addition of energy will cause a small portion of the liquid to convert to vapor. Likewise, along the saturated vapor line, any removal of energy will cause some of the vapor to condense back into a liquid, producing a mixture. When a substance reaches the saturated liquid line, it is commonly said to be at its boiling point. The temperature will remain constant while it is at constant pressure underneath the saturation dome (boiling water stays at a constant of 212 °F) until it reaches the saturated vapor line. This line is where the mixture has converted completely to vapor. Further heating of the saturated vapor will result in a superheated vapor state. This is because the vapor will be at a temperature higher than the saturation temperature (212 °F for water) for a given pressure.

Fig. 4.18 *PVT* in a three-dimensional diagram

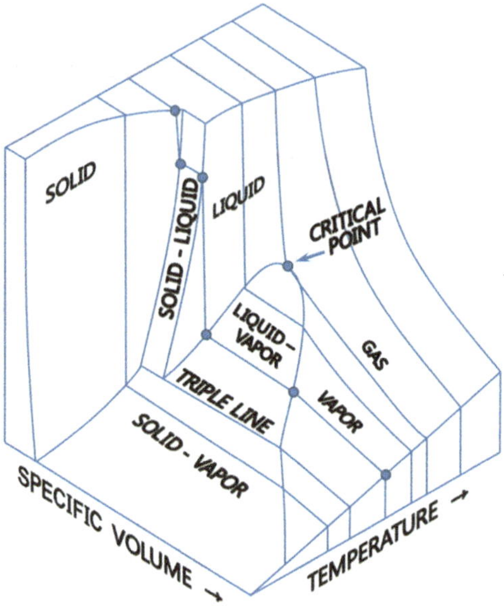

Then, the vapor quality refers to the vapor-liquid mixture that is contained underneath the dome. This quality is defined as the fraction of the total mixture, which is vapor, based on mass [11]. A fully saturated vapor has a quality of 100%, while a saturated liquid has a quality of 0%. Quality can be estimated graphically as it is related to the specific volume or how far horizontally across the dome the point exists. At the saturated liquid state, the specific volume is denoted as υ_f, while at the saturated vapor stage, it is denoted as υ_g.

However, quality can be calculated by Eq. 4.5 as:

$$x = \frac{\upsilon - \upsilon_f}{\upsilon_g - \upsilon_f} \tag{4.5}$$

Now that we have the understanding of saturation dome thermodynamics, using Fig. 4.19, we can easily see where the current existing nuclear power plant of Generation III (GEN-III) is suffering as drawback of this dome.

Figure 4.19 shows comparison of typical Rankine cycle for a typical LWR versus coal plant.

As it can be seen from Fig. 4.19, the following points are driving the drawback thermal efficiencies of GEN-III NPPs and extend footprint for coal power plants (CPPs):

- LWRs are trapped in the vapor dome for the steam water system.
- Coal plants have a separate heat exchanger called a superheater that allows them to go to higher temperatures and better efficiency.
- This has been tried with nuclear plants by adding a combustion superheater and by designing another heat exchanger in the core. Neither worked very well.

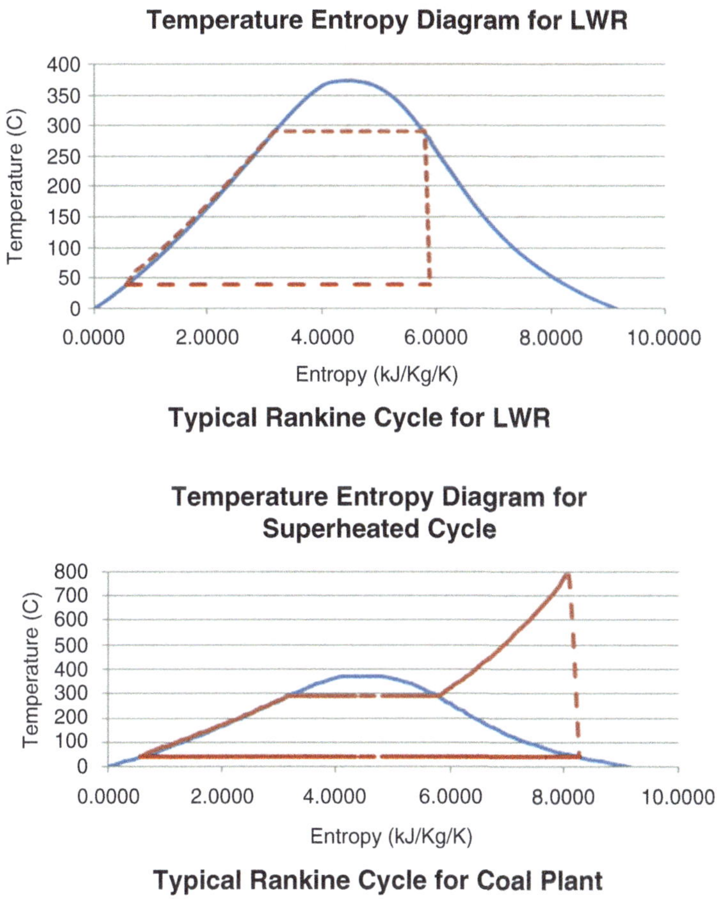

Fig. 4.19 Rankine cycle comparison between LWR and coal power plants

- One Generation IV concept is to go to supercritical water at ~25 Mpa, but this will require significant in-core development.

These above issues are very simple; yet it is very important to look into innovative combined cycle design approach, when it comes to the future of Generation IV (GEN-IV) NNPs.

Typical combined cycle suggested by this author based on his recent research and presented in various publications of journals and books is depicted here, and the computer code developed by him and McDaniel was modeled and developed according to these drawings. See Fig. 4.20.

Bear in mind in order to model the combined cycle as part of overall steam generator and electrical turbine generator components, compact heat exchangers play a

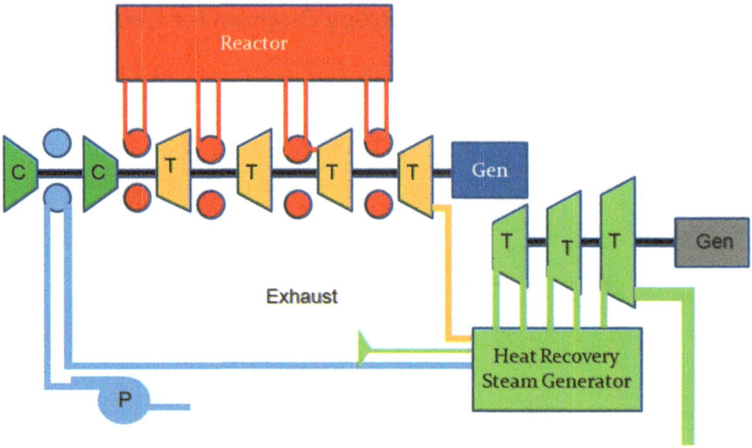

Fig. 4.20 Brayton system diagram

big role. The model and FORTAN code, considered types of compact heat exchangers that are categorized as follows:

- Tubular heat exchanger
- Fin-plate heat exchanger
- Tube-fin heat exchanger
- Plate-frame heat exchanger
- Regenerative heat exchanger

Designing this type of heat exchangers are an easy task and the reasoning to argue why this is not that simple, can be summarized below:

- Perform the required heat transfer AND
- Minimize size and weight
- Minimize pressure drop
- Meet required life
- Minimize cost

In order, to achieve all the required in above in need of combined cycle, another important factor for designing these types of heat exchangers (i.e., Compact) particularly, in their applications for high temperature combined cycle for next generation (i.e., Gen-IV) nuclear power plant, is pinch point as illustrated in Fig. 4.21, and imposed golden rules by this point [11].

Outcomes of this innovative approach going forward are summarized here:

- Adapt advanced open cycle Brayton systems to Gen IV concepts.
- Develop general simulation tools to study performance and optimization of heat power systems.
- Analyze adapted systems in enough detail to accurately predict overall efficiencies to within 1–2%.

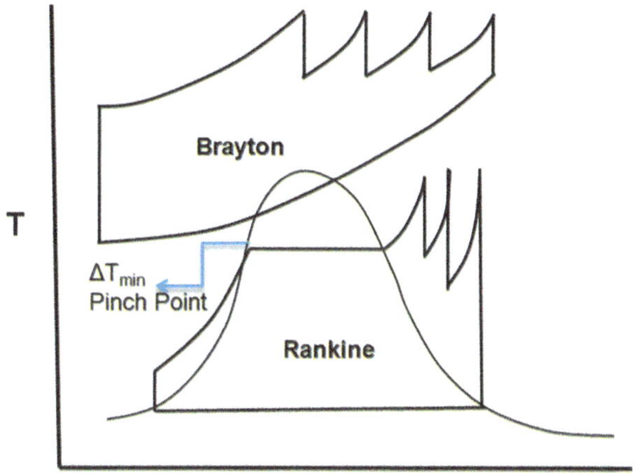

Fig. 4.21 Model of typical pinch point between Brayton and Rankine cycles

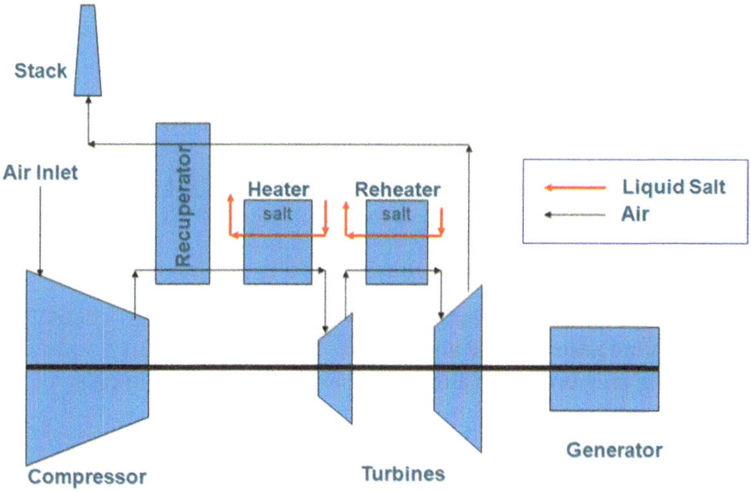

Fig. 4.22 Open-air recuperated Brayton liquid-salt power conversion cycle

- Optimize system configurations to achieve greatest overall efficiencies.
- Estimate power conversion system sizes as a function of power level.

See Figs. 4.22 and 4.23 for a more detail layout of open-air Brayton-type power plants.

Whoever is interested to have marching steps toward this innovation should plan on the following activities as this author can see:

- Model high-temperature Brayton open cycle systems for Next Generation Nuclear Plant:
  - Power conversion systems for high-temperature systems

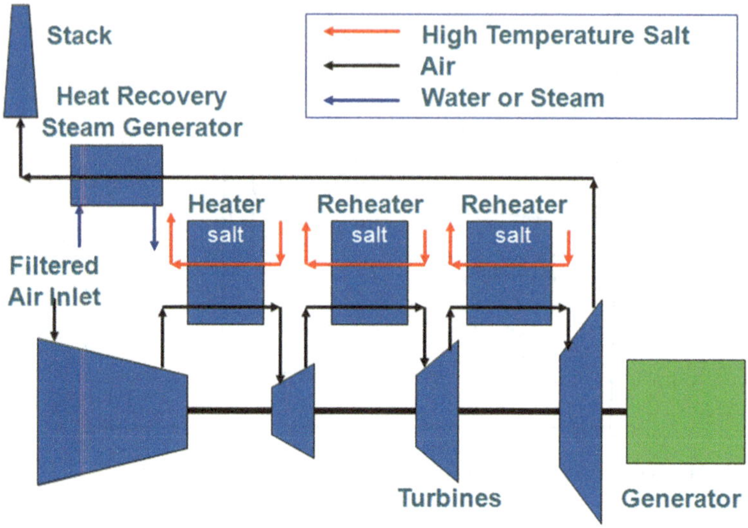

Fig. 4.23 Open-air Brayton liquid-salt combined cycle power conversion

- Develop computer codes, modeled on transient not steady-state analyses as this author started, in order to improve:
  - Brayton-Rankine combined cycle (BRCC) and Rankine bottoming cycle (RBC) computer codes
- Optimize both configurations for Brayton open cycle systems:
  - Number of reheat gas turbines
  - Temperature ratios for gas turbines
  - Operating pressures for bottoming cycle in combined cycle case

4.14 End of Cheap Oil and Future of Nuclear Power

Dusk of cheap oil is in rise, while the dawn of new generation of power plant (i.e., GEN-IV) is in horizon. Global production of conventional oil will begin to decline sooner than most people think, probably within 10 years. As we recall, two sudden price increases took place in 1973 and 1979 and rudely impacted the industrial world and made it to recognize its dependency on cheap crude oil. The first event in 1973 that caused oil price increase took place in response to an Arab embargo during Arab and Israel war when the price tripled and then nearly doubled again when Iran Shah was dethroned, sending the major economies into spin. Although emotional and political reaction of most analysts predicting a shortage of crude oil in the world due to these types of crises and not having enough underground reservoir for exploration of oil will put the future survival of world economy into a critical path, even at the time, oil experts knew that they had no scientific basis. Just a few years earlier, oil explorers had discovered enormous new oil province on the North Slope of Alaska and below the North Sea off the coast of Europe.

The five Middle Eastern nations who are members of Organization of Petroleum Exporting Countries (OPEC) were able to hike price of crude oil not because oil was growing short but because they managed to control 36% of international market. Later, due to pumping oil flow from Alaska and North Sea, the demand for crude oil sagged, then prices of oil dropped, and the OPEC's control of prices collapsed. The next oil crunch will not be so temporary; this is because the exploration and discovery of oil field as well as production of it around the world suggest that within the next decade, the supply of conventional oil will not support and cannot keep up with the demand. Whether this conclusion is in contradiction with what oil companies are reporting is a question. Distributing today's oil production rate of about 23.6 Gbo (giga barrel oil) per year may suggest a cheap crude oil for the next 43 more years based on the official charts shows the reserves are growing. But there are three critical errors:

- *First*, it relies on distorted estimates of reserves.
- *A second mistake* is to pretend that production will remain constant.
- *Third*, and most important, conventional wisdom erroneously assumes that the last bucket of oil can be pumped from the ground just as quickly as the barrels of oil gushing from wells today.

In fact, the rate at which any well – or any country – can produce oil always rises to a maximum and then, when about half the oil is gone, begins falling gradually back to zero.

4.15 The Future of Energy

Energy in the future will be very different. For all the uncertainties highlighted in various reports by experts in the field, we can be certain that the energy world will look a lot different in 2030 than it does today. The world energy system will be transformed, but not necessarily in the way we would like to see. We can be confident of some of the trends highlighted in reports on current global trends in energy supply and consumption, environmentally, economically, and socially. But that can – and must – be altered when there is still time to change the road we are on. The growing weight of China, India, the Middle East, and other non-OECD (Organization for Economic Co-operation and Development) regions in energy markets and in CO_2 emissions is something we need to take under consideration in order to deal with global warming. The rapidly increasing dominance of national oil companies and the emergence of low-carbon energy technologies seem to be one necessary solution to the problem in hand, but not sufficient enough. And while market imbalances could temporarily cause prices to fall back, it is becoming increasingly apparent that the era of cheap oil is over. But many of the key policy drivers (not to mention other external factors) remain

Fig. 4.24 Typical nuclear plant in our backyard

in doubt. It is within the power of all governments, of producing and consuming countries alike, acting alone or together, to steer the world toward a cleaner, cleverer, and more competitive energy system. Time is running out and the time to act is now. So what we need to ask is that "Can nuclear power compete?"

A variety of companies that are in energy production business say the answer may be yes. Manufacturers have submitted new designs to the Nuclear Regulatory Commission's safety engineers, and that agency has already approved some as ready for construction if they are built on a previously approved site. Utilities, reactor manufacturers, and architecture/engineering firms have formed partnerships to build plants, pending final approvals. Swarms of students are enrolling in college-level nuclear engineering programs, and a rosy projection from industry and government predict a surge in construction.

Like another moon shot, the launch of new reactors after a 35-year hiatus in orders is certainly possible, though not a sure bet. It would be easier this time, the experts say, because of technological progress over the intervening decades. But as with a project as large as a moon landing, there is another question: Would it be worthwhile?

Figure 4.24 is an illustration of typical nuclear power plant in our backyard and farms, where we live today.

In order to answer this question, we need to at least satisfy the four unresolved problems associated with nuclear power plant that was brought up by MIT report [1] and they were mentioned at the beginning of this write-up. In order to argue the first point which is the cost of producing a nuclear power plant with its modern and today's technologies from total ownership and return on the investments, we need to understand the nature of the beast from the day was born in basement of University of Chicago and later was shown to the world as first nuclear explosion.

4.16 Nuclear Power in the World Today and Time for Change

The world faces serious difficulties in creating enough energy that will be meeting supply and demand for the next decades to face population growth. One of the challenges facing these difficulties, beside obtaining that will be needed in coming decades for a growing population, especially given the problem of climate change caused by fossil fuel use and as result the impacts of green effects in our day-to-day life around the globe.

The need for carbon-free energy option and reduction in cost of producing electricity from nuclear power of new generation (GEN-IV), due to the increase of their thermal output efficiency per Eq. 4.1 and innovative suggestion by different researchers, makes nuclear energy, either fission or fusion types, somewhat attractive [5–8].

Current arguments about possibilities to prevent global warming and mitigate the green effects due to fossil and gas fuels power plants have also opened up our vision to a potential revival of nuclear power, regardless of some countries such as Germany has decided to abandon all their nuclear power plants by shutting them down and off-course aftermath of few nuclear accidents in past and as recent one as Fukushima Daiichi nuclear power plant in Japan in 2011 are few obstacles that we need to overcome for our time change to go forward with nuclear as source of energy.

Generation IV innovative design by increasing their safety, life cycle and thermal efficiency of their existence suggest a time change for nuclear power in the world today, in order to meet the demand for electricity at lower cost for population growth and prevent global warming due to emissions for old fossil and gas fuel power plants and off course gasoline burning cars are additional parameter to add to our global warming effect.

As part of pro-argument for taking look at the GEN-IV reactors as a new generation and produce lower cost of electricity and source of renewable of energy from these nuclear power plants, seems to be one of the strongest point of this argument in favor of atomic energy either in near term such as fission process [8] or longer term process that is known to us as *Fusion Process* both Magnetic Confinement Fusion (MCF) [6] and Inertial Confinement Fusion (ICF) [7].

To determine the future cost of electricity from nuclear power source, the cost from currently operating power stations is taken into consideration, especially if we are looking at alternative sources of renewable energy going forward with time and increasing demand for electricity based on existing supply today. Although as alternative some scientists, researchers and engineers in energy filed suggest Solar and Wind energy as part of our solution, but we know that wind does always blow and sun does not always shine, yet a save nuclear power plant can produce energy 24 × 7 time interval.

Civil nuclear power can now boast more than 17,000 reactor years of experience, and nuclear power plants are operational in 30 countries worldwide. In fact, through regional transmission grids, many more countries depend in part on nuclear-generated

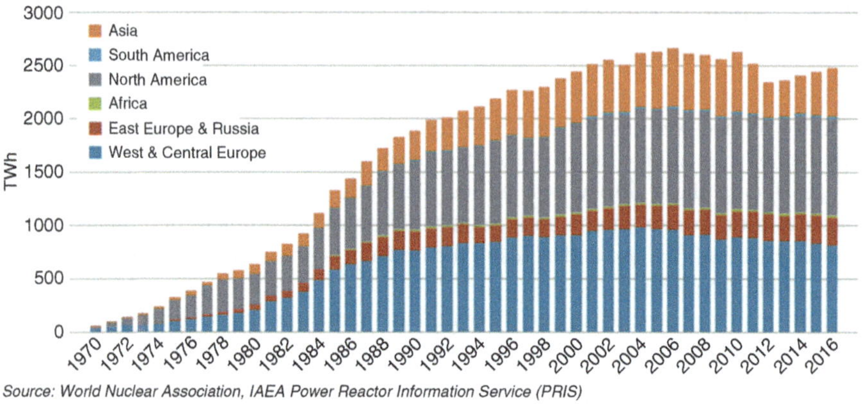

Fig. 4.25 Nuclear electricity production

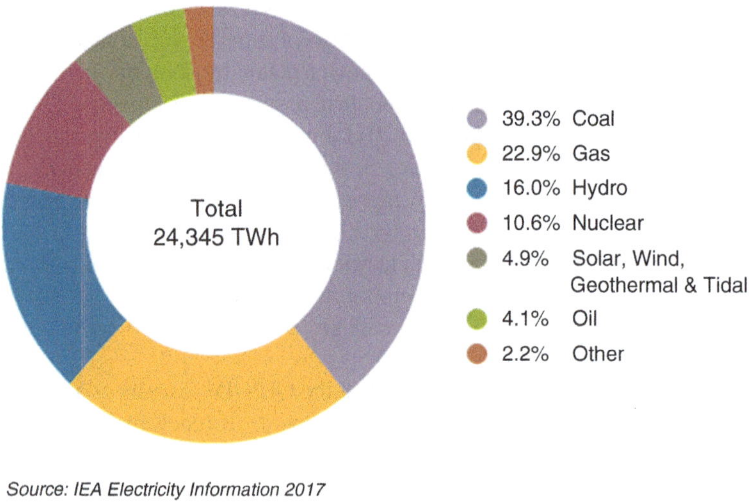

Fig. 4.26 World's electricity production by source, 2017

power; Italy and Denmark, for example, get almost 10% of their electricity from imported nuclear power [25].

Around 11% of the world's electricity is generated by about 450 nuclear power reactors. About 60 more reactors are under construction, equivalent to 16% of existing capacity, while an additional 150–160 are planned, equivalent to nearly half of existing capacity. See Fig. 4.25.

In 2016 nuclear plants supplied 2476 TWh of electricity, up from 2441 TWh in 2015. This is the fourth consecutive year that global nuclear generation has risen, with output of 130 TWh higher than in 2012. See Fig. 4.26.

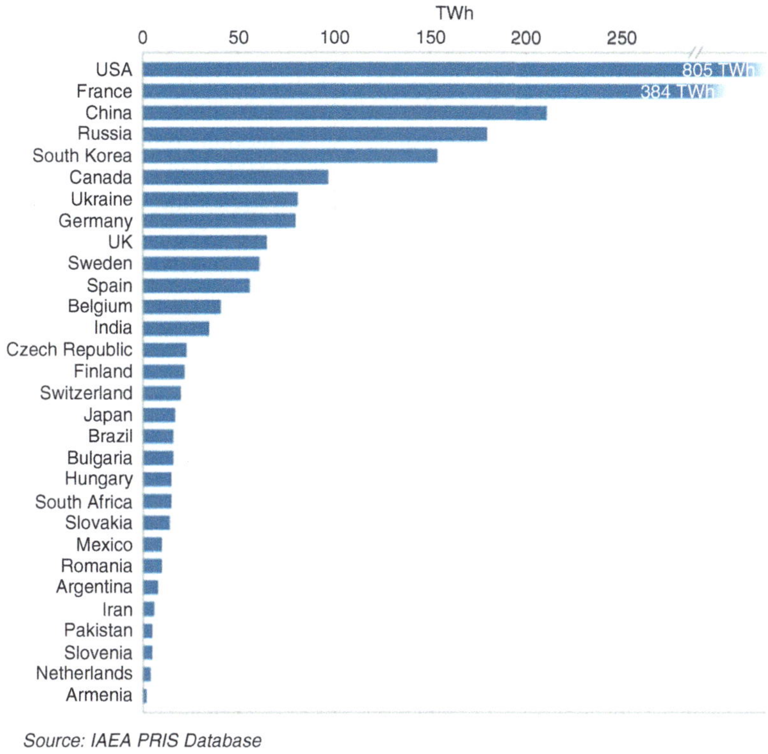

Fig. 4.27 Nuclear generation by country in 2016

Sixteen countries depend on nuclear power for at least one-quarter of their electricity (Fig. 4.27). France gets around three-quarters of its electricity from nuclear energy; Hungary, Slovakia, and Ukraine get more than half from nuclear, while Belgium, Czech Republic, Finland, Sweden, Switzerland, and Slovenia get one-third or more. South Korea and Bulgaria normally get more than 30% of their electricity from nuclear, while in the United States, United Kingdom, Spain, Romania, and Russia about one-fifth of electricity is from nuclear. Japan is used to relying on nuclear power for more than one-quarter of its electricity and is expected to return to somewhere near that level [25].

There is a clear need for new generating capacity around the world, both to replace old fossil-fuel units, especially coal-fired ones, which emit a lot of carbon dioxide, and to meet increased demand for electricity in many countries. In 2015, 66.0% of electricity was generated from the burning of fossil fuels. Despite the strong support for and growth in intermittent renewable electricity sources in recent years, the fossil fuel contribution to power generation has remained virtually unchanged in the last 10 years (66.5% in 2005) [25].

The Organization for Economic Co-operation and Development (OECD) International Energy Agency publishes annual scenarios related to energy. In its World Energy Outlook 2017, there is an ambitious "Sustainable Development

Scenario" which is consistent with the provision of clean and reliable energy and a reduction of air pollution, among other aims. In this decarbonization scenario, electricity generation from nuclear more than doubles by 2040, increasing to 5345 TWh, and capacity grows to 720 GWe. The World Nuclear Association has put forward a more ambitious scenario than this – the Harmony program proposes the addition of 1000 GWe of new nuclear capacity by 2050 to provide 25% of electricity then (10,000 TWh) from 1250 GWe of capacity (after allowing for 150 GWe retirements). This would require adding 25 GWe per year from 2021, escalating to 33 GWe per year, which is not much different from the 31 GWe added in 1984, or the overall record of 201 GWe in the 1980s. Providing one-quarter of the world's electricity through nuclear would substantially reduce carbon dioxide emissions and have a very positive effect on air quality [25].

As per World Nuclear Association statistical data, the countries involved in nuclear power development worldwide are outlined below by continent:

- **North America**

Canada has 19 operable nuclear reactors, with a combined net capacity of 13.5 GWe. In 2016, nuclear generated 16% of the country's electricity.

All but one of the country's 19 nuclear reactors is sited in Ontario. In the first part of 2016, the government signed major contracts for the refurbishment and operating lifetime extension of six reactors at the Bruce Generating Station. The program will extend the operating lifetimes by 30–35 years. Similar refurbishment work enabled Ontario to phase out coal in 2014, achieving one of the cleanest electricity mixes in the world.

Mexico has two operable nuclear reactors, with a combined net capacity of 1.6 GWe. In 2016, nuclear generated 6% of the country's electricity.

The **United States** has 99 operable nuclear reactors, with a combined net capacity of 99.6 GWe. In 2016, nuclear generated 20% of the country's electricity.

There had been four AP1000 reactors under construction, but two of these have been halted. One of the reasons for the hiatus in new build in the USA to date has been the extremely successful evolution in maintenance strategies. Over the last 15 years, improved operational performance has increased utilization of US nuclear power plants, with the increased output equivalent to 19 new 1000 MWe plants being built.

The year 2016 saw the first new nuclear power reactor enter operation in the country for 20 years. Despite this, the number of operable reactors has reduced in recent years, from a peak of 104 in 2012. Early closures have been brought on by a combination of factors including cheap natural gas, market liberalization, oversubsidy of renewable sources, and political campaigning.

- **South America**

Argentina has three reactors, with a combined net capacity of 1.6 GWe. In 2016, the country generated 6% of its electricity from nuclear.

Brazil has two reactors, with a combined net capacity of 1.9 GWe. In 2016, nuclear generated 3% of the country's electricity.

- **West and Central Europe**

Belgium has seven operable nuclear reactors, with a combined net capacity of 5.9 GWe. In 2016, nuclear generated 52% of the country's electricity.

Finland has four operable nuclear reactors, with a combined net capacity of 2.8 GWe. In 2016, nuclear generated 34% of the country's electricity. A fifth reactor – a 1720 MWe EPR – is under construction, and plans to build a Russian VVER-1200 unit at a new site (Hanhikivi) are well advanced.

France has 58 operable nuclear reactors, with a combined net capacity of 63.1 GWe. In 2016, nuclear generated 72% of the country's electricity.

A 2015 energy policy had aimed to reduce the country's share of nuclear generation to 50% by 2025. In November 2017, the French government postponed this target. The country's Energy Minister said that the target was not realistic and that it would increase the country's carbon dioxide emissions, endanger security of supply, and put jobs at risk.

One reactor is currently under construction in France – a 1750 MWe EPR at Flamanville.

In **Germany**, seven nuclear power reactors continue to operate, with a combined net capacity of 9.4 GWe. In 2016, nuclear generated 13% of the country's electricity.

Germany is phasing out nuclear generation by about 2022 as part of its Energiewende policy. Energiewende, widely identified as the most ambitious national climate change mitigation policy, has yet to deliver a meaningful reduction in carbon dioxide (CO^2) emissions. In 2011, the year after the policy was introduced, Germany emitted 731 Mt. CO^2 from fuel combustion; in 2015, the country emitted 730 Mt. CO^2 and remained the world's sixth biggest emitter of CO^2.

The **Netherlands** has a single operable nuclear reactor, with a net capacity of 0.5 GWe. In 2016, nuclear generated 3% of the country's electricity.

Spain has seven operable nuclear reactors, with a combined net capacity of 7.1 GWe. In 2016, nuclear generated 21% of the country's electricity.

Sweden has eight operable nuclear reactors, with a combined net capacity of 8.4 GWe. In 2016, nuclear generated 40% of the country's electricity.

The country is closing down some older reactors but has invested heavily in operating lifetime extensions and uprates.

Switzerland has five operable nuclear reactors, with a combined net capacity of 3.3 GWe. In 2016, nuclear generated 34% of the country's electricity.

The **United Kingdom** has 15 operable nuclear reactors, with a combined net capacity of 8.9 GWe. In 2016, nuclear generated 20% of the country's electricity.

The UK government energy paper in mid-2006 endorsed the replacement of the country's ageing fleet of nuclear reactors with new nuclear build. The government aims to have 16 GWe of new nuclear capacity operating by 2030. The placement of structural concrete at Hinkley Point C (two EPR units) has begun ahead of full construction.

- **Central and East Europe and Russia**

Armenia has a single nuclear power reactor with a net capacity of 0.4 GWe. In 2016, nuclear generated 31% of the country's electricity.

Belarus has its first nuclear power plant under construction and plans to have the first of the two Russian reactors operating by 2019. At present almost all of the country's electricity is produced from natural gas.

Bulgaria has two operable nuclear reactors, with a combined net capacity of 1.9 GWe. In 2016, nuclear generated 35% of the country's electricity.

The **Czech Republic** has six operable nuclear reactors, with a combined net capacity of 3.9 GWe. In 2016, nuclear generated 29% of the country's electricity.

Hungary has four operable nuclear reactors, with a combined net capacity of 1.9 GWe. In 2016 nuclear generated 50% of the country's electricity.

Romania has two operable nuclear reactors, with a combined net capacity of 1.3 GWe. In 2016, nuclear generated 17% of the country's electricity.

Russia has 35 operable nuclear reactors, with a combined net capacity of 26.9 GWe. In 2016, nuclear generated 17% of the country's electricity.

A government decree in 2016 specified construction of 11 nuclear power reactors by 2030 in addition to those already under construction. At the start of 2018, Russia had seven reactors under construction, with a combined capacity of 5.9 GWe.

The strength of Russia's nuclear industry is reflected in its dominance of export markets for new reactors. The country's national nuclear industry is currently involved in new reactor projects in Belarus, China, Hungary, India, Iran, and Turkey and to varying degrees as an investor in Algeria, Bangladesh, Bolivia, Indonesia, Jordan, Kazakhstan, Nigeria, South Africa, and Tajikistan among others.

Slovakia has four operable nuclear reactors, with a combined net capacity of 1.8 GWe. In 2016, nuclear generated 54% of the country's electricity. A further two units are under construction, with both due to enter commercial operation before the end of the decade.

Slovenia has a single operable nuclear reactor with a net capacity of 0.7 GWe. In 2016, Slovenia generated 35% of its electricity from nuclear.

Ukraine has 15 operable nuclear reactors, with a combined net capacity of 13.1 GWe. In 2016, nuclear generated 52% of the country's electricity.

- **Asia**

Bangladesh started construction on the first of two planned Russian VVER-1200 reactors in 2017. It plans to have the first unit in operation by 2023. The country currently produces virtually all of its electricity from fossil fuels.

China has 38 operable nuclear reactors, with a combined net capacity of 34.6 GWe. In 2016, nuclear generated 4% of the country's electricity.

The country continues to dominate the market for new nuclear build. At the start of 2018, 20 of the 58 reactors under construction globally were in China. These include the world's first Westinghouse AP1000 units and a demonstration of high-temperature gas-cooled reactor plant. China is commencing export marketing of the Hualong One, a largely indigenous reactor design.

The strong impetus for developing new nuclear power in China comes from the need to improve urban air quality and reduce greenhouse gas emissions. The government's stated long-term target, as outlined in its Energy Development Strategy Action Plan 2014–2020, is for 58 GWe capacity by 2020, with 30 GWe more under construction.

India has 22 operable nuclear reactors, with a combined net capacity of 6.2 GWe. In 2016, nuclear generated 3% of the country's electricity.

The Indian government is committed to growing its nuclear power capacity as part of its massive infrastructure development program. The government in 2010 set an ambitious target to have 14.6 GWe nuclear capacity online by 2024. At the start of 2018, six reactors were under construction in India, with a combined capacity of 4.4 GWe.

Japan has 42 operable nuclear reactors, with a combined net capacity of 40 GWe. At the start of 2018, only five reactors had been brought back online, with a further 21 in the process of restart approval following the Fukushima accident in 2011. In the past, 30% of the country's electricity has come from nuclear; in 2016, the figure was just 2%.

South Korea has 24 operable nuclear reactors, with a combined net capacity of 22.5 GWe. In 2016, nuclear generated 30% of the country's electricity.

South Korea has four new reactors under construction domestically as well as four in the United Arab Emirates. It plans for two more, after which energy policy is uncertain. It is also involved in intense research on future reactor designs.

Pakistan has five operable nuclear reactors, with a combined net capacity of 1.4 GWe. In 2016, nuclear generated 4% of the country's electricity. Pakistan has two Chinese Hualong One units under construction.

- **Africa**

South Africa has two operable nuclear reactors and is the only African country currently producing electricity from nuclear. In 2016, nuclear generated 7% of the country's electricity. South Africa remains committed to plans for further capacity, but financing constraints are significant.

- **Middle East**

Iran has a single operable nuclear reactor with a net capacity of 0.9 GWe. In 2016, nuclear generated 2% of the country's electricity.

The **United Arab Emirates** is building four 1450 MWe South Korean reactors at a cost of over $20 billion and is collaborating closely with the International Atomic Energy Agency and experienced international firms.

- **Emerging Nuclear Energy Countries**

As outlined above, Bangladesh, Belarus, and the United Arab Emirates are all constructing their first nuclear power plants. A number of other countries are moving toward use of nuclear energy for power production. For more information, see page on Emerging Nuclear Energy Countries.

4.17 Improved Performance from Existing Reactors

The performance of nuclear reactors has improved substantially over time. Over the last 40 years, the proportion of reactors reaching high-capacity factors has increased significantly. For example, 64% of reactors achieved a capacity factor higher than 80% in 2016, compared to 24% in 1976, whereas only 8% of reactors had a capacity factor lower than 50% in 2016, compared to 22% in 1976 [25].

Figure 4.28 is an illustration of long-term trends in capacity factors, while Fig. 4.29 is a plot presenting the median capacity factor from 2007 to 2016 of life cycle by age of reactor in operational existence today.

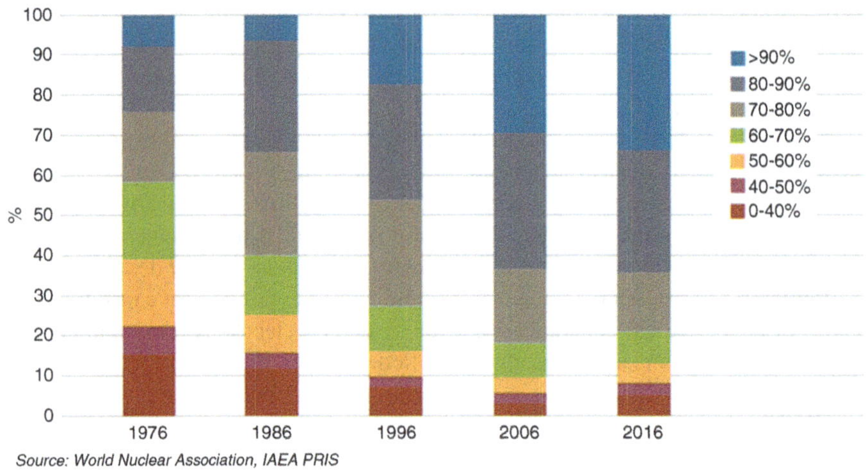

Fig. 4.28 Long-term trends in capacity factors

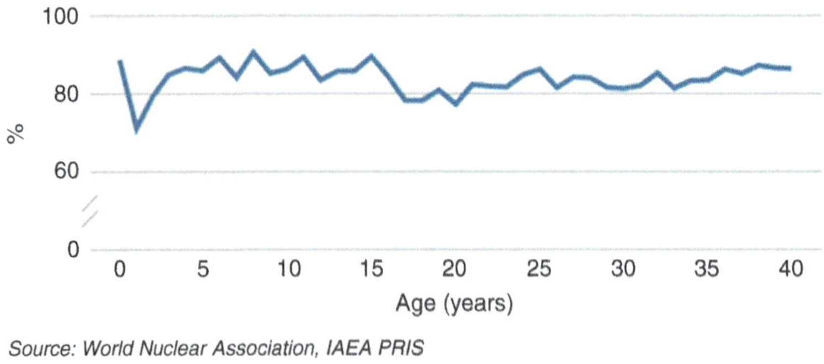

Fig. 4.29 Median capacity factor 2007–2016 by age of reactor

4.18 Other Nuclear Reactors

In addition to commercial nuclear power plants, there are about 225 research reactors operating in 50 countries, with more under construction. As well as being used for research and training, many of these reactors produce medical and industrial isotopes.

The use of reactors for marine propulsion is mostly confined to the major navies where it has played an important role for five decades, providing power for submarines and large surface vessels. At least 140 ships, mostly submarines, are propelled by some 180 nuclear reactors, and over 13,000 reactor years of experience have been gained with marine reactors. Russia and the USA have decommissioned many of their nuclear submarines from the Cold War era.

Russia also operates a fleet of four large nuclear-powered icebreakers and has three more under construction. It is also completing a floating nuclear power plant with two 40 MWe reactors adapted from those powering icebreakers for use in remote regions [25].

References

1. An Interdisciplinary MIT Study 'The Future of Nuclear Power', MIT, ISBN 0–615–12420-8, 2003
2. http://www.energystar.gov/ia/business/GBN_Energy_Strategy.pdf?7d10-c2f8.
3. http://www.iea.org/.
4. http://www.iea.org/Textbase/npsum/WEO2008SUM.pdf.
5. Zohuri, B. "Plasma Physics and Controlled Thermonuclear Reactions Driven Fusion Energy" 1st ed. 2017 Edition, Springer Publisher.
6. Zohuri, B. "Magnetic Confinement Fusion Driven Thermonuclear Energy" 1st ed. 2017 Edition, Springer Publisher.
7. Zohuri, B. "Inertial Confinement Fusion Driven Thermonuclear Energy" 1st ed. 2017 Edition, Springer Publisher.
8. Zohuri, B. "Combined Cycle Driven Efficiency for Next Generation Nuclear Power Plants: An Innovative Design Approach" 1st Edition 2015, Springer Publisher.
9. Patrick, McDaniel; Cassiano, de Oliviera; Bahman, Zohuri; James, Cole. "A Combined Cycle Power Conversion System for the Next Generation Nuclear Power Plant" American Nuclear Society Transactions, November 2012, San Diego, California.
10. Zohuri, B. "Thermal-Hydraulic Analysis of Nuclear Reactors" 2nd ed. 2017 Edition, Springer Publisher.
11. Zohuri, B; McDaniel, P. "Thermodynamics In Nuclear Power Plant Systems" 1st ed. 2015 Edition, Springer Publisher.
12. Zohuri, B. "Dimensional Analysis and Self-Similarity Methods for Engineers and Scientists" 1st ed., 2015 Edition, Springer Publisher.
13. Zohuri, B; McDaniel, P; de Oliviera, C. "A Comparison of a Recuperated Open Cycle (Air) Brayton Power Conversion System with the Traditional Steam Rankine Cycle for the Next Generation Nuclear Power Plant", ANS Transactions, June 2014.
14. Mattingly, JD. Elements of Gas Turbine Propulsion, McGraw-Hill, Inc., New York, 1996.
15. Fraas, AP. Heat Exchanger Design, 2nd Ed., John Wiley & Sons, New York, 1989.

16. Macpherson, RE; Amos, JC; Savage, HW. "Development Testing of Liquid Metal and Molten Salt Heat Exchangers", Nuclear Science and Engineering, 8, pp. 14–20, 1960.
17. Fraas, AP. "Design Precepts for High Temperature Heat Exchangers", Nuclear Science and Engineering, 8, pp. 21–31, 1960.
18. Yarosh, MM. "Evaluation of the Performance of Liquid Metal and Molten Salt Heat Exchangers", Nuclear Science and Engineering, 8, pp. 32–43, 1960.
19. Kays, WM; AL. London, Compact Heat Exchangers, 2nd Ed. McGraw-Hill Book Company, New York, 1964.
20. Zohuri, B; Patrick, McDaniel; Cassiano, De Oliviera. "Advanced Nuclear Open-Air Brayton Cycles for Highly Efficient Power Conversion", Nuclear Science Technology Journals, Jan 23rd, 2015.
21. M. M. EL-WAKIL, Powerplant Technology, McGraw-Hill (1984).
22. Wilson, DG; Korakianitis, T. The Design of High-Efficiency Turbomachinery and Gas Turbines, 2nd Ed., Prentice-Hall, Upper Saddle River NJ, 1998.
23. Haywood, RW. Analysis of Engineering Cycles, 4th Ed., Pergamon Press, Oxford, 1991.
24. Boss, M. Steam Turbines for STAG Combined-Cycle Power Systems, GE Power Systems
25. http://www.world-nuclear.org/information-library/current-and-future-generation/nuclear-power-in-the-world-today.aspx

Chapter 5
Small Modular Reactors, the Next Big Renewable Energy Source

Abstract Policy Exchange's Energy and Environment Unit conducts innovative and independent policy research into a wide range of environmental, infrastructure, and regulatory challenges. Our objectives are to influence policy making and to shape debate. We produce publications, organize events, and use the media to promote our findings and policy proposals. A key focus of our work is to identify ways to tackle environmental challenges effectively while minimizing adverse impact on living standards. We promote well-designed regulation to exploit the power of markets to achieve environmental outcomes innovatively and cost-effectively.

5.1 Introduction

The discovery of nuclear fission in 1939, following the Manhattan Project, was an event that opened the prospect of entirely new source of power utilizing the internal energy of the atom. Nuclear energy is an incredibly efficient method of producing and generating electricity that also enables low-carbon emissions. Currently, many nuclear power plants use Generation-III reactors to produce power.

In the United States today, there are 99 operating nuclear power reactors with a total installed capacity of about 100 GWe. This represents about 10% of US electrical generating resources. However, the high reliability and base-load status of these plants result in a contribution of about 20% of the actual US electricity generation.

More importantly, these 99 nuclear power reactors are the source of 63.3% of our clean-air electricity. Other sources include hydro at 21.2%, wind at 13%, geothermal at 1.3%, and solar at 0.7%. See Fig. 5.1 for US electric generation by energy source.

Nuclear power is a largely controversial topic in environmental physics as there are multiple pros and cons. It could be the future for our planet or just the same story as fossil fuels where it lasts for a couple hundred years or so then becomes high in demand and exceedingly expensive. Unfortunately, nuclear power is not renewable as it uses radioactive "heavy" metals such as uranium as its fuel which is not very abundant with about two to four parts per million in the earth's crust however still frothy times more abundant than silver. Also, for it to be economically viable

B. Zohuri, *Small Modular Reactors as Renewable Energy Sources*,
https://doi.org/10.1007/978-3-319-92594-3_5

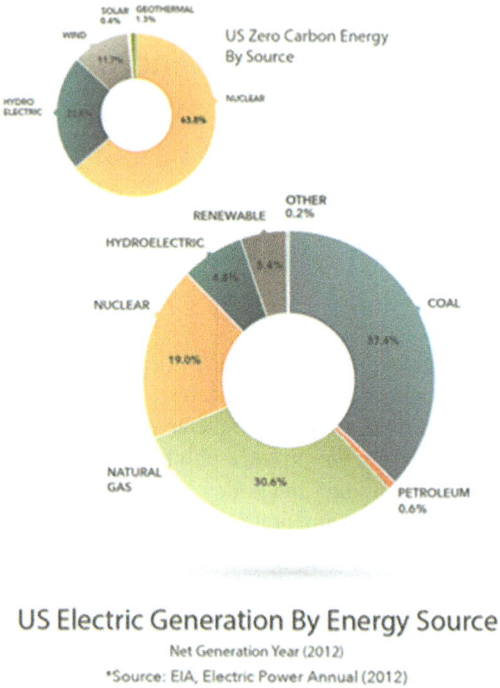

Fig. 5.1 Electric generation by energy source

to extract, it needs to be extracted in large quantities which can only be done in countries like Canada and Australia. Many countries with small amounts of uranium cannot extract it and make it a profitable proposition from return on investment (ROI) point of view.

The contribution of nuclear to clean, reliable electricity is threatened by the approaching, inevitable retirement of existing reactors. NRC licenses expire, and reactor retirements begin in 2029. All currently operating plants will retire by 2050.

Despite the advantages of high reliability, competitive generating costs, and low environmental impact, the potential for new builds in the United States of GWe scale, conventional plants is not very promising, especially for utilities in unregulated markets. This hesitance is due to inherently uncertain licensing and construction costs and durations and perceived health and environmental risks (Fig. 5.2).

However, new advanced generation of SMRs have key characteristics that answer many of the concerns raised by environmental communities as utilities and their owners are considering new nuclear power plants for their future generation portfolio to produce enough supply of electricity, to meet their demands due to population increase, where this demand is at 19% increase globally.

One of the key characteristic of small modular reactors are their efficiency of producing thermal output for generating, the electricity both for day-to-day consumption as well as being a new source for renewable energy while could get coupled in conjunction with plant producing hydrogen as fuel for the new generation of cars to prevent any additional carbon dioxide by existing engine in today's cars

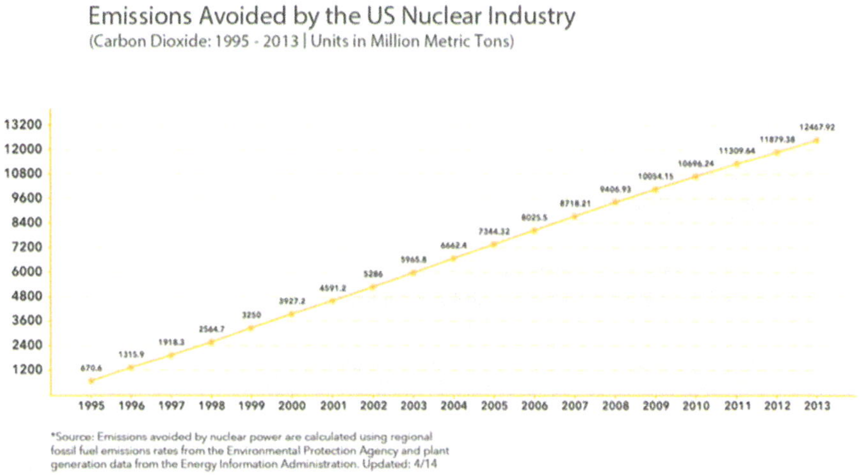

Fig. 5.2 Emission avoided by the US nuclear industry

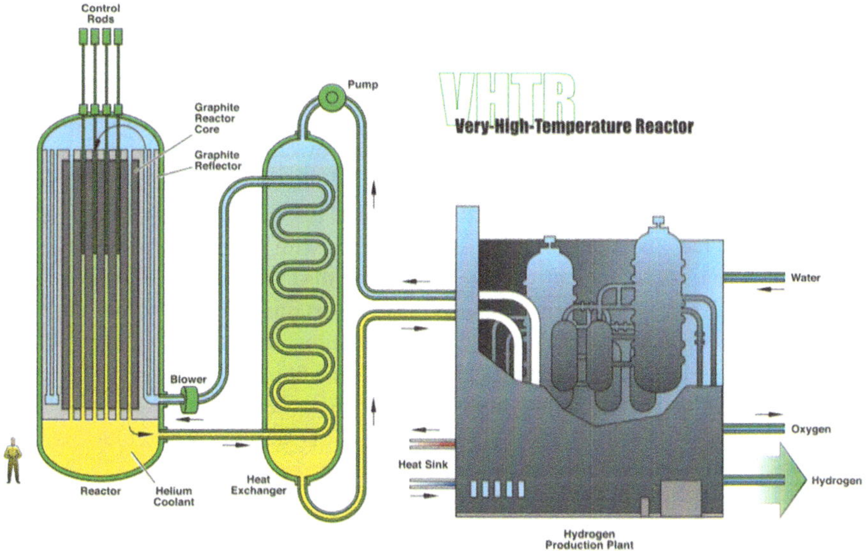

Fig. 5.3 Very-high-temperature reactor of Generation IV in conjunction with hydrogen power plant facility

running around [1]. Figure 5.3 is an illustration of such conceptual idea, by utilizing one of the six final designs of GEN IV that is considered to be very-high-temperature reactor (VHTR).

Decarbonizing our existing electricity system with 100% renewable energy would be possible, but unnecessarily expensive and perhaps unsustainable. The intermittent nature of solar and wind would mean that large amounts of underutilized backup capacity and storage would be required at great expense to the consumer/taxpayer. Biomass could be used to provide some backup power supply, but this is unlikely to be a sustainable solution for more than a small part of our electricity system [2].

Decarbonizing our whole *energy* system using renewable sources would test the limits of the possible. Electricity comprises just one fifth of the final energy demand in the United Kingdom. For example, so creating a 100% renewable energy economy would be an order of magnitude more difficult than the already challenging task of powering our existing electricity grid with 100% renewable sources.

Here in the United States under the Department of Energy watch, nuclear industries are also pushing the same agendas as the United Kingdom, when it comes to the new advanced generation of nuclear power plants.

There is much hope for Generation-IV reactors in terms of sustainability, safety, and price. While Generation-III reactors will likely remain popular as the choice technology for new reactors currently being built, Generation IV will be an opportunity to build more sustainable nuclear reactors for the longer-term future (by that meaning in the next 20–50 years) [3]. The technology is certainly in route for this timeline, but there are other political, strategic, and economic hurdles that Generation-IV prototypes will need to overcome in order to become part of the nuclear power infrastructure [3]. One of the biggest hurdles is the fact that there has been a decrease in funding for Generation-IV reactors, especially given the popularity of Generation-III reactors [3]. Though Generation IV shows quite a bit of promise, especially in developing applications outside of current nuclear power plants (NPPs), it is hard to move the current technology past the status quo [4]. To find a solution to this problem, many large international organizations, such as the Generation-IV International Forum (GIF), have been working to find synergies with other systems to promote the research and development (R&D) of Generation-IV reactors [3]. Major areas of R&D for Generation-IV reactors have been in sustainability and safety, two key measures that nuclear reactor development is measured against [4].

As we have discussed in previous chapters of this book, GEN-IV nuclear power plants offer an excellent safety margin and impose better safety criteria. 2014 marked a period of important progress in terms of safety design guidelines for Generation-IV reactors. There has been an established hierarchy of safety standards, starting with Safety Fundamentals, moving on to Safety Design Criteria and Guidelines, and ending with Technical Codes and Standard [3]. The safety criteria are also being explored for the variety of different systems that have been prototyped, including very-high-temperature reactor (VHTR), sodium fast reactor (SFR), supercritical water reactor (SCWR), gas fast reactor (GFR), lead fast reactor (LFR), and molten salt reactor (MSR) [5, 6]. These six are the main designs being explored, with modifications being made after the Fukushima accident [7]. The six designs fall under two general categories: thermal reactors and fast reactors [6]. Main safety designs being explored would allow reactors to avoid pressurized operations and have automatic reactor shutdowns in the case of emergency [4]. Many designs also seek to avoid the usage of water to cool the reactor (which would reduce the risks in the situation where water is lost through leaks or heat) [4].

The main advantages Generation IV seeks to provide are reducing the amount of time the waste remains radioactive for (on the magnitude of ten, reducing from millennia to centuries), improving the energy yield for the nuclear fuel, increasing the variety of fuels that can be used to power the reactor, and allowing for reactors to use already present nuclear waste in its operations [7]. Many of these goals are encompassed under the umbrella of sustainability, to enable nuclear reactors to become more sustainable and environmentally friendly [4].

In conclusion we can state that Generation-IV reactors have some very large goals ahead, but they are the ones that we should all be aiming for. Nuclear power has the potential to change the way energy is accessed on this planet and could provide an alternative that is efficient and sustainable.

5.2 The Limit of Renewable Energy and Small Modular Reactor

Historically, the coal-fired power plants have been generating electricity in the United Sates and globally to start with; then with progress in technology of power plant, we turned to fossil fuel; and furthermore by improvement of gas turbine, we start using gas fuel power plants to generate electricity for our needs. In 2015, coal plants generated 39% of the 3944 billion kilowatt-hours of electricity generated in the United States. However, coal's contribution has steadily eroded, down from 50% just a decade earlier.

Nuclear power is one of the most recent achievements in the long history of harnessing energy and one of the most controversial. A result of research originally done to produce the atomic bomb, nuclear energy takes advantage of the incredible potential energy within the atom in a productive instead of destructive way. As of 2011, nuclear energy provides nearly 20% of the electric power in the United States.

Aging infrastructure has made many older and smaller units uneconomical to operate. Nearly 70% of coal-fired generating units comprising more than 50% of the coal generating capacity are more than 40 years old. At the end of 2015, the coal-fired generating units in the United States totaled 286 gigawatts of capacity. In 2015 alone, 11.3 gigawatts of coal-fired capacity was retired. The US Energy Information Administration (EIA) projects that a total of 30 gigawatts of coal-fired generating capacity will retire by 2025, 87% of which by the end of 2020. See Fig. 5.4, where the coal plant shutdown is scheduled.

Tightening environmental regulations have accelerated the trend. New regulatory standards require reductions in emissions of mercury, acid gases, and toxic metals. These standards are scheduled to take effect in April 2015. The capital improvements required to reduce these emissions would make many coal plants uneconomical to operate resulting in the bubble of closures over the next few years. Extensions on compliance are being traded for pledges to close older, dirtier plants completely. The growing public concern with climate change and CO_2 emissions further increases the pressure for coal plant closures.

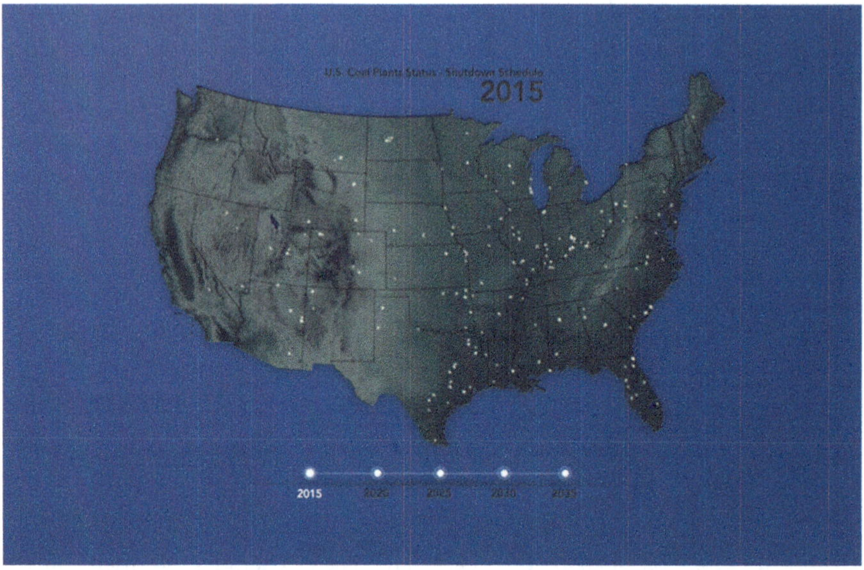

Fig. 5.4 Coal plant shutdown schedule

When it comes to replacing retiring generation, coal now faces greater competition from another fossil fuel, natural gas. Due to technological advances, accessible natural gas reserves have increased dramatically. Natural gas is now available in greater quantities and at low prices. While cleaner and, for the moment, cheaper, natural gas still produces substantial CO_2, emissions and fuel prices are volatile.

Along with the circumstance above, what comes to play was the nuclear energy that stated for peace around the 1950 time frame, where it went through so many generations (see Chap. 1 of this book), and now Generation III of these power plants is opening room for new and advanced Generation IV.

If we compare each source of energy to nuclear one as illustrated in Fig. 5.5, we obviously can see that nuclear goes a long way.

However, one may argue if we consider nuclear power energy as a big revolution and evolution in our life since Manhattan project took place, thus the argument may continue on the issue of the bigger is not necessarily the better energy resource.

However, it is very clear that nuclear energy can play a very significant long-term role for meeting the world's increasing supply and demand for energy, based on growth in population globally, while simultaneously addressing challenges associated with global climate and environmental impact.

Today's need for electricity and new source of renewable energy in a very cost-efficient way has sent many countries/nations of the world, particularly the Asia/Pacific Rim countries into quest of new and innovative source of energy beyond what we have from our past technologies, and they all are actively engaged in a major expansion of their nuclear energy complex. This is to the degree to which nuclear energy can address long-term energy solution needs, either globally or

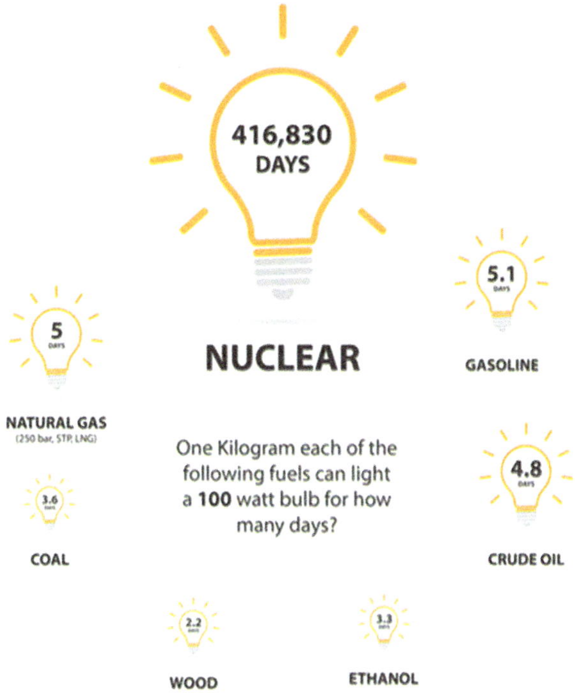

Fig. 5.5 Energy fuel density

regionally, where such resource of energy will be dictated by the peace and adequacy of technical and policy solutions for safety, security, waste management, nonproliferation, and finally greenhouse effects issues, as well as the capital cost of construction, where also energy efficiency has to be challenged. See Fig. 5.6.

Although the chart in Fig. 5.6 is an indication of wind energy to be the most efficient way of producing electrical energy, we have to keep in our mind that the wind is not always blowing 24 x 7 around the year and probably there are some regions that the wind blowing is not as energetic as we need it to be in order to meet the demand for the electricity.

In addition, here, we briefly describe each means of producing electrical energy from each source that is mentioned in Fig. 5.6 as:

1. *Biomass*

   Everything from crops left in the field to weedy trees, from animal waste to humans' garbage, can be recycled and transformed into usable energy. Biomass is a very broad term covering a wide variety of materials that can be used as energy resources. Since the sun's energy is absorbed by all living things, humans, animals, and especially plants, a lot of materials we see as leftovers are storehouses of energy. For example, a tree uses photosynthesis to store energy in its leaves and trunk. The tree is biomass. The tree can be burned to release the energy in the form of heat.

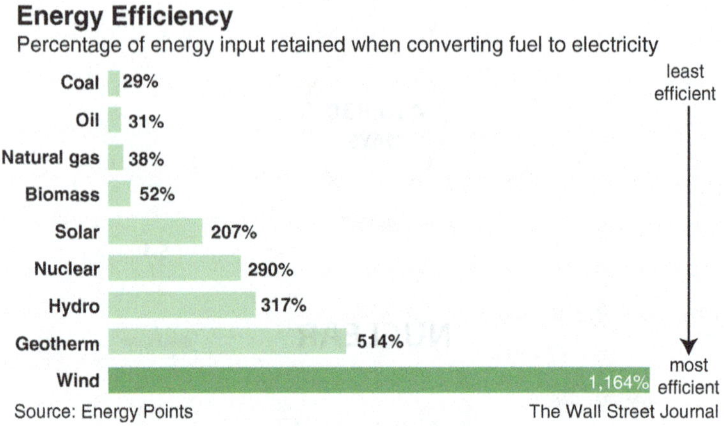

Fig. 5.6 Energy efficiency chart

2. *Geothermal Power*

   Geothermal energy is energy that is generated and stored within the earth. It takes advantage of the Earth's core heat to produce usable energy.

3. *Hydropower*

   Water's raw power provides the energy to produce enough electricity for 28 million Americans every year and, as of 2011, creates nearly 10% of all electricity in the United States. Worldwide, hydropower generates more than 2.3 trillion Kilowatt (kW) hours of electricity per year, the energy equivalent of 3.6 billion barrels of oil.

4. *Wind Power*

   One of the most important alternative energy resources can't be seen or touched, but its power is obvious to anyone who's ever weathered a hurricane, a tornado, or even a strong storm: wind. At its worst, wind can wreak havoc, destroying everything in its path. At its best, it's a source of clean, efficient, inexpensive energy; but as of 2011, it provides less than 3% of all the electricity in the United States.

5. *Solar Power*

   The sun is primarily a source of light and heat. But can it be our primary source of energy? Solar panels or thin films designed to collect sunlight are integral parts of the process to generate electricity by way of the sun. The sun is our most impressive source of energy. More than 1 million times larger than the earth, the sun gives us ten times more energy than is stored in all the world's reserves of coal and oil every year. Despite this, as of 2011, solar power accounts for less than 1% of all the electricity generated in the United States.

6. *Fossil Fuels*

   Fossil fuels (coal, oil, and natural gas) provide the energy that powers our lifestyles and our economy. Fossil fuels power everything from the planes in the sky to the cars on the road. They heat our homes and light up the night. They're the bedrock we base our energy mix on. But they are a limited resource.

7. *Hydrogen Power*
   Hydrogen can be used as a way of storing or transporting energy.
8. *Energy Basics*
   Electrical power is produced and distributed through three simple steps:
   (a) Generation
   (b) Transmission
   (c) Distribution

It is also clear that in meeting our low-carbon energy needs, nuclear power should play a crucial role. The energy density of nuclear fission means that just a few plants can provide a large percentage of our electricity requirements. In Western liberalized economies, however, traditional large nuclear power plants are not thriving. Struggling utility companies now have difficulty financing projects that can cost upwards of 10 billion pounds and reactor vendors do not have a good record in reducing costs or bringing new plants online on schedule.

Small Modular Reactors (SMRs) Could Be a Solution Each unit would require a smaller investment than large reactors, and their modular nature means that they can be built in a controlled factory environment where, with increased deployment, costs can be brought down over time through improved manufacturing processes and economies of volume. This learning-by-doing effect has helped the offshore wind industry achieve impressive cost reductions, and the nuclear industry could replicate their success.

Furthermore, the advanced and innovative SMRs could successfully address several of these issues and offer simpler, standardized, and safer modular design by being factory built, requiring smaller initial capital investment per power plant by virtue of modularity and having shorter construction time periods. The SMRs also could be small enough to be portable by means of transportation and occupy smaller real estate due to smaller footprint. It could be implemented in an isolated location without even accessing to the water as coolant media and advanced infrastructure and with no access to power grid (i.e., remote military bases overseas) or could be clustered in a single site to provide a multi-module, large capacity power plant.

To emphasize our argument here in defense of SMR technology, we can express what Matt Rooney [2] is saying, and that is, "SMRs could offer a number of advantages in a flexible power system, including the potential for dual output, producing other useful services in addition to electricity, like hydrogen or heat. SMRs could, for example, provide a demand/grid management solution by redirecting the power from an SMR to hydrogen production when renewable output is high."

A new fleet of small modular reactors (MSRs) would also provide a large quantity of secure low-carbon energy, thus reducing reliance on imports of natural gas, electricity via interconnectors, and biomass. Uranium, the main source of nuclear fuel, is an inexpensive commodity traded worldwide, and the pioneers in this implementation of source of energy have the capability to both enrich uranium and manufacture its own nuclear fuel. As we know by now, the nuclear power reduces import dependency from multi-angle point of views.

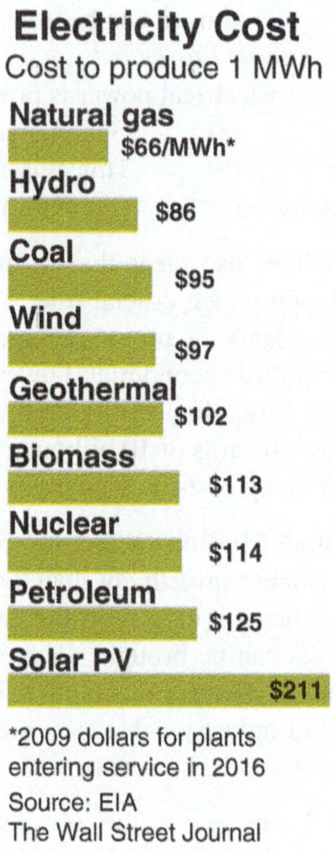

Fig. 5.7 Cost of electricity per 1 MWh

5.3 Small Modular Reactor-Driven Renewable and Sustainable Energy

In order to address this subject within this section, we need to ask ourselves the question: What is the most efficient source of energy?

The answer falls into the following fact that the true cost of electricity is difficult to pin down. That's because a number of inputs comprise it: the cost of fuel itself, the cost of production, as well as the cost of dealing with the damage that fuel does to the environment.

Energy Points, a company that does energy analysis for business, factors in these myriad values in terms of what percentage of the energy input – fossil fuel energy, plus energy for production and energy for environmental mitigation – will become usable electricity.

The chart above in Fig. 5.7 shows that fossil fuels yield, on a national average, only a portion of their original energy when converted into electricity. That's because they are fossil fuels that require other fossil fuels to make the conversion into electricity; their emissions, such as carbon dioxide, also require a lot of energy to be mitigated. Renewables, however, have energy sources that are not fossil fuel and their only other energy inputs are production and mitigating the waste from that production. That actually results in more energy produced than fossil fuels put in. Wind, the most efficient fuel for electricity, creates 1164% of its original energy inputs when converted into electricity; on the other end of the efficiency spectrum, coal retains just 29% of its original energy.

These are national averages, meaning that, for example, solar might be more efficient in a place such as Arizona with lots of infrastructure and direct sunlight than it is across the whole nation. Thus, a scenario such as solar farm technology may very well be suited in such environment, and arguably source of fresh water shortage as a coolant media for fossil, gas, and nuclear power plant may also enhance the solar power plant farm as the only choice of solution to generate electricity as well renewable energy approach, which may very well be cost-efficient for such production.

However, no matter what, in any given area, electricity might come from a number of different sources, including oil, coal, gas, wind, hydropower, and solar. Each has its own set of costs, both internal and external. From Energy Points:

Energy Points' methodology measures environmental externalities and calculates the energy it takes to mitigate them. For example, it quantifies the greenhouse gas (GHG) emissions that result from turning coal and natural gas into electricity and then calculates the energy it would take to mitigate those emissions through carbon capture and sequestration. Water scarcity and contamination are quantified as the energy that is required to durably supply water to that area. And in the case of solar or wind energy, Energy Points incorporates the life-cycle impact of manufacturing and shipping the panels.

This metric is a more rounded calculation than merely cost or carbon footprint. For example, hydroelectricity has the lowest carbon footprint of 4 gCO^2/kWh, but when Energy Points factors in the full life cycle of the different fuels, wind is the most efficient. Additionally, natural gas is the cheapest fuel to produce electricity, according to levelized cost data from the Environmental Protection Agency, which measured the total cost of building and operating a generating plant over an assumed financial life and duty cycle. Though it's cheap, it's not very efficient if you factor in its production and emissions.

5.4 Small Modular Reactor-Driven Hydrogen Energy for Renewable Energy Source

Research is going forward to produce hydrogen based on nuclear energy. Hydrogen production processes necessitate high temperatures that can be reached in the fourth-generation nuclear reactors (i.e., small modular reactors). Technological studies are now underway in order to define and qualify components that in the future will enable us to retrieve and transfer heat produced by these reactors. Hydrogen combustion turbine power could be one of the solutions to our future energy needs particularly in on-peak demand for electricity, but until recently the problem with hydrogen power was its production for use as an energy source.

Although hydrogen is the most common element in the known universe to human being, actually capturing it for energy use is a process which itself usually requires some form of fuel or energy [1].

Germany is to take a drastic measurement to revise their nuclear energy policy that had long heralded nuclear power plants as its main source of energy. For example, while Germany decided to abandon all of their atomic power plants, the new energy policy that is announced by Japan is taking steps to decrease its dependency on nuclear as much as possible while increasing and enhancing their research and development (R&D) to quest for an alternative renewable energy source. Also in parallel effort, the government is promoting for a "Hydrogen Society" and use of hydrogen as source of energy to pave their way to such goal, by making, for example, fuel cell vehicle (FCV), where Fuel Cell and Hydrogen Technology Group at the New Energy and Industrial Technology Development Organization (NEDO) is in charge of such R&D.

Burning hydrogen in a combustion form does not emit any carbon dioxide (CO_2), so it is considered as source of clean energy that can greatly help reduce the greenhouse gases effects. Although expectations are set so high, it comes with the technical challenges and cost of ownership as well as return on investment of such research and development toward full production of such source of energy as part of renewable form. As an example, setting up expensive hydrogen stations for FCVs, securing sufficient supplies of the gas, and coming up with ways to produce it without emitting carbon dioxide are just a few of those challenges and hurdles [1].

Another industrial application of hydrogen is in oil refinery, where it is used to process crude oil into refined fuel, such as gasoline and diesel, and for removing contaminants, such as sulfur, from these fuels. See Fig. 5.8.

The total hydrogen consumption in oil refineries is estimated at 12.4 billion standard cubic feet per day, which equates to an average hydrogen consumption of 100–200 standard cubic feet per barrel of oil processed. Hydrogen consumption in the oil refining industry grew at a compound annual growth rate of 4% from 2000 to 2003, and growth in consumption is expected to increase to between 5% and 10% through to 2010 [Oil & Gas Journal, CryoGas International]. See Fig. 5.9. [14].

The principal drivers of this growth in refinery hydrogen demand are:

- Low sulfur in diesel fuel regulations – hydrogen is used in refineries to remove sulfur from fuels such as diesel.
- Increased consumption of low-quality "heavy" crude oil, which requires more hydrogen to refine.
- Increased oil consumption in developing economies such as China and India.

Approximately 75% of hydrogen currently consumed worldwide by oil refineries is supplied by large hydrogen plants that generate hydrogen from natural gas or other hydrocarbon fuels, with the balance being recovered from hydrogen-containing streams generated in the refinery process.

Pressure swing adsorption (PSA) (see Fig. 5.10a, b) technology is used in both hydrogen generation plants and for hydrogen recovery.

Fig. 5.8 A typical oil refinery plant

Fig. 5.9 Hydrogen PSA unit – HYDROSWING. (Courtesy of Mahler Advanced Gas Systems)

Hydrogen is used in a range of other industries, including chemical production, metal refining, food processing, and electronics manufacturing. Hydrogen is either delivered to customers in these industries as compressed or liquid hydrogen or generated on-site from water using a process known as electrolysis or from natural gas using a process called reforming. In certain applications, there is a gradual shift toward on-site generation to replace delivered compressed or liquid hydrogen, largely based on the lower cost of new on-site hydrogen generation technologies when compared to delivered hydrogen.

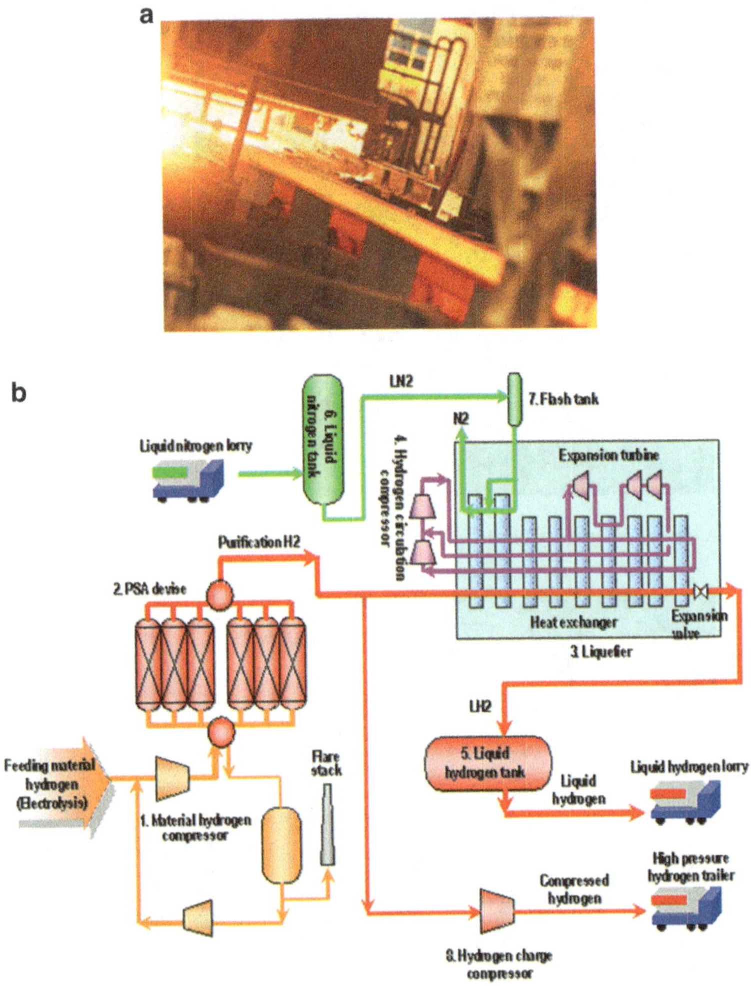

Fig. 5.10 (**a**) A typical metal refining plant. (**b**) Flow of liquid hydrogen production facilities

Other applications of hydrogen in industry that are worth to mention are listed below:

1. Weather balloons in meteorologist, where these balloons are fitted with equipment to record information necessary to study the climates.
2. Hydrogen is used in fertilizer and paint industries.
3. Food industries, where in food it is used as element to make hydrogenated vegetable oils, while using nickel as a catalyst, solid fat substances are produced.
4. Welding companies use the hydrogen as part of welding torches element. These torches are utilized for steel melting.
5. Chemical industries use them for metal extraction. For example, hydrogen is needed to treat mined tungsten to make them pure.

6. In home uses, hydrogen peroxide can be used in nonmedical ways. Other applications include a pest controller in gardens, removing stains on clothing, and functioning as a bleaching agent for cleaning homes.

As we can see, hydrogen is an important utility for numerous applications in multiple industries. Users in a wide range of industries can benefit from operating a cost-effective hydrogen plant and reduce their production costs significantly. See Fig. 5.10a, b.

As we know from our knowledge of chemistry, hydrogen is the lightest and most common element in the cosmos. Its atomic number is 1. In its elemental state, hydrogen is rare. But it is one of the components of water and vital to life. Hydrogen alone does not exist as a natural resource, and it needs to be produced by separating from other elements and molecules, such water as we have waste oceans sounding us. By far the most common method of producing hydrogen in industry currently is due to stripping hydrogen from natural gas using a process known as steam reforming.

Another way of producing hydrogen is through electric hydrolysis as an alternate to steam reforming approach, and both methods were mentioned as below.

Currently, fossil fuels, including naphtha, natural gas, and coal, are the main sources of hydrogen, which is generated by "steam reforming" method, in which steam is added to methane to yield hydrogen. A huge amount of hydrogen is also produced as a by-product from the production of caustic soda plants and from coke ovens.

In contrast electric hydrolysis is a relatively simple process and method in which production of hydrogen takes place in any high school chemistry laboratory course, where two electrodes, one with positive charge known as anode and other negatively charged know as cathode by a battery are placed into water. Result of such induced electric current through water splits the hydrogen ion from oxygen with positive hydrogen ion being attracted to cathode and negative oxygen ion going toward anode. Once the ions touch the electrodes, the hydrogen gains and electron while oxygen loses one, and they are creating fully fledged atoms of hydrogen and oxygen, which then rise in the water, and they can be collected separately at the top of water container.

The Japanese organization, NEDO, published a white paper on hydrogen energy in July 2012 that states the importance of promoting hydrogen-related products, which in Japan are expected to develop into a market worth ¥1 trillion by 2030 and ¥8 trillion by 2050 [15].

5.4.1 *Hydrogen Energy for Renewable Energy Source*

Hydrogen is considered to be the main choice for storing renewable energy, as it is easily transported and stored in large amounts. A major advantage of hydrolysis is that it produces hydrogen in a clean way. The problem is that it needs a source of the electrical power in the first place. A new working project in Prenzlau, 75 miles North of Berlin in Germany, aims to combine the advantages of wind power and hydrogen to overcome the problems of both forms of power production and to embrace the advantages [16].

The three wind turbines at the Prenzlau project can each produce 2 Megawatts of power. Based on an estimated assumption that 1 megawatt can supply enough power for 240 to 500 households, WindUpBattery® estimates that the three turbines together should therefore be able to supply enough energy between them to power 720 to 1500 homes, assuming a constant wind. As we all know, the wind isn't constant and reliable though, so when it comes to the future power companies working out how many homes should be powered by a certain number of wind turbines, the calculations need to take into consideration periods of low wind. This is combined with the fact that people do not use a set amount of electricity constantly but instead vary their demands during the day; this means that there is always the possibility of differences between supply and demand.

Interestingly, according to Vattenfall, one of the energy companies participating in the Prenzlau project, currently there is no system within the renewable energy sector which is designed to compensate for supply and demand differences. Wind farms produce energy which is fed directly into the power grid, but the challenge is to develop and employ a way of storing the energy when wind power production is greater than demand. The Prenzlau project enables a balance to be found in the renewable energy production system. Operating on the 6 megawatt that provides, this is the first full-scale European hybrid power plant to convert the variable nature of wind power to hydrogen for a reliable long-term renewable energy source for general electrical usage and for use in hydrogen fuel cell-powered cars and other transport, as well as for heat to co-fire the power plant itself [17].

From the three 2 megawatt wind turbines, the electrical power can either be fed straight into the power grid or to an electrolysis unit which is used to produce the hydrogen. This is stored in the plant's hydrogen storage tanks. From there the hydrogen can then be fed along the system and mixed with biogas from a separate unit at the project, and this mixture can be used to run two combined heat and power plants to produce both electricity for the power grid and district heating at times when high demand coincides with low wind power production. See Fig. 5.11.

Alternatively, the hydrogen from the storage tanks can be used at vehicle filling stations in chosen location and facilities which can run by the energy-related companies to supply fuel cell cars and other vehicles.

Hydrogen is a significant chemical product where about half of the annual production is being used in making nitrogen fertilizers and about half to convert low-grade crude oils (especially those from tar sands) into transport fuels [18]. Both uses are increasing significantly. Some are used for other chemical processes. World consumption in 2009 was about 70 million tones [19] per year, growing at about 7% pa. There is a lot of experience handling hydrogen on a large scale, though it is not entirely straightforward. About 96% of hydrogen is made from fossil fuels: half

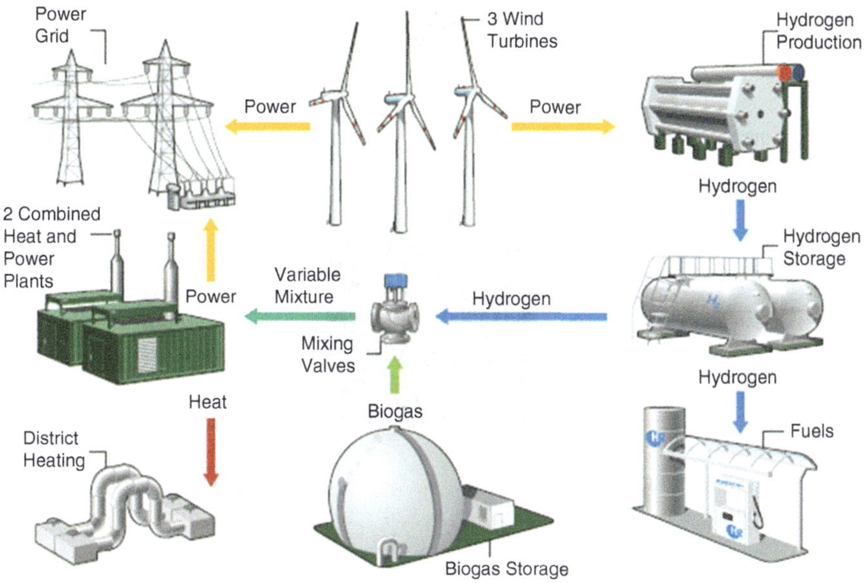

Fig. 5.11 Hydrogen production plant based on wind energy. (Courtesy of Prenzlau Corporation)

from natural gas, 30% from liquid hydrocarbons, and 18% from coal. This gives rise to quantities of carbon dioxide emissions – each ton produced gives rise to 11 tons of CO_2. Electrolysis accounts for only 4%.

At rate that we are burning fossil fuel due to increasing demand on electricity driven by growth of world population, there exists a looming specter of the exhaustion of fossil resources; therefore a lot of industrial countries that are in quest of more electrical energy are vigorously conducting research and development on various kinds of new energies. Recent global environmental and green effect problem has imposed great importance on utilizing of clean energy as a solution to this problem. Hydrogen seems as an attractive source of energy when it goes through any combustion process resulting in the form of water. Up to now, almost all practical-use hydrogen has been produced by fossil fuels [1].

Figure 5.12 depicts all the characteristics of hydrogen, and if hydrogen could be produced by renewable energies then, energy use will truly become harmonized with the environment.

As a matter of fact, Iwatani Corporation with its headquarters in Osaka and Tokyo in Japan with a capital of 20 billion yen was the first plant to produce liquid hydrogen at the Gases Corporations' facility in Ichihara, Chiba, as a renewable source of energy to meet the increasing demand for liquid hydrogen nationwide and to improve the stability of the supply chain system.

This plant has been in operation since July of 2009, and it is the first liquid hydrogen plant in Eastern Japan along with Hydro Edge's facilities in Sakai and Osaka which started its operation since April 2006 and establishes a hydrogen

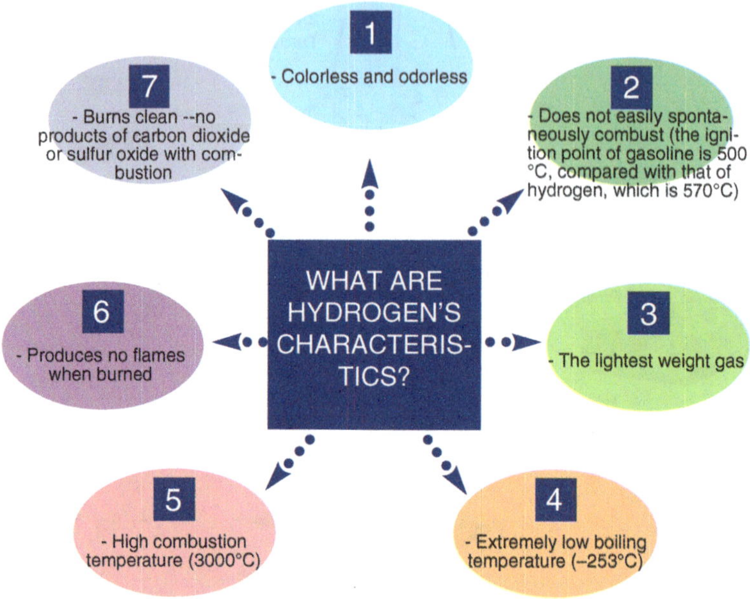

Fig. 5.12 Hydrogen characteristics

Fig. 5.13 Liquid hydrogen facility in Japan. (Courtesy of Iwatani Corporation)

supply/chain system in order to improve manufacturing efficiency and supply stability, logistic efficiency, and back-up system with two bases located in each eastern and western front of Japan as part of their continuous research and development program to go forward with hydrogen production plants (Fig. 5.13).

Companies like Iwatani with its plant which uses by-product hydrogen produced by the adjacent factory in the industrial complex as material, with a liquid hydrogen manufacturing capability of 3000 Liter/Hour (L/H) and a storage capacity of 300 Kilo/Liter (KL).

5.5 The Future of Nuclear Power

Small modular reactors (SMRs) are the latest "new" technology that nuclear advocates preach about as the game changer that will overcome previous economic and safety failure associated to previous generation nuclear power plants of past. The debate over SMRs has been particularly intense by pioneer manufacturers such as NuScale, because of the rapid failure of large "nuclear renaissance" reactors in market economies, the urgent need to address climate change, and the dramatic success of alternative, decentralized resources in lowering costs and increasing deployment. Therefore, assessing the prospects for SMR technology from three perspectives is in order, and they can be implemented as:

1. The implementations of the history of cost escalation in nuclear reactor construction for learning economies of scale and other processes that small modular reactor (SMR) advocates claim will lower the cost of construction and its trend into production market by the demand on energy by virtue of modulization
2. The challenges SMR technology faces in terms of high costs resulting from lost economies of scale, long lead time needed to develop a new design, the size of the task to create assembly lines for modular reactors, and intense concerns about safety
3. The cost and other characteristics – e.g., scalability, speed to market, flexibility, etc. – of available alternatives compared to SMR technology and last but not the least operational licensing

One of the important elements of surprise as part of "nuclear renaissance" that needs to be observed is the recent decision of the major vendors such as Westinghouse and Babcock & Wilcox at least in the United States, which dramatically reduces small modular reactor development efforts reflecting the severe disadvantages that SMR technology faces in the next several decades, yet on the other hand companies such as NuScale is expanding on SMR technology and going forward at full speed with design and implementation of the advanced SMRs into production and is on virtue of growing as a new company into being a major player in the nuclear energy industry [8].

The analysis that is shown by Cooper [8] has identified the four factors that are creating the conditions for a "nuclear renaissance" and the dozen characteristics that suggest small modular technology would play a large part in that renaissance have turned sharply against nuclear power energy and its future of existence. In fact, he argues that, looking at the long history of commercial nuclear power, this outcome is not merely repetitive, and it is endemic and possibly inevitable. Thus, the failure of nuclear economics (see next section of this chapter) is not just bad luck; however nuclear energy being inherently uneconomic is due to the fact that it relies on a catastrophically dangerous resource that is vulnerable to human frailties and the vicissitudes of "mother nature."

Cooper also argues that the severe threats to public safety posed by nuclear power and the evolving demands of safety result in an extremely complex technology that

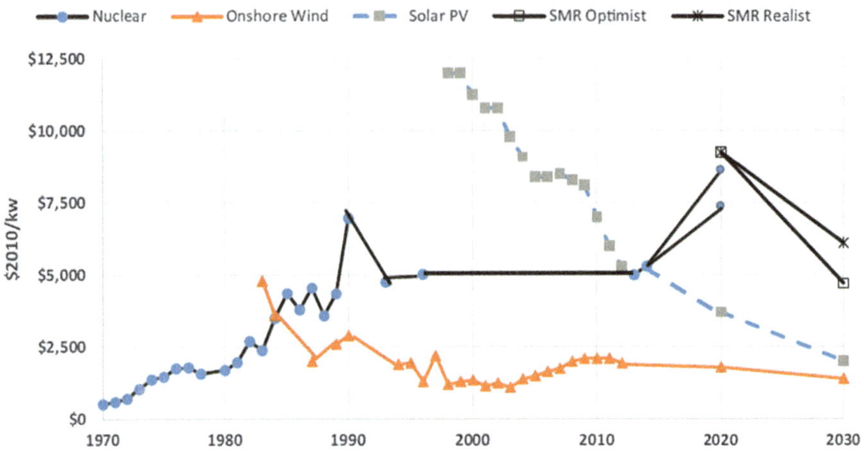

Fig. 5.14 Illustration of cost trend for nuclear, winds, and solar. (Complement of US Department of Energy, August 2013)

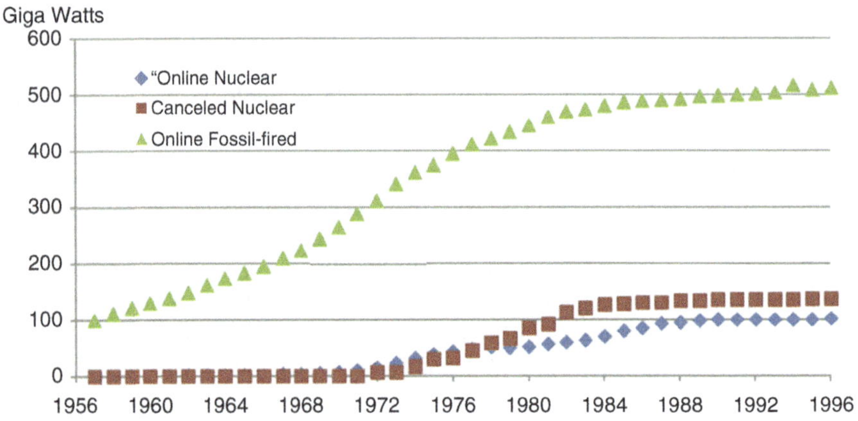

Fig. 5.15 Illustration of completed and canceled nuclear capacity compared to fossil-fired capacity. (Complement of Bull Atomic Science 2013)

requires long lead times and large sunk capital cost. The technology suffers constant stream of cost escalation and does not exhibit cost reduction processes that are observed in other industries due to mass production involved with mass consumption by consumers, which is the main driver for cost reduction of a good as well as competition by other manufacturers playing the same domain and space, where the particular good is sold.

Such fluctuation of roller-coaster motion can be observed by the following two charts, Figs. 5.14 and 5.15.

Figure 5.15 is a completed and canceled nuclear energy capacity compared to fossil-fired capacity. In this figure we notice the data between 1970 and 1996 time frame based on the argument that, was the Three Mile Island accident in 1979 the main cause of US nuclear power's woes?

Therefore, any nation that claims to have the wherewithal (technical expertise and economic resources) to build a "safe" nuclear reactor will have the wherewithal to meet its needs for electricity with alternatives that are less costly and less risky. Thus, at present and for the foreseeable future, it is a virtual certainty that nuclear power is not going to be the least cost option or close to it, even in a low-carbon utility sector [9].

Considering this dismal picture of the prospects for nuclear technology, illustrated by Figs. 5.8 and 5.9, large and small, it is not surprising that SMR technology has stumbled getting into the starting gate, with a dramatic reduction in interest from two of the leading developers, such as Babcock & Wilcox (B&W) and Westinghouse in the United States for the time being, unless the future picture of need for nuclear power energy drastically would change and force the major developers to pay their attention back to it, given the economical perspective in favor of such source of energy.

Babcock & Wilcox, one of the major firms that had received a federal SMR subsidy, stepped back from the development of SMR technology because of the failure "to secure significant additional investors or customer engineering, procurement, and construction contracts to provide the financial support necessary to develop and deploy reactors, and this move cut a lot of advocates of this source of energy by surprise, where the spending from $80 million per year got reduced to around $60 million per year [10, 11].

By the same talking, Westinghouse, also one of the major US developers in SMR space technology and the leading vendor supplying the design for the large-scale nuclear power and presently handling projects under construction with their AP-1000 nuclear power plant in the United States, also announced a similar move that was made by B&W. This also was stepping back from development of small modular nuclear technology. The reason for the decision: Westinghouse could find no customers. Instead of pushing ahead to build SMRs, Westinghouse said it would focus on decommissioning of existing reactors [12, 13].

However, countries like the United Kingdom are strong advocates of the nuclear power based on technology of advanced SMRs and have a very innovative strategy for the deployment of the GEN-IV package. The United Kingdom is swiftly pursuing the development of at least one GEN-III small modular reactor design. The initial support will involve providing funding for front end engineering design studies of this generation of nuclear power plants. There is a good argument for enabling the development of more than one design at this stage as this approach will assist the reduction of development risk, where one project could fail to meet expectations, but this must be balanced against constraints on the overall research and development (R&D) budget and the capability of the UK nuclear industry to concurrently develop more than one technology, possibly among those six selected SMRs of Generation IV (GIF). This will include the ability of the Office for Nuclear

Regulation to process multiple Generic Design Assessment (GDA) for small modular reactors alongside their current work in processing the designs of larger reactors to meet the demand for electricity both tactically and strategically based on near-term and long-terms scenario.

Finally, the UK government should ask potential SMR vendors to take into consideration our evolving energy system. With increased penetration of intermittent renewable technologies, like solar and wind, nuclear power plants will need to find new and inventive ways of adding value to the system. Small modular reactor (SMR) design studies should at least be asked to consider how they could contribute to such a system, be it for hydrogen production, energy storage, or combined heat and power. The long-term future of nuclear power will depend on its ability to adapt to the new world of flexible power systems and low marginal cost renewable electricity [8].

5.6 Key Conclusion

Cost reductions of solar and wind have been impressive and should not be downplayed. Biomass can play a role in reducing emissions from electricity generation and even more so in heat. Battery developments are welcome, and much of our electricity system will become distributed and smart. These are promising and exciting developments. But the limits of these technologies are such that all options should be kept on the table. Carbon capture and storage will be a vital technology in the second half of the twenty-first century.

Development of large-scale storage would be welcome. There are many good ideas in this field, and any additional large-scale, low-cost energy storage that can be added to our energy system should be welcomed, while the door should be also kept open to large nuclear, but vendors need to find ways to bring down cost, either through better project management, technological learning, or innovative financing.

However, taken together, the limitations of all the above technologies and the scale of the challenge we face mean that we may need more. The gap in our energy needs could be met by small modular reactors. Regardless of some major nuclear player firms in the United States pulling out of SMR business, the United Kingdom is seeking companies to lead on the design and build of new British small modular reactors. Given the ongoing problems of financing and constructing large reactors across the western world, this is a sensible strategy.

In the short-to-medium term, the most important aim should be to bring down the cost of nuclear power and add as much low-cost, low-carbon electricity to the grid as possible to meet our obligations under the Climate Change Act.

The government should choose at least one Gen-III+ design SMR to take forward through detailed design to demonstration. The metrics on which to judge the best SMR should be simplicity of design, potential for cost reductions, and the speed of deployment.

The focus at this stage should be on simple designs that reduce costs through replication, while not compromising on safety. The chosen project(s) should be assessed against measurable deliverables, most importantly if they can be built on time and to budget. Some extra expense should be allowed due to the fact that all of these projects will be first-of-a-kind, but on completion of the first SMR, the project leaders should be asked to produce plans for how they intend to reduce costs in the future. Any support for future projects should be based on the credibility of these plans.

An extra consideration is the potential for using the SMR for other services, including district heat, hydrogen production, or even in the manufacture of isotopes for medical imaging and treatment. In the future it is clear that nuclear power increasingly will not be able to continuously provide maximum power to the grid at all times of operation ("base-load" power). Periods of high wind and solar output will displace nuclear power from the grid and reduce revenues. As the marginal cost of nuclear power is so low, it makes little sense to load follow, however. The electricity produced at these times should be stored in some way in order to be sold later at times of higher demand. In order to maintain the financial viability of nuclear power in the future, reactor vendors should design SMRs with this in mind.

References

1. Zohuri, B. "Nuclear Energy for Hydrogen Generation through Intermediate Heat Exchangers: A Renewable Source of Energy" Springer Publishing Company, Jul 15, 2016
2. Rooney Matt, "Small Modular Reactors, The next big thing in energy?", Policy Exchange 2017, www.policyechange.org.uk
3. "Annual Report 2014," Gen IV International Forum, Nuclear Energy Agency, 2014.
4. "A Technology Roadmap for Generation IV Nuclear Energy Systems," U.S. Department of Energy, GIF-002-00, December 2002
5. B. Kallman, "The Very High Temperature Reactor," Physics 241, Stanford University, Winter 2013.
6. G. Roberts, "Nuclear Reactor Basics and Designs for the Future," Physics 241, Winter 2013.
7. G. Locatelli, M. Mancini, and N. Todeschini, "Generation IV nuclear reactors: current status and future prospects." Energy Policy **61**, 1503 (2013).
8. Cooper, Mark, "Small modular reactors and the future of nuclear power in the United States", published in Elsevier at www.elsevier.com/locate/erss
9. Cooper M. Public risk, private profit, ratepayer cost, utility imprudence: advanced cost recovery for reactor construction creates another nuclear fiasco, not a renaissance; 2013. March.
10. Downey J. Westinghouse slows small reactor development. Charlotte Bus J2014. February.
11. Electric Energy Online. B&W announces restructuring of small modular reac-tor program 2014. April 14
12. Akikur RK, et al. Comparative study of stand-alone and hybrid solar energy systems suitable for off-grid rural electrification: a review. Renew Sustain Energy Rev. 2013:27
13. Litvak A. Westinghouse backs off small nuclear plants. Pittsburgh Post-Gazette 2014. February 1.
14. http://www.xebecinc.com/applications-industrial-hydrogen.php
15. http://www.japantimes.co.jp/news/2014/10/12/national/japan-rises-challenge-becoming-hydrogen-society

16. http://www.windupbattery.com/wind/hydrogen-hybrid-prenzlau.htm
17. http://www.windupbattery.com/wind/hydrogen-hybrid-prenzlau.htm
18. Eg (CH)n tar sands or (CH1.5)n heavy crude to (CH2)n transport fuel. Upgrading heavy crude oil and tar sands requires 3 to 4 kilograms of hydrogen per barrel (159 liters) of product.
19. In thermal terms @ 121 MJ/kg: 8470 PJ, equivalent to all of US nuclear electricity

Chapter 6
Safety and Nonproliferation Aspect of GEN-IV Reactors

Abstract The Generation IV International Forum (GIF) was initiated by the US Department of Energy in 2000 and formally chartered in mid-2001. It is an international collective representing governments of 14 countries where nuclear energy is significant now and also seen as vital for the future. Most are committed to joint development of the next generation of nuclear technology. The original charter members of GIF are Argentina, Brazil, Canada, France, Japan, South Korea, South Africa, the United Kingdom, and the United States. They have been joined by Switzerland, the EU (Euratom), China, Russia, and Australia. The purpose of GIF is to share R&D rather than build reactors. Any future decision to go ahead with the industrial construction of Generation IV nuclear power plants among the countries, involved with this type of power plant, will be guided by a number of strategic considerations, among which safety performance and nonproliferation aspect of it, will obviously be foremost.

6.1 Introduction

Several numbers of countries are currently conducting research to develop GEN-IV reactors using various fuels and cooling systems. France, as one of the lead countries in developing GEN-IV reactors, in particular has concentrated its research efforts on the construction of a new sodium-cooled fast reactor. Within this context, in Chap. 1, we considered various six conceptual GEN-IV reactor designs. In this chapter, we should discuss briefly each concept to a thorough review from a purely safety perspective. The review is based on the premise that future reactors should match, if not outperform, last-generation pressurized water reactors in terms of safety performance, in accordance with general specifications and taking into account feedback from the Fukushima accident, under guideline by the Nuclear Regulatory Commission (NRC), Department of Energy (DOE) in the United States, and other government agencies within other countries such as the French Institute for Radiological Protection and Nuclear Safety (IRSN) as well as French Nuclear Safety Authority (ASN), in the United Kingdom, under the Office for Nuclear Regulation (ONR). All these efforts were taken into account in relative on transparency, nuclear safety, and nanofiltration.

B. Zohuri, *Small Modular Reactors as Renewable Energy Sources*,
https://doi.org/10.1007/978-3-319-92594-3_6

All these agencies globally, with over few thousand specialists such as engineers, researchers, doctors, agronomists, veterinarians, technicians, and experts in nuclear and radiation risks, are involved to perform expert assessments and conduct research in the following fields in respect to upcoming future of the Next Generation Nuclear Plant (NGNP) to meet the uprising demand for electricity around the world with growth of population.

The list of these fields, in general, that are constantly under review and subject to updating the regulations is:

- Nuclear safety
- Safety relative to the transportation of radioactive and fissile materials
- Protection of human health and the environment from ionizing radiation
- Protection and control of nuclear materials
- Protection of facilities and transports dealing with radioactive and fissile materials against malicious acts
- Proliferation and nonproliferation of these new generations during their life cycle in operation

It is clear from this general review that whatever the reactor system, among those NGNP, are considered, and are not withstanding the intrinsic advantages of each one, significant technological progress is required before any claim can be made that expected safety levels have been met.

Many energy future studies foresee uranium shortages in the twenty-first century, including some that assume effective energy demand management, drastic limitation of greenhouse gas emissions, and strong "renewable energy" policies.

In this context, the US Department of Energy (DOE) launched the Generation IV International Forum (GIF) in the year 2000. Currently, the GIF has 13 members. Its first action was to select the six "GEN-IV" nuclear reactor technologies (concepts) which are considered the most promising, both in terms of conserving uranium resources and with respect to the following criteria:

- Reduction in radioactive waste, especially long-lived high-level waste
- Safety improvements
- Robustness with respect to nonproliferation and malicious acts
- Economic competitiveness

Many reactor types were considered initially; however, the list was downsized to focus on the most promising technologies and those that could most likely meet the goals of the GEN-IV initiative. At present the focuses are on six different types of reactors.

The six chosen reactor concepts are as follows:

1. Sodium-cooled fast reactors (SFR)
2. Gas-cooled fast reactors (GFR)
3. Lead-cooled fast reactors (LFR)
4. Molten salt reactors (MSR)
5. High- or very-high-temperature reactors (V/HTR)
6. Supercritical water-cooled reactors (SCWRs)

These Generation IV reactors are a set of theoretical nuclear reactor designs currently being researched. Most of these designs are generally not expected to be available for commercial construction before 2030. Demonstration of concepts are however foreseen around 2015–2020. Current reactors in operation around the world are generally considered second- or third-generation systems, with the first-generation systems having been retired some time ago. Relative to current nuclear power plant technology, the claimed benefits for fourth-generation reactors include:

1. Sustainability
2. Increased safety
3. Better use of the nuclear fuel
4. High efficiency and better economics
5. Minimal waste production
6. The ability to consume existing nuclear waste in the production of electricity
7. Increased proliferation resistance

According to the studies cited above, industrial deployment of these new designs could occur toward the middle of the twenty-first century, following initial operating experience on experimental reactors, demonstrators, or prototypes.

Most of these are party to the 2005 Framework Agreement, which formally commits the nations involved with NFNP Research and Development (R&D), and are participating in the development of one or more Generation IV systems selected by GIF for further improvements. Argentina, Australia, and Brazil did not sign the Framework Agreement, and the United Kingdom withdrew from it; accordingly, within GIF, these four are designated as "inactive members." Russia formalized its accession to the Framework Agreement in August 2009 as its tenth member, with Rosatom as implementing agent. In 2011 the 13 members decided to modify and extend the GIF charter indefinitely. (Australia joined as the 14th GIF member in June 2016.) In February 2015 the Framework Agreement was extended for 10 years, with Rosatom signing for the extension in June and Euratom in November 2016.

In Fig. 6.1, here is presentation timeline for generations of nuclear power systems from historical viewpoint.

As we mentioned in previous chapters, the key steps for this roadmap that was signed off by participant countries involved with the Next Generation of Nuclear Plant (NGNP) in general are listed below as:

- Identify potential areas of multilateral collaborations on Generation IV nuclear energy systems.
- Foster collaborative R&D projects.
- Establish guidelines for collaboration and reporting of their results (review, recommendations, etc.).
- Define technology goals for Generation IV.
- Identify concepts with potential.
- Evaluate concepts with a common and consistently applied methodology.
- Identify R&D gaps and needs.
- Roadmap issued in December 2002.

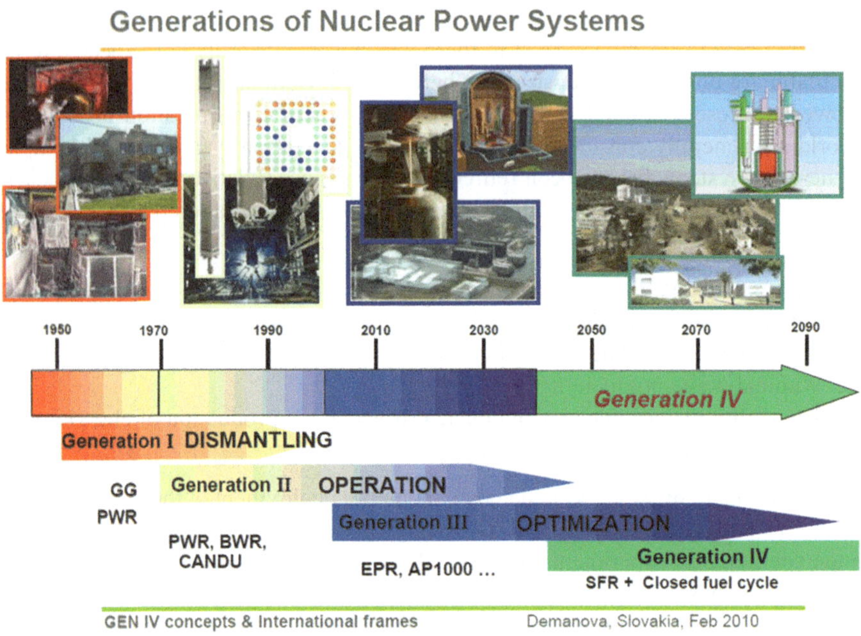

Fig. 6.1 Generations of nuclear power systems

In addition, the rational for the selection, in general, was based on the following high-level points and they are listed as:

- Scoring for potential, significant advances in
  - Economics
  - Safety and reliability
  - Sustainability
  - Security and nonproliferation
- Offering various energy applications
  - Electricity generation
  - Hydrogen, clean water (desalination), heat
- Composite and robust set of concepts
- Innovation, leverage effect of R&D needs
- Time sequencing and opportunities for development of GEN-IV vs. INTD list

However, expanding on the criteria of selection for the NGNP, and first level goals for innovative systems were based on the following list as:

- *Sustaiability-1*
- Generation IV nuclear energy systems will provide sustainable energy generation that meets clean air objectives and promotes long-term availability of systems and effective fuel utilization for worldwide energy production.

- *Sustaiability-2*
- Generation IV nuclear energy systems will minimize and manage their nuclear waste and notably reduce the long-term stewardship burden in the future, thereby improving protection for the public health and the environment.
- *Economics-1*
- Generation IV nuclear energy systems will have a clear life-cycle cost advantage over other energy sources.
- *Economics-2*
- Generation IV nuclear energy systems will have a level of financial risk comparable to other energy projects.
- *Safety and Reliability-1*
- Generation IV nuclear energy systems operations will excel in safety and reliability.
- *Safety and Reliability-2*
- Generation IV nuclear energy systems will have a very low likelihood and degree of reactor core damage.
- *Safety and Reliability-3*
- Generation IV nuclear energy systems will eliminate the need for off-site emergency response.
- *Proliferation Resistance and Physical Protection*
- Generation IV nuclear energy systems will increase the assurance that they are a very unattractive and least desirable route for diversion or theft of weapons-usable materials and provide increased physical protection against acts of terrorism.

The overall Generation IV roadmap organization and involvement with the rational decision-making for the selection were depicted in Fig. 6.2.

The given criteria were laid out in bolt format based on performance expectation for these new generations of nuclear plant, and Fig. 6.3 lays them down as performance indicator based on 4 objectives and 15 criteria as well.

Furthermore, the goals of safety and reliability point of view are given in the next section.

6.2 Safety and Reliability Goals

One of the most important questions in this section is that: Are nuclear reactors safe? And if we think they are safe, are they safe enough though? And finally, what do we mean by the word "safe" and how safe is safe for present and future nuclear power plant? After all we are advocating a push toward the new generation of these power plants that are known as "the Next Generation of Power Plant (GNPP)," or we know them by now as Generation IV International Forum (GIF) that besides the United States through firms such as NuScale leading the way to the future operation of these reactors, yet many other countries worldwide under IAEA organization are

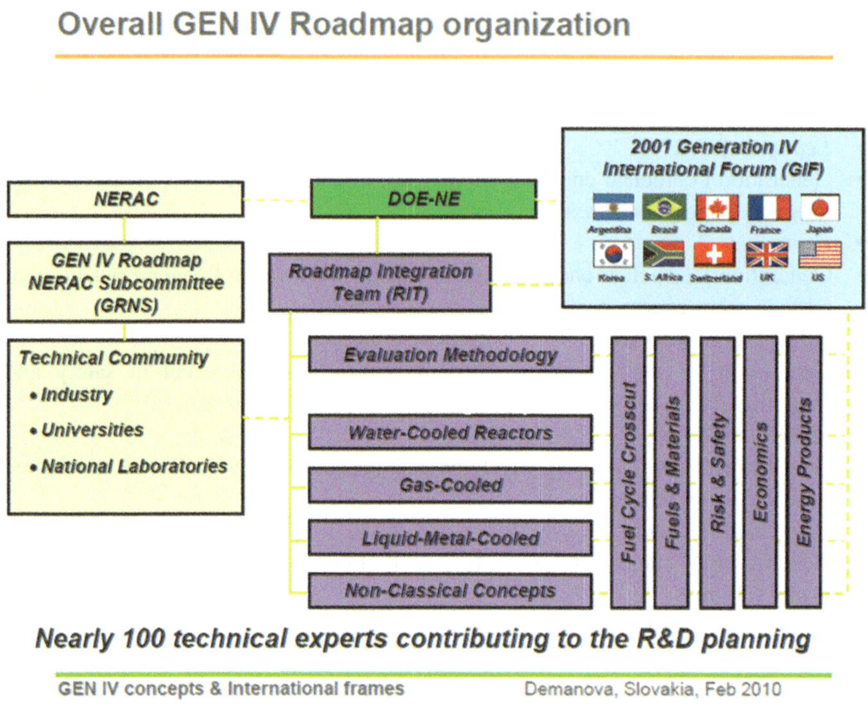

Fig. 6.2 Overall Generation IV roadmap organization

building trust and momentum toward such advanced nuclear technology and one their banner is perching the safety and reliability of these generations.

Historically, soon after the 1942 nuclear experiments at Chicago Stagg Field, Enrico Fermi expressed a concern that remains even today, whenever nuclear reactor safety comes to play and is discussed, and that is: "The public may not accept an energy source that is encumbered by vast amounts of radioactivity, and that produces a nuclear explosive, which might fall into hostile hands" [1].

The public's safety concerns were exacerbated by 1979 Three Mile Island (TMI) and 1986 Chernobyl and then 2011 episode of Fukushima Daiichi nuclear power plant accident and aftermath disaster and recovery from it, due to natural made tsunami, and even today Japan still hasn't fully recovered from such disaster.

The TMI accident destroyed the reactor and did not cause any fatality and physical harm from small leakage of radioactive material into local environment surrounding TMI in Pennsylvania. But it traumatized many thousands though. The Chernobyl accident destroyed the entire nuclear reactor as well as shut down the site and led to a few dozen near-term deaths of the first responders. It released a large amount of radioactive materials, and release may cause thousands of the long-term side effects and death [2].

The 2011 Fukushima Daiichi was adding salt to injury, which caused public concern, but the long-term implications are far from certain of the future of

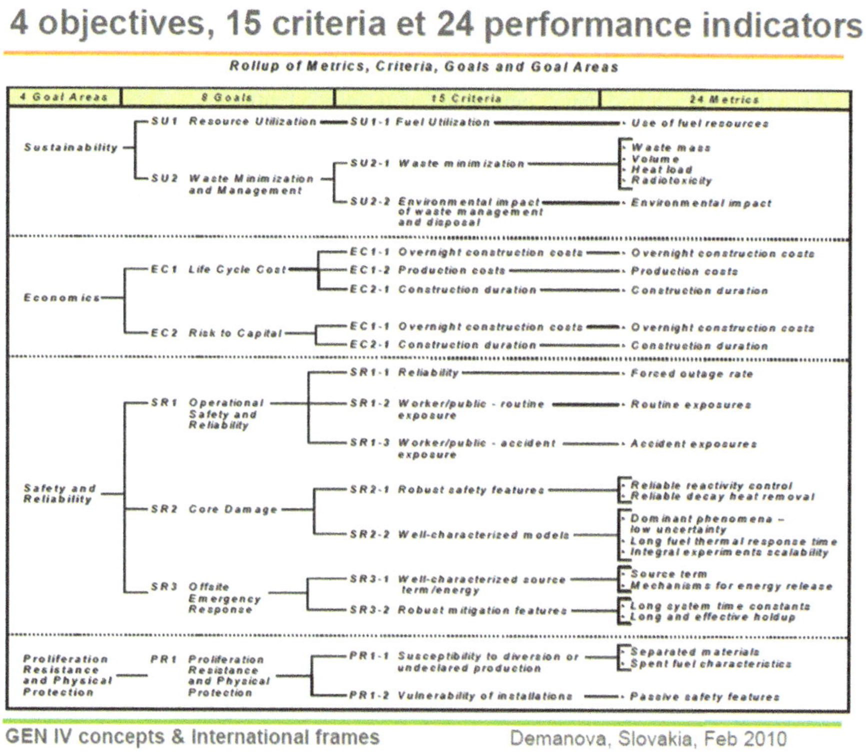

Fig. 6.3 Illustration of 4 objectives, 15 criteria, and 24 performance indicators

nuclear power plants in the United Sates and globally involved nations in an effort toward the innovative GIF and, technologically, more advanced version of these Generation IV International Forum (GIF) reactors that we call them small modular reactors (SMRs).

According to Remy Carle, a senior executive with Electricité de France: Nuclear safety encompasses all the technical and organizational measures to be taken to ensure that operation of a nuclear installation has no harmful consequences for public health and the environment.

Nuclear safety is based on an approach known as "defense in depth" which involves:

- Accident prevention, from the initial design stage, through careful sizing of all installations, the taking into account of possible equipment failures and human error, the taking into account of external hazards, the implementation of safety systems, and the quality control of the design and execution of equipment and work
- Continuous monitoring during operation, according to procedures monitored by national authorities
- Implementation of safety systems to maintain the cooling of nuclear fuel and prevent the release of radioactive products in the event of abnormal operation

- Lastly definition of emergency planning and procedures to deal with the highly improbably event of a serious accident [3]

As will be described later, new, enhanced safety features are incorporated in current reactor designs and will continue to be important components of future reactors. The industry has been concerned about what to call these new plants. Should they be "inherently safe," "passively safe," "transparently safe," or some other term. "Naturally safe" received the votes of 44%. One term did receive more: 49% thought favorably of the term "safer"[4].

As part of safety and reliability goals of Generation IV nuclear power plants, we are relying on energy resilience system, while naturally we are also concerned with decarbonation of our environment when we talk about deployment of these types of reactors with their advanced technologies in mind.

Energy resilience is the ability to prepare for and recover from energy disruptions such as extreme weather events, physical attacks, cyber-attacks, and electromagnetic interference with safety and reliability goals in mind.

Resilient power systems minimize the effect of such failures, in which DOE estimates cost of US business to be approximately $150 billion per year [5]. Having a resilient power source during a power outage can save billions of dollars, maintain critical services, and protect lives.

Governmental entities are increasingly focusing on energy resilience through the passage of legislation, issuance of executive directives, the purchasing of resilient energy sources, and the commissioning of research like the DOE's August 2017 Staff Report on Electricity Markets and Reliability and Safety of this source of energy driven by advance technology of small modular reactor (SMR) as new generation of nuclear power plant (NPP).

Stating with the above argument around resilience energy which is the fundamental driving factor for safety issue, the responsible agencies within governmental entities of the nations involved in R&D of Generation IV International Forum (GIF) are required to provide an updated overview of specific safety and radiological protection issues for all the reactor concepts adopted by the GIF, independent of their advantages or disadvantages in terms of resource optimization or long-lived waste reduction.

In particular, this new document attempts to bring out the advantages and disadvantages of each concept in terms of safety, taking into account the Western European Nuclear Regulators' Association (WENRA) statement concerning "safety objectives for new nuclear power plants" issued in November 2010, Reference Document [6].

Although the WENRA objectives target third-generation nuclear reactors (and should be revised before 2020 as stated in) [6], it still seems useful to take these objectives into consideration given the lack of specific documents for fourth-generation reactors.

Section 5 of International Nuclear Safety Advisory Group Report (NSAG) 10 (INSAG 10), issued by the IAEA, covers defense in depth for future nuclear power plant projects, including "advanced reactors," which correspond to innovative reactors, but it dates from 1996. INSAG 12 ("75-INSAG-3 Rev. 1"), from 1999, refers to "future nuclear power plants" in Section 3.3.10, but with primary emphasis being placed on water-cooled nuclear reactors. In 2008, the GIF issued the document cited as Reference Document [7].

It should be noted that, at the current (early) stage of considerations on the safety requirements for fourth-generation reactors, it is generally stated that fourth-generation reactors must be at least as safe as third-generation reactors and should therefore at minimum comply with the objectives in [6], where these objectives are relevant. Using an identical framework for each reactor concept, this summary report provides some general conclusions regarding their safety and radiological protection issues, inspired by WENRA's safety objectives and on the basis of available information. Initial lessons drawn from the events at the Fukushima Daiichi nuclear power plant in March 2011 have also been taken into account in IRSN's analysis of each reactor concept.

In general, security aspects have not been covered in this document. Some general considerations have been mentioned in Reference Document [8], and other very specific comments have been added in this document. Further information can be found in document that is published by "IRSN, INSTITUT DE Radioprotection Et DE SURETE NUCLEAIRE, Enhancing Nuclear Safety" under the title of Overview of Generation IV (Gen IV) Reactor Designs: Safety and Radiological Protection Considerations, published on September 24, 2012.

6.3 Potential Benefits Offered by Small Modular Reactors

As we said in previous sections of this chapter, many reactor types were considered initially; however, the list was downsized to focus on the most promising technologies and those that could most likely meet the goals of the GEN-IV initiative. At present one focuses on six different types of reactors, and here all benefits of these types in Table 6.1 are pointed out, from top-view not granular level though.

These potential benefits that are offered by small modular reactors are considered as part of the roadmap described in Chap. 1, Sect. 1.19 of this book, and are driving factor for development of Generation IV International Forum (GIF), which is one of the combination solutions for demand on energy due to population growth as well as a way for storage and renewable energy, thus hydrogen production for hydrogen-driven cars of the near future.

6.4 Generation IV International Forum (GIF) Focus

After some 2 years' deliberation and review of about 100 concepts, late in 2002 GIF (then representing 10 countries) announced the selection of 6 reactor technologies which they believe represent the future shape of nuclear energy. These were selected on the basis of being clean, safe, and cost-effective means of meeting

Table 6.1 Potential benefits offered by SMRs

Potential benefits offered by SMRs	
Carbon-free baseload power	SMRs provide carbon-free baseload power
Enhanced safety	SMRs may be safer than conventional reactors since they will be built below ground, thus making them better protected from human and natural risks Passive safety systems allow for improved accident avoidance and tolerance Transportation of fuel may be minimized since the reactors can be fueled when manufactured in a factory
Modularity	As major components can be manufactured off-site and shipped to the point of use, SMRs allow for the centralization of manufacturing expertise Limited on-site construction is required, as work is concentrated in the manufacturing stage Individual factories could fabricate components for multiple SMRs, increasing fleet-wide design consistency and standardization. By manufacturing multiple reactors of smaller size at centralized facilities, manufacturers are likely to experience rapid learning curves Modularity and standardized designs can also increase the safety and efficiency of plant operations, as they eliminate idiosyncratic design features between plants and streamline operating and maintenance procedures
Low cost	The cost of an SMR has been estimated to be between $800 million and $3 billion per unit, whereas a large reactor typically costs between $10 billion and $12 billion per unit The smaller size of SMRs should translate to each reactor being less capital intensive; costs associated with manufacturing and construction are reduced as less material is required. Factory fabrication can mean quicker on-site construction, which reduces the cost of labor and shortens the interval between construction of the reactor and when the reactor begins to generate electricity
Scalability	SMRs have a responsive, adaptable site capacity and can provide power for a range of applications In developing countries or rural communities that lack the transmission infrastructure to support a large nuclear plant, SMRs provide a way for utilities to still have baseload power on the grid Nuclear plant operators can gradually scale up the number of SMRs at a single plant location as demand grows, distributing cost evenly throughout the lifetime of a nuclear power plant Utilities could use SMRs as on-site replacement for aging fossil fuel plants ¡V taking advantage of existing transmission infrastructure
Improved energy security	Having an SMR located on-site may provide long-term energy security to the federal agency, rather than relying on a separate grid that is outside the control of the federal agency By providing 2 years of fuel on-site, vulnerabilities relating to fuel transportation disruptions are minimized
Integration of renewables	NuScale's SMR design allows for output to be varied over days, hours, or minutes. This can allow SMRs to adjust their output in response to changes in electricity output from intermittent renewable generation
Siting flexibility	The small size of SMRs may allow them to be sited in places where a large baseload plant is not feasible or not needed. For example, SMRs have been considered as a power source for remote mines in Canada which cannot access the grid

(continued)

Table 6.1 (continued)

Potential benefits offered by SMRs	
Small land requirements	SMRs require significant less land than would power plants with the same output which use wind, solar, biomass, or hydropower. NuScale estimates that SMRs require only 1% of the land area required for similar generation by other technologies
Process heat	SMRs heat water in the process of producing electricity. Some SMR designs may be useful for producing process heat for desalination and other industrial activities
International export opportunities	United States companies that produce SMRs or sell related goods or services may have opportunities to sell to foreign markets. The Energy Information Administration (EIA) estimates that global electricity generation will increase by 69% from 2012 to 2040
Reduced fuel risk	SMRs can help diversify a generation portfolio and reduce fuel risk. The price of electricity from SMRs, especially under a long-term contract, should be relatively stable and predictable Natural gas prices have historically been very volatile, although they have been low in recent years. Higher natural gas prices would significantly increase the price of electricity produced from natural gas

increased energy demands on a sustainable basis, while being resistant to diversion of materials for weapon proliferation and security from terrorist attacks. They are the subject of further development internationally, with expenditure of about $6 billion over 15 years. About 80% of the cost is being met by the United States, Japan, and France.

In addition to selecting these six concepts for deployment between 2010 and 2030, GIF recognized a number of International Near-Term Deployment advanced reactors available before 2015. See also information page on Advanced Nuclear Power Reactors.

Most of the six systems employ a closed fuel cycle to maximize the resource base and minimize high-level wastes to be sent to a repository. Three of the six are fast-neutron reactors and one can be built as a fast reactor, one is described as epithermal, and only two operate with slow neutrons like today's plants.

Only one is cooled by light water, two are helium-cooled, and the others have lead-bismuth, sodium, or fluoride salt coolant. The latter three operate at low pressure, with significant safety advantage. The last has the uranium fuel dissolved in the circulating coolant. Temperatures range from 510 °C to 1000 °C, compared with less than 330 °C for today's light-water reactors, and this means that four of them can be used for thermochemical hydrogen production.

The 2014 GIF Roadmap says that a 600 MWt very-high-temperature reactor (VHTR) dedicated to hydrogen production could yield over 2 million normal cubic meters per day. An R&D priority is qualification of tristructural-isotropic (TRISO) fuel for operation up to 1250 °C and 200 GWd/t burnups, though US development has attained this, and also its robustness for hundreds of hours at 1600, 1700, and 1800 °C. However, in the short term, electricity production and industrial processes based on high-temperature steam that require outlet temperatures of 700–850 °C hold most potential.

Gas-cooled fast reactors	fast	helium	850	high	U-238 +	closed, on site	1200	electricity & hydrogen
Lead-cooled fast reactors	fast	lead or Pb-Bi	480-570	low	U-238 +	closed, regional	20-180** 300-1200 600-1000	electricity & hydrogen
Molten salt fast reactors	fast	fluoride salts	700-800	low	UF in salt	closed	1000	electricity & hydrogen
Molten salt reactor - **advanced high-temperature reactors**	thermal	fluoride salts	750-1000		UO_2 particles in prism	open	1000-1500	hydrogen
Sodium-cooled fast reactors	fast	sodium	500-550	low	U-238 & MOX	closed	50-150 600-1500	electricity
Supercritical water-cooled reactors	thermal or fast	water	510-625	very high	UO_2	open (thermal) closed (fast)	300-700 1000-1500	electricity
Very high temperature gas reactors	thermal	helium	900-1000	high	UO_2 prism or pebbles	open	250-300	hydrogen & electricity

* high = 7-15 MPa
+ = with some U-235 or Pu-239
** 'battery' model with long cassette core life (15-20 yr) or replaceable reactor module.

Fig. 6.4 Generation IV reactor designs under development by GIF. (Courtesy of World Nuclear)

One question to be resolved regards used fuel. Some synergy with used LWR fuel is possible, and the graphite may be recycled, but much remains to be worked out.

Euratom, France, Japan, China, Korea, Switzerland, and the United States originally signed on to the system arrangement for the VHTR under the Framework Agreement, and Australia joined in 2017. Two Project Arrangements (PA) were signed within the VHTR system: the Fuel and Fuel Cycle PA and the Hydrogen Production PA.

The HTR-PM demonstration unit in China is under construction at Shidaowan and will pave the way for a commercial version of the VHTR.

The main research needs, according to the China Academy of Sciences, are fuels, materials and hydrogen production.

Again, here in Fig. 6.4, we have summarized all the Generation IV International Forum (GIF) reactor designs under development by GIF.

Closely related to GIF, but more focused on Generation III immediately, is the Multinational Design Evaluation Program (MDEP) set up by the regulators. It was launched in 2006 by the US Nuclear Regulatory Commission (NRC) and the French Nuclear Safety Authority (ASN) to develop innovative approaches to leverage the resources and knowledge of national regulatory authorities reviewing new reactor designs. It involves the IAEA and 14 countries, and its secretariat is with the OECD Nuclear Energy Agency. Ultimately it aims to develop multinational regulatory standards for design of Generation IV reactors. The US Nuclear Regulatory Commission (NRC) has proposed a three-stage process culminating in international design certification for new reactor types, notably Generation IV types. Twelve countries are involved so far, Canada, China, Finland, France, India (from 2012), Japan, Korea, Russia, South Africa, Sweden (from 2013), the United Kingdom, and the United States, and others which have or are likely to have firm commitments to building new nuclear plants may be admitted – the UAE is an associate member.

The MDEP pools the resources of its member nuclear regulatory authorities for the purpose of (i) cooperating on safety reviews of designs of nuclear reactors that are under construction and undergoing licensing in several countries and (ii) exploring opportunities and potential for harmonization of regulatory requirements and practices. It also produces reports and guidance documents that are shared internationally beyond the MDEP membership. It has five design-specific working groups (EPR, AP1000, APR1400, VVER, ABWR) and three issue-specific ones (digital instrumentation and control (I&C), mechanical codes and standards, and vendor inspection cooperation). The Instrumentation and Control (I&C) Systems in Nuclear Power. Plants: A Time of Transition. I&C systems are the nervous system of a nuclear power plant. They monitor all aspects of the plant's health and help respond with the care and adjustments needed [9].

In relation to Generation IV reactors, the NRC has called for countries involved in their development to develop common design requirements so that regulatory standards can be harmonized. The NRC has published its draft design requirements.

Meanwhile the MDEP is being used to share information among countries engaged in certifying particular new reactor designs, notably the EPR and AP1000, but with APR1400, VVER designs, and ABWR-ESBWR following. MDEP members are looking at different design codes to ensure that consistently high safety standards are achieved in different countries. Harmonization of design requirements will ultimately assist this, in the same way as has been achieved in civil aviation.

6.5 Associated Ongoing Programs

While Russia was not initially part of GIF, one design corresponds with the BREST reactor being developed there, and Russia is now the main operator of the sodium-cooled fast reactor for electricity – another of the technologies put forward by GIF.

A major project relevant to several Generation IV designs is investigating the use of actinide-laden fuel assemblies in fast reactors as part of the sodium-cooled fast reactor program. The Global Actinide Cycle International Demonstration (GACID) is being undertaken by France's Atomic Energy Commission (CEA), Japan's

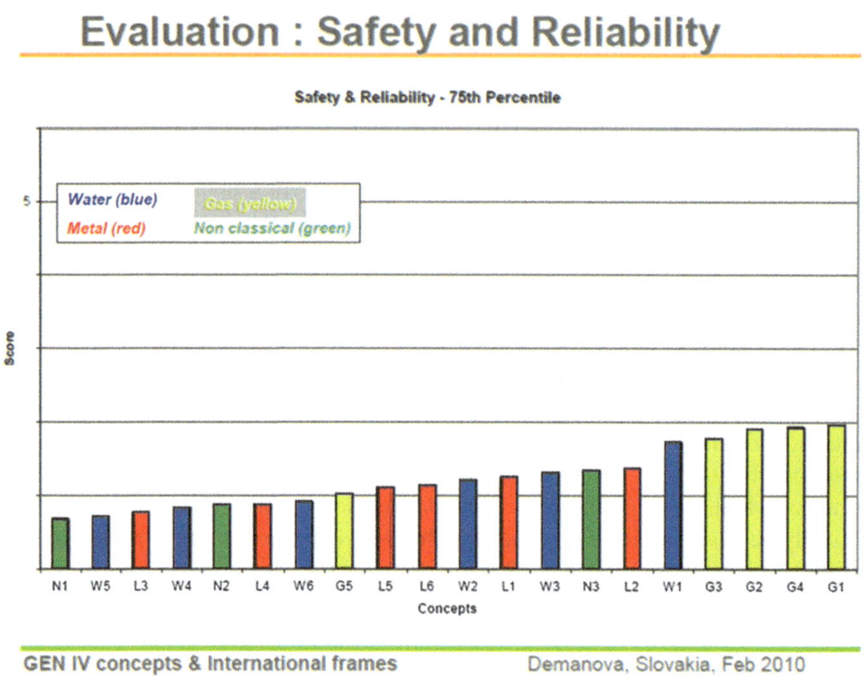

Fig. 6.5 Evolution of safety and reliability

Atomic Energy Agency (JAEA), and the US Department of Energy (DOE) under the US Advanced Fuel Cycle Initiative (AFCI). The first stage will lead to demonstration fuel containing minor actinides being used in Japan's Monju reactor [9].

In summary fuel issues and selections in conjunction with GEN-IV are part of safety feature of these generations of nuclear power plant (NPP) and very well thought in advance will open up a better and smoother roadmap gaining forward with idea of using GIF technology as part of energy renewable market share, while these reactors are offering a solution to decarbonization of our surrounding air.

6.6 Summary

In summary, we can illustrate the evaluation of Safety and Reliability charts as depicted in Fig. 6.5 here as:

In parallel and associated to safety and reliability, we can present depiction of Economics and Competitiveness for SMRs by Concepts as it is illustrated in Fig. 6.6.

Finally, the global evaluation for all the concepts for SMRs is illustrated in Fig. 7.

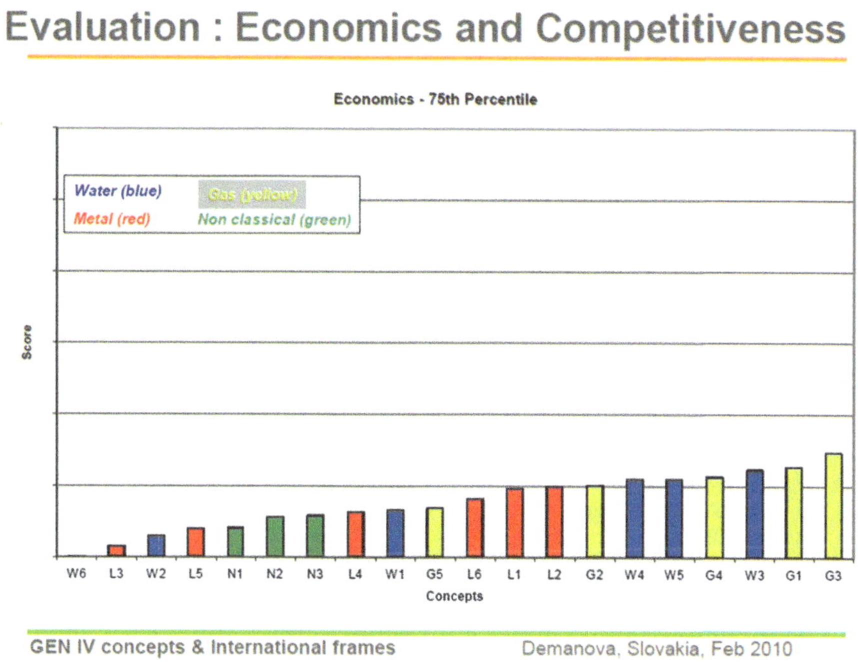

Fig. 6.6 Evolution of economics and competitiveness

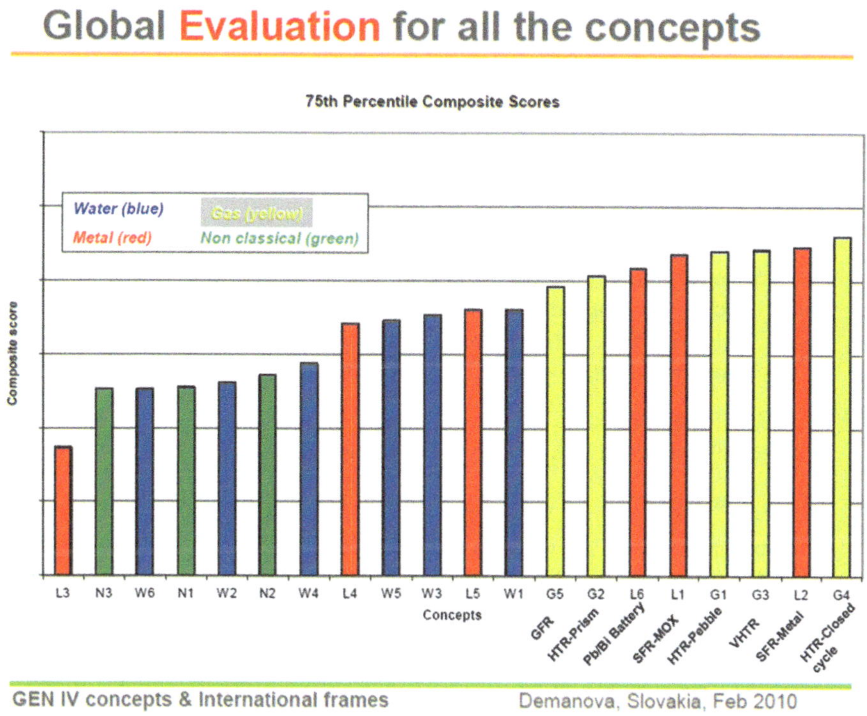

Fig. 6.7 Global evaluation for all the concepts

Fig. 6.8 Strategic Research Agenda

As conclusion of this chapter, we may put the summary of dusk of past generation of nuclear power plant to dawn of new generation of nuclear energy systems in the following perspective as:

- Nuclear energy is a vital component of the world energy mix.
- Sustainable nuclear energy should be based on fissile material regeneration.
  - Several thousand years of resources will be so available.
- Fast-neutron systems can achieve the goal.
- Future nuclear systems are studied at an international level (Generation IV Forum, IAEA-INPRO, European SNE-TP, etc.).
- Fast-neutron systems have been selected worldwide to be developed by 2040:
  - SFR, GFR and LFR.
  - Together with two more prospective systems: MSR and SCWR
- This is in indeed vital:
  - To progress toward a low-carbon future.
  - To be more influential on design features and international standards for future nuclear systems.

We can wrap up this chapter and summary by Fig. 6.8 that shows Strategic Research Agenda, which is at least open to all European members of the platform that are about 50 of them for time being.

References

1. Quoted in letter from A. M. Weinberg John Gibbons, Assistant to the President for Science and Technology, August 19, 1997
2. Assessment of the delayed (latent) fatalities associated with the exposure of radioactive material released by the Chernobyl accident indicates numbers up to 33,000 over the next 70 years assuming a linear non-threshold effect of radiation….on this basis, natural background radiation would result in 1500 times as many deaths…over the same timescale so these additional fatalities, if the occur, would be very difficult to observe. *Comparing Nuclear Accident Risks with Those from Other Energy Sources*, Nuclear Energy Agency No. 6861, OECD 2010
3. Remy Carl, Nuclear Power (Presses Universitaires de France, 1994)
4. *Post TMI: What Have We Learned*, J.F. Ahearne in Nuclear Safety 1989 Conference Proceedings
5. Distributed Generation: Cleaner, Cheaper, Stronger - Industrial Efficiency in the Changing Utility Landscape," Pew Charitable Trusts, October 2015, p. 6, http://www.pewtrusts.org/~/media/assets/2015/10/cleanercheaperstronger_finalweb.pdf?la=en.
6. "WENRA statement on safety objectives for new power plants", November 2010 (document faisant suite au rapport Safety objectives for new power reactors – WENRA Reactor Harmonization Group, December 2009). www.wenra.org
7. "Basis for the safety approach for design and assessment of generation IV nuclear systems", GEN IV International Forum, GIF/ RSWG/2007002, revision 1, November 24, 2008. www.gen-4.org
8. IRSN summary report on GEN IV reactors, 21 November 2007. www.irsn.fr
9. http://www.world-nuclear.org/information-library/nuclear-fuel-cycle/nuclear-power-reactors/generation-iv-nuclear-reactors.aspx#ECSArticleLink2

Chapter 7
Electricity Production and Renewable Source of Energy, Economics

Abstract Most of the electricity in the United States today is produced using steam turbines. A turbine converts the kinetic energy of a moving fluid (liquid or gas) to mechanical energy. In a steam turbine, steam is forced against a series of blades mounted on a shaft. The steam rotates the shaft connected to a generator. The generator, in turn, converts its mechanical energy to electrical energy based on the relationship between magnetism and electricity. In steam turbines powered by fossil fuels (coal, natural gas, and petroleum), the fuel is burned in a furnace to heat water in a boiler to produce steam. However, considering the rise on the demand for electricity going forward toward the future industrial world at the fast pace that the world is going, is the gas turbine generated electricity enough source to meet this demand? Or even in conjunction with solar and wind resource, is it the uphill battle for our demand? In this chapter we assess the situation and justify the future faith of nuclear power plant in form of advanced research and development of the Generation IV International Forum (GIF).

7.1 Introduction

From the point of view of electricity energy providers and utility owners, many of the industry challenges, where they supply electricity to, are being converted into opportunities by these industries that are considered as their clients. These utility owners are seeing changes in how they generate electricity and manage distribution to the grid as well as maintain the grid and in how they generate innovation and new ideas for this energy generation, while they are cost-effective that provide profitable return on their investment that goes into research and development (R&D) for these innovative ideas that form total cost of ownership (TCO) make it feasible for their stockholders.

They are also on the cusp of generation change, and a new generation of industry leaders is emerging and their demand for electricity constantly varies. For these reasons, these utility owners see this issue as "generation to generation" as an *energy evolution*.

B. Zohuri, *Small Modular Reactors as Renewable Energy Sources*,
https://doi.org/10.1007/978-3-319-92594-3_7

For them to get this energy evolution right is that many questions which require answers and are needed to be addressed are the following:

- Which technologies will co-exist? And How?
- Which generation technologies should remain valued and valuable?
- What is on the horizon? And finally, will new technologies fill the gap for power technology "generations" that recede but still provide the energy, performance attributes, and social goods that people want?
- What is the proper role of markets, federal government, state and local governments, and others with an interest in getting this generational shift right?

Getting the signals right, the central station generation mix is changing, as solar-, wind-, and gas-fired generation continue to be added, with significantly more expected and distributed energy resources, while still a small proportion of overall generation are growing very rapidly.

The fate and future of energy driven by the nuclear power as a source of production is a big unknown, as some units already have been retired, as well as more retirements due to end of their life cycle have been announced, and construction of new nuclear power plant units in the United States has hit a rough patch. But, wholesale price formation is under review, recognition, and compensation for their carbon-free emission characteristics are being discussed and implemented by some states, and small modular reactors (SMRs) are garnering interest.

Seeking to advance the renewables shift, communities in some jurisdictions, especially California, are aggregating load and buying their own power. The projections for growth of this are extremely high, but "stranded" costs are an unresolved issue. However, the Federal Energy Regulatory Commission (FERC) continues to seek an optimal balance between power generation characteristics in bid-based power markets, while seeking to preserve reliability and reasonable rates. At the same time, the Department of Energy has laid down a marker and an aggressive timeframe for market reform aimed at preserving baseload generation.

Furthermore, if we are considering that electricity generation from supply and demand prospection, this matter is *evolution not revolution*. New supply and demand options slowly begin to share the spotlight. Energy efficiency, for example, has significant potential, enhanced by improved interactive, Internet of things technology, but program designers must still deal with behavioral responses to get the most savings possible. An old storage technology – pumped hydroelectric storage – is gaining attention as new concepts get tested, improved turbines provide more flexible operating options, and utility-scale renewables emerge as a potential energy resource for pumping. The solar industry is raising awareness of increasing amounts of "flexible" solar power, which could provide grid services beyond energy and, when teamed with energy storage, could someday provide a trifecta of characteristics: good grid citizenship, energy when needed (not just when the sun shines), and cost competitiveness with gas-fired generation.

Parallel to the above concerns, preparing the grid for a multi-generation resource mix is another evolution we need to deal with in order to meet the supply and demand scenario for electricity, considering the growth of new population around

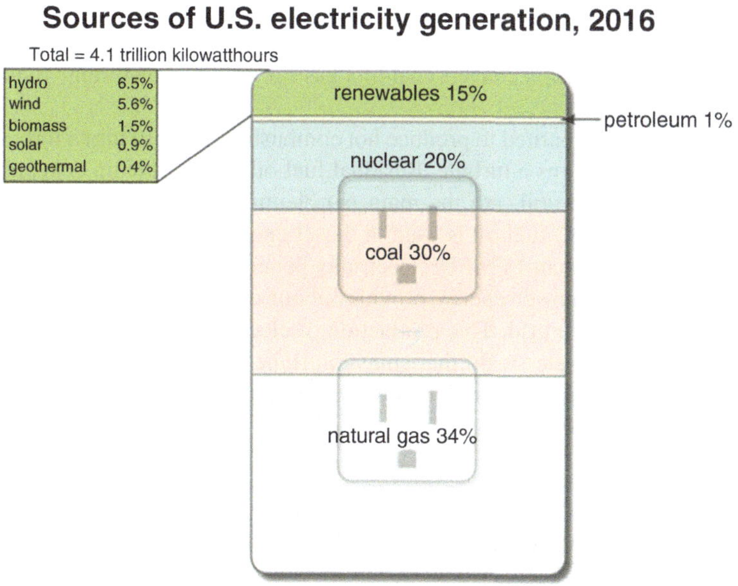

Fig. 7.1 US source of electricity generation. (Note: Electricity generation from utility-scale facilities. Sources: U.S. Energy Information Administration, *Electric Power Monthly*, February 2017, preliminary data for 2016)

the globe, in particular countries such as China that is at virtue of industrial renaissance. As generations of resources begin to coexist on the grid – both "old" and "young" – utilities continue to invest. But they are adapting grid architecture to provide a more flexible backbone to enable two-way power flows and more variable supply and demand. One example of a grid that is changing to accommodate distributed resources is that of Illinois, which has implemented grid modernization efforts as well as change incentive mechanisms to encourage distributed energy resource development.

7.2 Electricity Production in the United States

Most of United States electricity is generated using fossil fuels, such as coal, gas, and oil form or compounds. In 2016, natural gas was the largest energy source for the 4 trillion kilowatt-hours of electricity generated in the United States, and these types of fuels are broken down by their usage to produce energy in form of electricity, and they are listed below, as well as depicted in Fig. 7.1.

- *Natural gas:* was the source of about 34% of US electricity generation in 2016. In addition to burning natural gas to heat water for steam, it is also burned to produce hot combustion gases that pass through a gas turbine, spinning the turbine's blades to generate electricity.

- *Coal:* was the second largest energy source for US electricity generation in 2016 – about 30%. Nearly all coal-fired power plants use steam turbines. A few coal-fired power plants convert coal to a gas for use in a gas turbine to generate electricity.
- *Petroleum:* can be burned to produce hot combustion gases to turn a turbine or to make steam that turns a turbine. Residual fuel oil and petroleum coke, products from refining crude oil, are the main petroleum fuels used in steam turbines. Distillate (or diesel) fuel oil is used in diesel-engine generators. Petroleum was the source of less than 1% of US electricity generation in 2016.
- *Nuclear:* nuclear power energy provides about one-fifth of United States electricity energy to the grid. This production of electricity comes with nuclear fission process to create steam that spins a turbine to generate electricity energy. Nuclear power plant in operation now was the source of about 20% US electricity generation in 2016.

7.3 Energy Supply, Demand, and Market

The macroeconomics of subject matter in this section can be described as follows:

Unlike power transmission and distribution, power generation is not a natural monopoly, meaning a competitive market could, in theory, produce efficient allocation of resources and lower energy costs. But it was thought that prices would only go down and we would never run out. In most competitive markets, prices can spike, and there are stockouts; therefore, to prevent that, electricity markets were tweaked with administrative overlays multiple times, e.g., price caps and floors in lieu of scarcity prices. But some claim that as a result of these tweaks, current centralized, administered markets lack some of the market mechanisms that would provide the right price signals to resources for market entry and exit.

Competitive electricity markets are also criticized for not accounting for certain social costs and benefits, such as fuel diversity, resilience, and environmental attributes, and some believe that participants are instead given to maximize near-term individual benefit rather than system-wide, long-term benefit.

However, in response to the microeconomies, states are subsidizing supply to reflect their economic development, emissions, and other goals, adding supply to already over-supplied markets. This supply is often at zero short-run marginal cost (SRMC), just because markets were born out of the dispatch system; SRMC is their "center of gravity." But microeconomics tells us that long-run marginal cost (LRMC) is the proper basis for entry/exit price signals for long-lived assets; therefore the price signal may already be too low based on microeconomic theory, and it is depressed further by low natural gas cost. Pouring zero SRMC generation into an over-supplied market exacerbates this, lowering both energy and capacity costs.

Market players with higher fixed costs, many of which provide useful grid services and fuel diversity, find themselves disadvantaged by this set of circumstances, incurring operating.

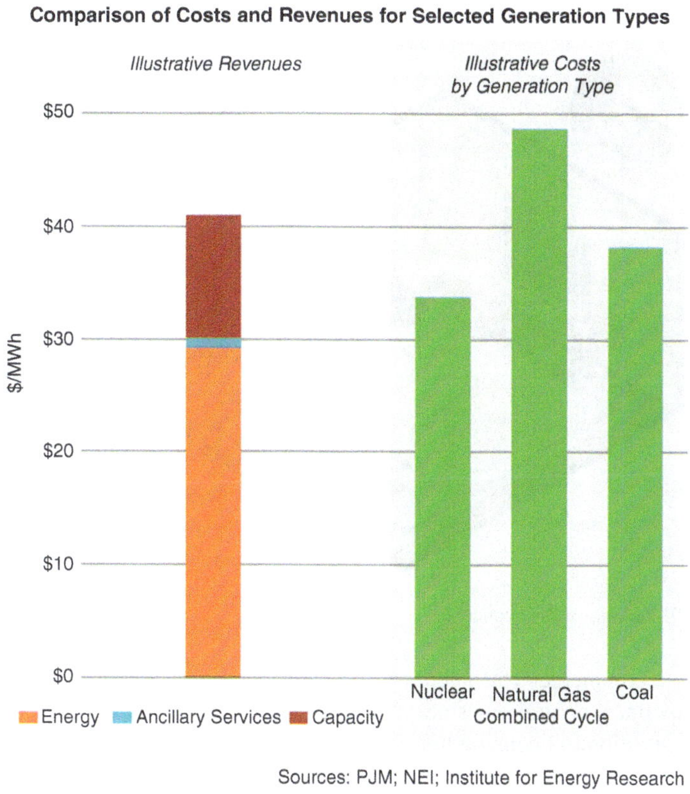

Fig. 7.2 Comparison of cost and revenues for selected electricity generation types

Fig. 7.3 A recent history of tweaking the market construction

losses that are becoming unsustainable; thus, the Federal Energy Regulatory Commission (FERC) finds itself looking to solve the issue of the "missing money." Figure 7.2 shows an illustration of the comparison of cost and revenues for selected electricity generation types, when we are taking into consideration the competition at crossroads, while Fig. 7.3 is an indication of tweaking the market construct based on recent history.

Keep in mind that Federal Energy Regulatory Commission is a US federal agency that regulates the transmission and wholesales of sales of electricity and natural gas in interstate commerce and regulates the transportation of oil by pipeline

Fig. 7.4 PJM administrative concept

in interstate commerce. This agency also reviews proposals to build interstate natural gas pipelines, natural gas storage projects, and liquified natural gas (LNG) terminals in addition to licensing non-federal hydropower projects.

Figure 7.3 also is indicator of Pennsylvania, Jersey, and Maryland (PJM), where this organization plays a role on an electric bill; electricity usage is commonly reported in kilowatt-hours. Ten 100 watt light bulbs left for 1-h usage and 1 kWh of electricity and at an electric rate of 11.5 cents per kWh.

PJM organization works quietly behind the scenes to coordinate and direct the flow of electricity to keep the lights on for 65 million people in 13 states and the District of Columbia. PJM does not own power lines or generators. Instead, it is a neutral, regulated organization that directs the operation of power lines and generators for many different owners. PJM acts as an agent to provide fair access to the transmission system for competing suppliers and electricity users. See Fig. 7.4.

PJM also administers competitive wholesale markets for large blocks of electricity – similar to the way the stock market works. In addition, PJM conducts a long-range planning process that identifies what changes and additions to the grid are needed to ensure reliability and the successful operation of the wholesale markets.

However, PJM is different from a consumer's local utility company in several ways, as illustrated in Fig. 7.5.

Operating the electric system is a balancing act that requires continuous monitoring of the system. PJM reacts instantly to changes in demand, equipment problems, weather conditions, and other factors to maintain safe and reliable service while meeting customer needs for electricity when and where it's needed.

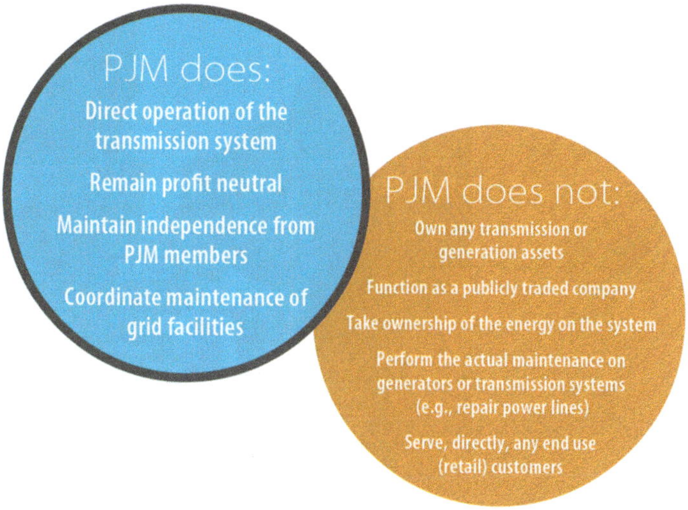

Fig. 7.5 PJM modes of operations

As part of PJM compliance, this organization is committed to compliance with various government entities in order to ensure the reliability of the power grid. Within this Compliance section, you can browse general compliance questions and select information that is tailored to generation owners, transmission owners, and more. Visit the Compliance section of pjm.com for additional compliance information.

PJM is regulated by the Federal Energy Regulatory Commission. The FERC is an independent agency that regulates the interstate transmission of electricity, natural gas, and oil. The FERC also regulates hydropower and natural gas projects.

PJM also is committed to full compliance with all applicable North American Electric Reliability Corporation and the North American Energy Standards Board business standards as well as Reliability First standards. PJM provides ongoing communication with its members on the status of compliance and compliance-related activities through the Reliability Standards and Compliance Subcommittee.

PJM is also compliant with the SERC Reliability Corporation, a nonprofit corporation responsible for promoting and improving the reliability, adequacy, and critical infrastructure of the bulk power supply systems in all or portions of 16 central and southeastern states. See Fig. 7.6, where the map the eight regional entities that have agreements with the North American Electric Reliability Corporation to monitor and enforce compliance.

The eight reginal states that are involved in infrastructure of the bulk power supply systems are listed below as:

- Western Electricity Coordinating Council (WECC)
- Midwest Reliability Organization (MRO)
- Southwest Power Pool (SPP)

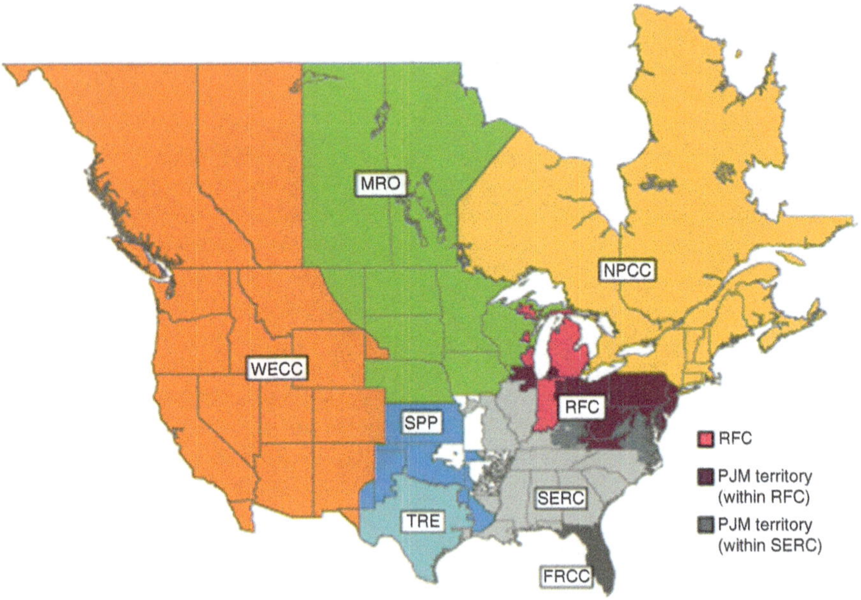

Fig. 7.6 Map of eight regional areas

- Texas Reliability Entity (TRE)
- Northeast Power Coordinating Council (NPCC)
- Reliability First (RFC)
- SERC Reliability Corporation (SERC)
- Florida Reliability Coordinating Council (FRCC)

Also, keep in mind that, the SERC Reliability Corporation (SERC) is one of eight regional electric reliability councils under North American Electric Reliability Corporation (NERC) authority. NERC and the regional reliability councils were formed following the Northeast Blackout of 1965. SERC was formed on April 29, 2005, as the successor to the Southeast Electric Reliability Council (also known as SERC).

The original SERC was formed January 14, 1970, by the functional merger of four smaller reliability entities: the CARVA Pool, Tennessee Valley Authority (TVA), Southern Company (SOCO), and the Florida Electric Power Coordinating Group (FEPCG). On September 16, 1996, the SERC member companies formerly represented by FEPCG formed the Florida Reliability Coordinating Council (FRCC) and separated from SERC. SERC's offices are located in Charlotte, NC.

As we can see in Fig. 7.7, it shows the three major and two minor NERC Interconnections, and the nine NERC Regional Reliability Councils.

The SERC region lies within the Eastern Interconnection, occupying most of Southeast North America, serving all of the States of Missouri, Alabama, Tennessee, North Carolina, South Carolina, Georgia, Mississippi, and portions of Iowa, Illinois,

Fig. 7.7 Illustration of North America regional reliability councils

Kentucky, Virginia, Oklahoma, Arkansas, Louisiana, Texas, and Florida. SERC members include investor-owned, municipal, cooperative, state, and federal systems, Reginal Transmission Organizations/Independent System Operators (RTOs/ISOs), Merchant Electricity Generators, and power marketers.

7.4 What Is a Capacity Market?

A difficult economic problem that evades a simple solution forces that Federal Energy Regulatory Commission (FERC) and market participants have long recognized that least marginal cost hourly dispatch did not necessarily ensure a diverse set of capacity resources and long-run total cost recovery of long-lived assets – the so-called "missing money" problem. Capacity markets in some competitive wholesale markets were formed to solve this problem, while not purely market based, capacity markets are administrative constructs to provide "guard rails" to ensure proper monetary incentives exist for resource adequacy. Various mechanisms have been used to manage unintended price impacts and distortions, particularly the FERC-approved Minimum Offer Price Rule (MOPR), which applies to RTO auction markets [1].

To serve their customers, electric power utilities provide not only electric power on a continuous basis but related services, including "capacity." Capacity is the maximum amount of electricity that a power plant can produce, usually stated in megawatts (MW). Customers who agree to curtail their use of electric power when called upon (known as demand response) also provide capacity.

In the context of wholesale electric markets, a capacity market is a mechanism to provide revenue to a power plant owner to stand ready to supply power when needed. An electric power utility or other Load-Serving Entity (LSE) purchases or owns capacity to ensure a reliable supply of power during peaks in demand (generally the hottest and coldest times of the year). The LSE needs to have in place sufficient capacity to meet the projected peak demand plus a reserve margin, as determined by regional reliability entities, which help ensure that the regional grid can "keep the lights on."

In several regions of the country, large bureaucratic, quasi-governmental entities, known as Regional Transmission Organizations (RTOs), operate and oversee the bulk power grid and operate wholesale markets for energy, capacity, and other "ancillary services" that help provide reliability to the bulk power grid. Three of the nation's seven RTOs operate mandatory capacity markets. In these mandatory markets, capacity must be bought and sold through the RTO market. Capacity that is owned or contracted for bilaterally by an LSE still must be offered into (and meet the market clearing price set by) the capacity auctions. The price paid for capacity purchased through the auction is set by the RTO. The three RTOs with mandatory capacity markets are the Independent System Operator (ISO) New England (ISO-NE), PJM Interconnection (PJM) (covering the mid-Atlantic states, Ohio, and Northern Illinois), and New York ISO (NYISO). PJM and ISO-NE both operate a "forward" market where capacity is procured 3 years in advance for a 1-year period. The capacity auctions in NYISO are shorter term and are procured close to the period when the capacity will be needed [2].

The Midcontinent ISO's (MISO) capacity market is voluntary, and LSEs can choose whether to participate. MISO proposed in November 2016 to develop a capacity market that would be mandatory just for the utilities that have been restructured (which means they no longer own generation to serve their customer's load). Public power and consumer representatives opposed the proposal out of concern that it will lead to a mandatory capacity market in the remainder of MISO. FERC rejected MISO's proposal out of concern that such a bifurcated market would result in inefficient and volatile pricing. Neither the California ISO nor the Southwest Power Pool operates a capacity market. The Electric Reliability Council of Texas (ERCOT), which functions as an RTO, but is not under FERC's jurisdiction because of the intrastate nature of its grid, does not operate a capacity market.

While RTO markets for capacity are described as "competitive," they are highly mechanized, centrally administered constructs governed by thousands of pages of complex rules. Transactions in these markets are opaque, with little meaningful data available to the public. RTO-operated wholesale electricity markets are regulated by the Federal Energy Regulatory Commission (FERC).

Considering what we have seen so far, the question is: *What is different*, and *what is the concern*?

There are an increasing number of "outside-of-market" mechanisms to address state policy preferences not solely driven by economic efficiency or reliability, e.g., state solicitations of renewables, zero emissions credits, and baseload capacity support. Subsidized entry can lead to imbalance (artificial surplus), changing market outcomes for other (including existing) resources, and power markets are not yet designed for difficult multivariate optimization, including all attributes valued by states. A key concern is that, at some point, the amount of capacity priced using "outside-of-market" mechanisms might reach a tipping point as listed below:

- Bid-based markets effectively become bilateral markets.
- Capacity markets then become residual markets with distorted price signals and ultimately cease yielding "just and reasonable" prices.

However, in May 2017, the FERC held a technical conference to discuss state policies and wholesale markets operated by the ISO New England, New York Independent System Operator, and PJM Interconnection. The FERC has considered a range of options (see next page), as the Department of Energy is calling for the end of study on price formation and for expeditious action by the FERC.

7.5 Renewable and Nonrenewable Energy Sources

When people use electricity in their homes, the electrical power was probably generated by burning coal, by a nuclear reaction, or by a hydroelectric plant on a river, to name just a few sources. Therefore, coal, nuclear, and hydro are called energy sources. When people fill up a gas tank, the source might be petroleum refined from crude oil or ethanol made by growing and processing corn.

Energy sources are divided into two groups:

1. Renewable (an energy source that can be easily replenished)
2. Nonrenewable (an energy source that cannot be easily replenished)

Renewable and nonrenewable energy sources can be used as primary energy sources to produce useful energy such as heat or used to produce secondary energy sources such as electricity.

When people use electricity in their homes, the electrical power was probably generated from burning coal or natural gas, a nuclear reaction, or a hydroelectric plant on a river, to name a few possible energy sources. The gasoline people use to

Fig. 7.8 Wind and solar energy source

fuel their cars is made from crude oil (nonrenewable energy) and may contain a biofuel (renewable energy) like ethanol, which is made from processed corn.

Unlike fossil fuels, which are finite, renewable energy sources regenerate. Renewable energy, clean energy, and green energy – all of these terms are becoming more and more common and pretty much in energy sector; you hear them, day in and day out, as part of the energy expansion to meet the challenge of supply and demand scenario due to global growth of population.

We are aware that wind energy and solar energy are a means of renewable source of energy Fig. 7.8), and we arguably presented advance technology and innovative approaches to design small modular reactors (SMRs), which also are ready to play a part of the family of these new renewable sources of energy as well.

Other than wind, solar, and nuclear, there are actually a ton of renewable energy sources. Basically, anything we can use to create energy that will quickly renew is a form or source of renewable energy. But, to get more specific, here's a rundown of five commonly used most popular renewable energy sources:

1. Biomass, which includes:
   - Wood and wood waste
   - Municipal solid waste
   - Landfill gas and biogas
   - Ethanol
   - Biodiesel
2. Hydropower
3. Geothermal
4. Wind
5. Solar

More details of each of these major sources of renewal energy can easily be found on the Internet or references such as the US Energy Information Administration website [3]; however, Fig. 7.9 is illustration of US primary energy production by

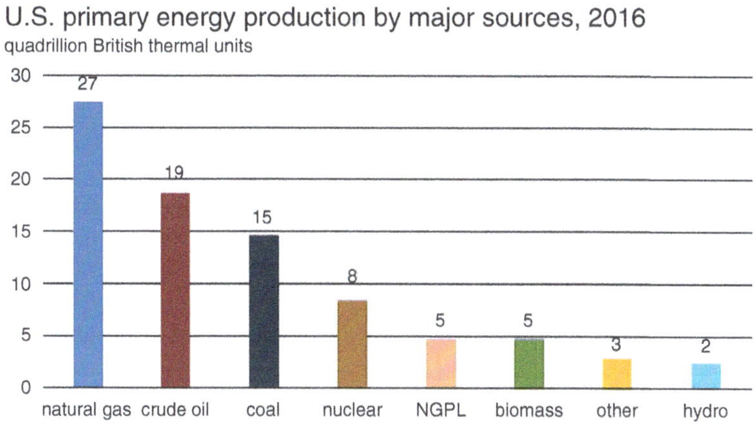

Fig. 7.9 US energy production by source of energy. (Courtesy of US Energy Information Administration)

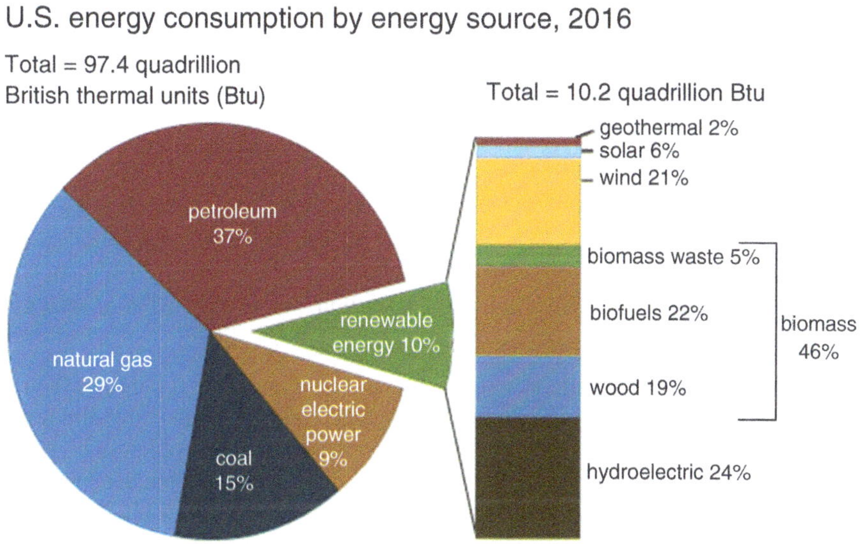

Fig. 7.10 US energy consumption by source of energy. (Note: Sum of components may not equal 100% because of independent rounding. Courtesy of US Energy Information Administration)

major energy sources in 2016, while Fig. 7.10 is the depiction of the US energy consumption by energy source.

Note that, in Fig. 7.9, the natural gas is dry gas, and NGPL is designation for Natural Gas Plant Liquids, other in the chart is presentation of Geothermal, Solar, and Wind, and Hydro is Conventional Hydroelectric.

However, renewable energy sources provide 15% of US electricity and as part of source of production, and they can be briefly summarized here as:

- Hydropower, the source of about 7% of total US electricity generation (about 44% of electricity generation from renewable energy) in 2016, is a process in which flowing water spins a turbine connected to a generator. Most hydropower production is at large facilities built by the federal government, such as the Grand Coulee Dam. Many of the largest hydroelectric dams are in the western United States, but many hydropower facilities operate around the country.
- Wind power is produced by converting wind energy into electricity with wind turbines. Electricity generation from wind has increased significantly in the United States since 1970. In 2016, wind power provided almost 6% of US electricity generation (about 37% of electricity generation from renewable energy).
- Biomass is material derived from plants or animals and includes lumber and paper mill wastes and the food scraps, grass, leaves, paper, and wood in municipal solid waste (garbage). Biomass also includes forestry and agricultural residues such as wood chips, corn cobs, and wheat straw. These materials can be burned directly in steam-electric power plants, or they can be converted to a gas that can be burned in steam generators, gas turbines, or internal combustion engine generators. Biomass accounted for about 2% of total US electricity generation in 2016.
- Solar power is produced with energy from the sun. Photovoltaic (PV) and solar-thermal power are the two main types of technologies used to convert solar energy to electricity. PV conversion produces electricity directly from sunlight in a photovoltaic (solar) cell. Solar-thermal power generators concentrate solar energy to heat a fluid and produce steam to drive turbines. In 2016, about 1% of US electricity generation was from solar energy.
- Geothermal power comes from heat energy beneath the surface of the earth. In some areas of the United States, geothermal energy is close enough to the earth's surface to heat underground water into steam, which is tapped for use at steam-turbine plants. Geothermal electricity generation was less than 1% of total US electricity generation in 2016.

Most of the energy consumed in the United States comes from fossil fuels (petroleum, coal, and natural gas). These fossil fuels and crude oil-based petroleum products are the major sources of energy used in the United States. Renewable energy resources currently supply about 10% of total US energy consumption. See Fig. 7.11.

The United States began using nuclear energy to generate electricity in the late 1950s. Nuclear energy supplies about 20% of total US electricity.

Along with renewable source of energy, there exist sources of energy that are not renewable, and they are considered as *nonrenewable energy sources*.

Energy sources are classified as nonrenewable because they do not form or replenish in a short period of time. Renewable energy sources such as solar and wind replenish naturally in a short period of time.

The four major nonrenewable energy sources are:

- Crude oil
- Natural gas
- Coal
- Uranium (nuclear energy)

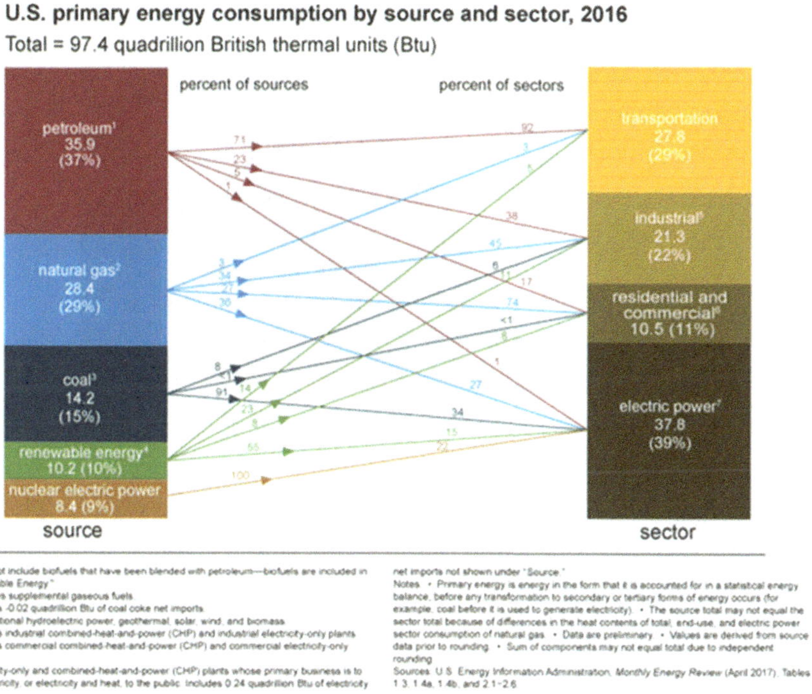

Fig. 7.11 US primary energy consumption by source and sector. (Courtesy of US Energy Information Administration)

Nonrenewable energy sources come out of the ground as liquids, gases, and solids. We use crude oil to make liquid petroleum products such as gasoline, diesel fuel, and heating oil. Propane and other hydrocarbon gas liquids, such as butane and ethane, are found in natural gas and crude oil.

All fossil fuels are nonrenewable, but not all nonrenewable energy sources are fossil fuels; coal, crude oil, and natural gas are all considered fossil fuels because they were formed from the buried remains of plants and animals that lived millions of years ago.

Uranium ore, a solid, is mined and converted to a fuel used at nuclear power plants. Uranium is not a fossil fuel, but it is classified as a nonrenewable fuel.

7.6 Role of Renewable Energy

As part of this section, we consider the fact that:

What role does renewable energy play in the United States?

Up until the mid-1800s, wood supplied nearly all of the nation's energy needs. As more consumers began using coal, petroleum, and natural gas, the United States

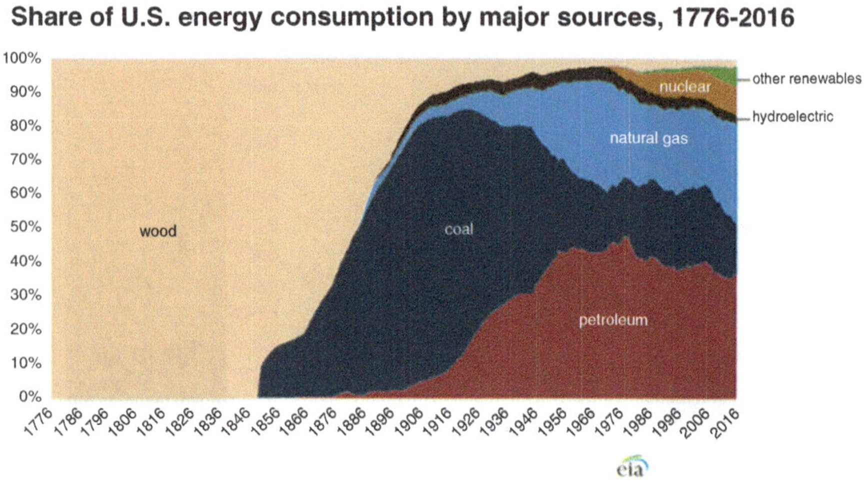

Fig. 7.12 Share of the US energy consumption by major source. (Source: U.S. Energy information Administration, *Monthly Energy Review*, April 2017, preliminary data for 2016. Courtesy of US Energy Information Administration)

relied less on wood as an energy source. Today, the use of renewable energy sources is increasing, especially biofuels, solar, and wind.

In 2016, about 10% of total US energy consumption was from renewable energy sources (or about 10.2 quadrillion British thermal units (Btu) – 1 quadrillion is the number 1 followed by 15 zeros). About 55% of US renewable energy use is by the electric power sector for producing electricity, and about 15% of US electricity generation was from renewable energy sources in 2016.

Renewable energy plays an important role in reducing greenhouse gas emissions. When renewable energy sources are used, the demand for fossil fuels is reduced. Unlike fossil fuels, non-biomass renewable sources of energy (hydropower, geothermal, wind, and solar) do not directly emit greenhouse gases.

The consumption of biofuels and other nonhydroelectric renewable energy sources more than doubled from 2000 to 2016, mainly because of state and federal government mandates and incentives for renewable energy. The US Energy Information Administration (EIA) projects that the use of renewable energy in the United States will continue to grow through 2040. See Fig. 7.12.

The usage of renewable energy from present to future time is becoming an inevitable situation that we as consumer cannot ignore.

In general, renewable energy is more expensive to produce and to use than fossil fuel energy. Favorable renewable resources are often located in remote areas, and it can be expensive to build power lines from the renewable energy sources to the cities that need the electricity.

In addition, renewable sources are not always available:

- Clouds reduce electricity from solar power plants.
- Days with low wind reduce electricity from wind farms.
- Droughts reduce the water available for hydropower.

7.7 Frequently Asked Questions

As summary of this chapter, we include this section as end of this chapter and ask:

What is the efficiency of different types of power plants that are generating electricity?

Heat rate is one measure of the efficiency of a generator or power plant that converts a fuel into heat and into electricity. The heat rate is the amount of energy used by an electrical generator or power plant to generate 1 kilowatt-hour (kWh) of electricity. The US Energy Information Administration (EIA) expresses heat rates in British thermal units (Btu) per net kWh generated. Net generation is the amount of electricity a power plant (or generator) supplies to the power transmission line connected to the power plant. Net generation accounts for all the electricity that the power plant consumes to operate the generator(s) and other equipment, such as fuel feeding systems, boiler water pumps, cooling equipment, and pollution control devices.

To express the efficiency of a generator or power plant as a percentage, divide the equivalent Btu content of a kWh of electricity (3412 Btu) by the heat rate. For example, if the heat rate is 10,500 Btu, the efficiency is 33%. If the heat rate is 7500 Btu, the efficiency is 45%.

The US Energy Information Administration (EIA) only publishes heat rates for fossil fuel-fired generators and nuclear power plants. There is a discussion about the method that EIA uses to estimate the amount of energy consumed to generate electricity with renewable energy sources in Alternatives for Estimating Energy Consumption, which includes a table with estimates for the conversion efficiencies of noncombustible renewable energy sources (geothermal, hydroelectric, solar, and wind energy) [4].

References

1. Generation to Generation, An Energy Evolution, The ScottMadden Energy Industry Update, Volume 17 – Issue 2
2. https://www.publicpower.org/policy/rto-capacity-markets-and-their-impacts-consumers-and-public-power
3. https://www.eia.gov/energyexplained/?page=renewable_home
4. https://www.eia.gov/tools/faqs/faq.php?id=107&t=3

Chapter 8
Energy Storage Technologies and Their Role in Renewable Integration

Abstract Today's world is at a turning point. Resources are running low, pollution is increasing, and the climate is changing. As we are about to run out of fossil fuels in the next few decades, we are keen to find substitutes that will guarantee our acquired wealth and further growth on a long-term basis. Modern technology is already providing us with such alternatives like wind turbines, photovoltaic cells, biomass plants, and more. But these technologies have flaws. Compared to traditional power plants, they produce much smaller amounts of electricity, and even more problematic is the inconsistency of the production. The global demand for electricity is huge, and it's growing by approximately 3.6% annually, but the sun isn't always shining nor is the wind always blowing. For technical reasons, however, the amount of electricity fed into the power grid must always remain on the same level as demanded by the consumers to prevent blackouts and damage to the grid. It leads to situations where the production is higher than the consumption or vice versa. This is where storage technologies come into play — they are the key element to balance out these flaws. (http://www.engineeringnews.co.za/article/electricity-consumption-to-increase-to-over-30-116-b-kwh-globally-in-2030-2009-04-17).

8.1 Introduction

With the growing importance of renewable energy sources, scientist and engineers are anxious to enhance efficiencies and to lower the costs of these technologies. Yet, there seems to be only a handful of technologies available that are efficient enough and also economical. Storing energy isn't an easy task, as most of us know. Our smartphone battery only lasts for about a day and laptops only a few hours; the range for electric cars is limited to only little more than 100 km; and these are only examples for comparatively small devices. Now imagine the problem of storing energy at the level of hundreds to thousands of wind turbines and photovoltaic cells [1].

The way we handle the fluctuating energy demand today works fine – for now. But, as we approach the point of peak oil faster and faster, and as we are trying hard

B. Zohuri, *Small Modular Reactors as Renewable Energy Sources*,
https://doi.org/10.1007/978-3-319-92594-3_8

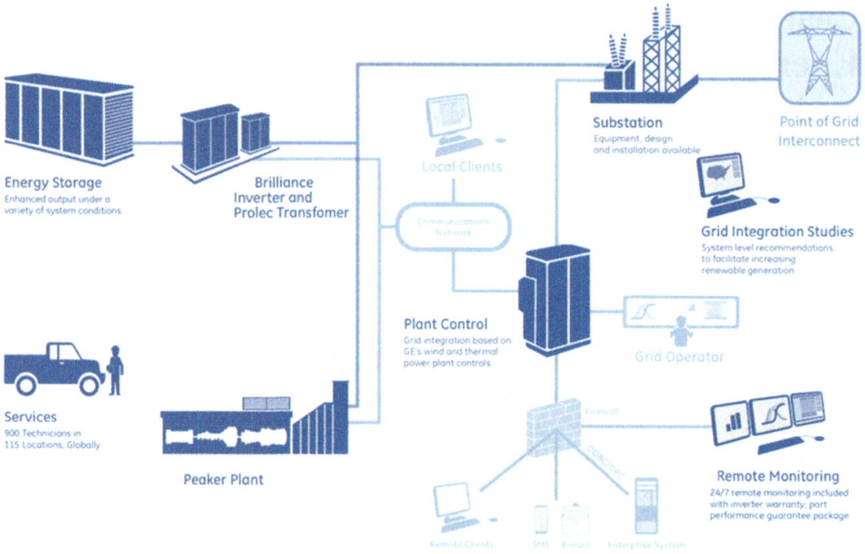

Fig. 8.1 Energy storage stages

to replace these conventional plants with regenerative energy sources, the grid changes, whereas the demand will remain about the same. With renewable energy, the production is fluctuating in a way that is hardly predictable. We may be able to predict the weather for the next few days, but as we all know, the weather forecast isn't always right, and even then, a few days isn't enough to calculate in the context of a national or even transnational power grid to guarantee a secure energy supply. Also, when the wind stops, it stops, foreseeing it won't change the fact that wind turbines won't produce the energy we need. So, there is a need to find ways to compensate for this fluctuation, to save the energy in times of sunny and windy days and use it for cloudy and windless days. The technology to do so exists, and we even use them today, but its capacity is not enough by a long shot – not if we're planning to go green and sustainable. The problem emerging is that we can't just simply build more of the existing storage technologies as each technology has its own flaws. For example, pumped hydro storage, the most reliable and so far the only economical storage technology available, is extremely limited by few potential sites and strict laws on nature conservation.

In the following chapters, we will be introducing some basic knowledge of power grids, the most important storage technologies so far, as well as a critical observation of their benefits, problems, and possible impacts in the future; and a small glance at promising technologies still in their development and pilot phases.

As we have stated in past chapters, energy storage is the capture of energy produced at one time for use at a later time. A device that stores energy is sometimes called an accumulator or battery as it is illustrated in Fig. 8.1.

Fig. 8.2 Chain of bees

To wrap up this introduction, it is notable that, "when energy storage joins the block-chain the entire energy community benefits" as it stated by Tim Larrison CFO Green Charge. Bees are an excellent example of joint community as it can be seen in Fig. 8.2.

Answering the call for increasing energy self-reliance, a grassroots electricity-sharing model is emerging. "Community microgrids," comprising community-owned or subscribed solar PV and other renewable energy sources, offer participants and surrounding consumers the security of energy resilience in times of grid failure, and protection from energy price increases driven by volatile energy markets. They also give energy producers/consumers (aka "prosumers") more control over the renewable energy they generate. Figure 8.3 is a presentation of blockchain-based transaction, where two buildings with individual loads agree to a transaction of energy credits.

One of the most intriguing of such projects in the United States is LO3's Brooklyn Microgrid (BMG), where residents with rooftop solar PV sell their excess energy to their neighbors, instead of relying on net metering to sell it back to their utility, ConEdison. Although the law prohibits energy consumers from selling energy to one another, BMG members are able to execute these peer-to-peer energy transactions using credits in a blockchain ledger. Every BMG prosumer connects to the microgrid through a dual-purpose meter called a TransActive Grid Element (TAG-e). The TAG-e both measures the participant's energy production and consumption and communicates with other TAG-e devices to record transactions in the blockchain.

It isn't too much of a stretch to imagine future energy storage systems, incorporating open-source, microgrid-sensing software that could participate in blockchain-based transactions (see figure below). If, on a sunny day, there were community members with a glut of excess solar energy, they could use energy storage to keep the excess for later use or sale depending on the going rate for energy.

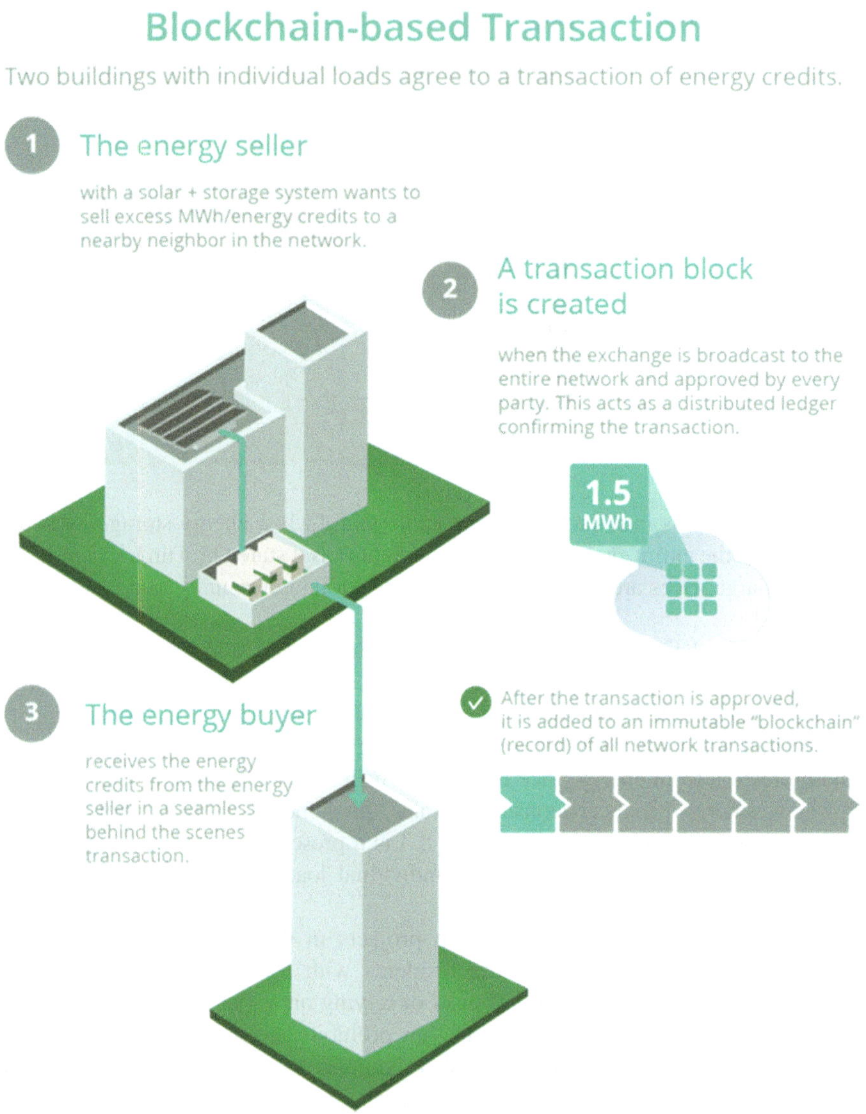

Fig. 8.3 Blockchain-based transaction illustration. (Courtesy of Green Charge)

8.2 The Electric Grid

An electrical grid is an interconnected network for delivering electricity from producers to consumers. It consists of generating stations that produce electrical power, high-voltage transmission lines that carry power from distant sources to demand centers, and distribution lines that connect individual customers.

Power stations may be located near a fuel source, at a dam site, to take advantage of renewable energy sources, and are often located away from heavily populated areas. They are usually quite large to take advantage of economies of scale.

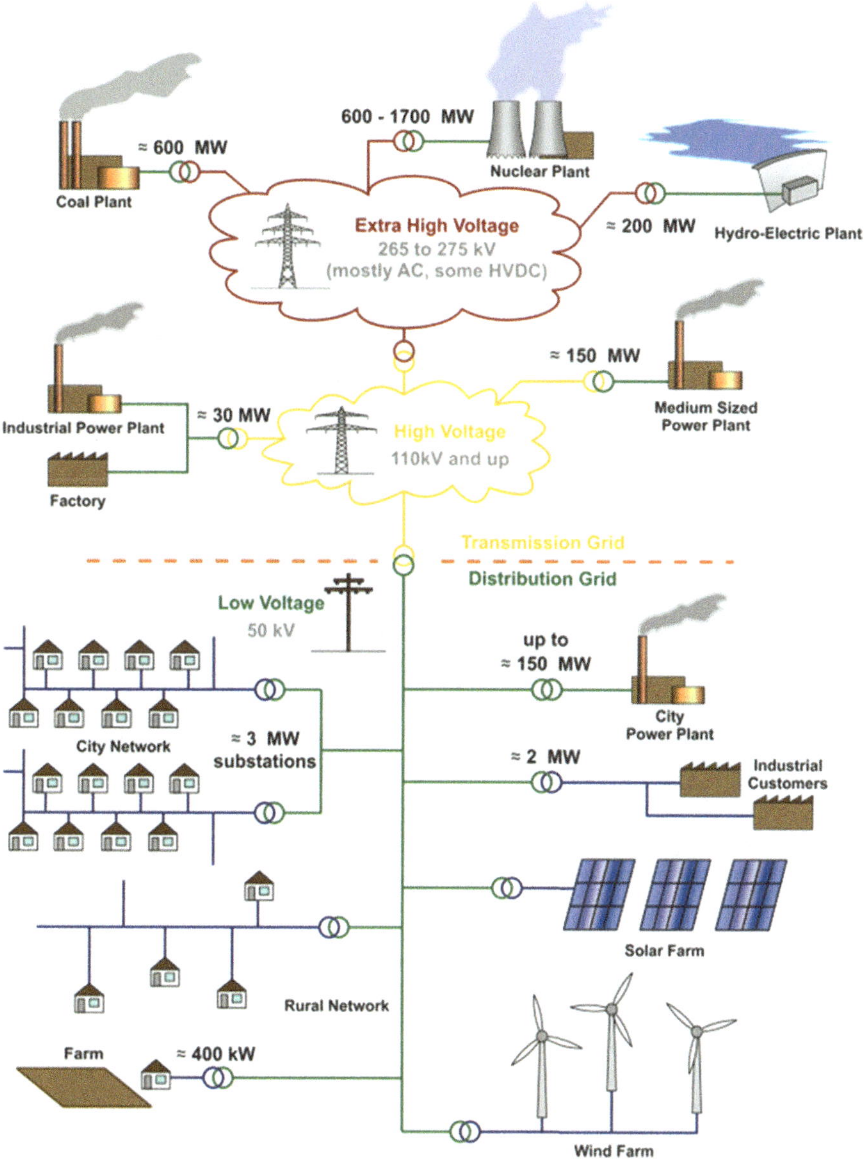

Fig. 8.4 General layout of electricity networks. (Courtesy of Wikipedia)

The electric power which is generated is stepped up to a higher voltage at which it connects to the electric power transmission network (see Fig. 8.4).

Note that voltages and depictions of electrical lines are typical for Germany and other European systems.

The bulk power transmission network will move the power long distances, sometimes across international boundaries, until it reaches its wholesale customer (usually the company that owns the local electric power distribution network).

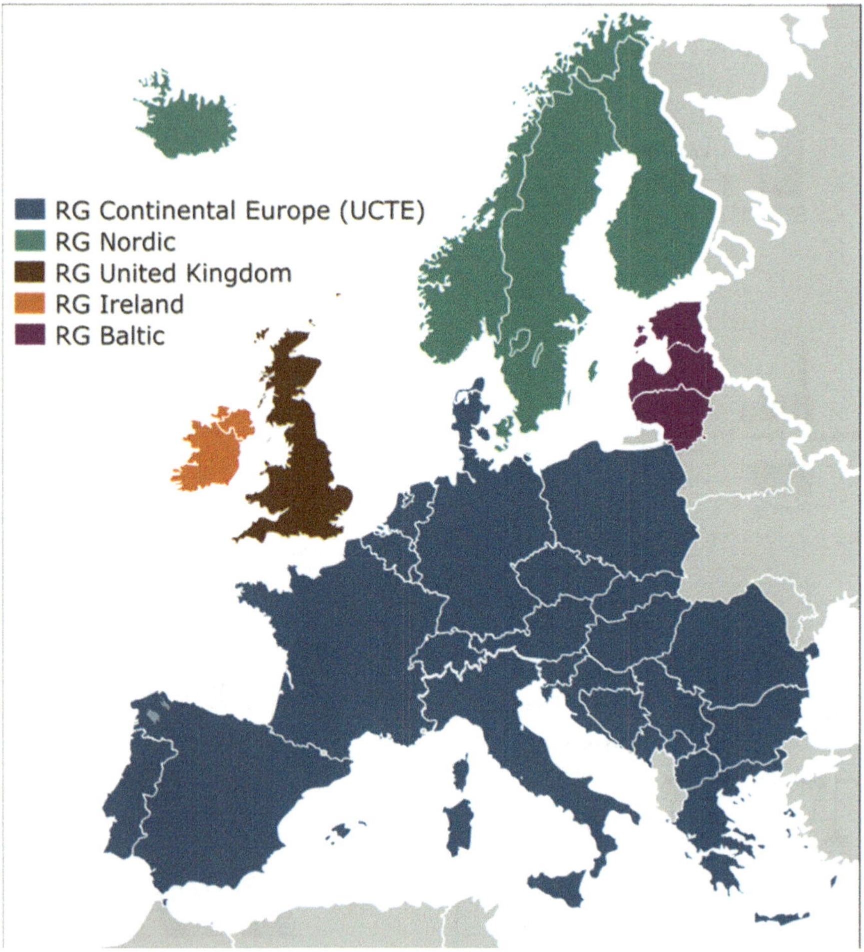

Fig. 8.5 Members of the European Transmission System Operators Association. (Courtesy of Wikipedia)

On arrival at a substation, the power will be stepped down from a transmission level voltage to a distribution level voltage. As it exits the substation, it enters the distribution wiring. Finally, upon arrival at the service location, the power is stepped down again from the distribution voltage to the required service voltage(s).

Electrical grids vary in size from covering a single building through national grids which cover whole countries to transnational grids which can cross continents.

Grids are designed to supply voltages at largely constant amplitudes. This has to be achieved with varying demand, variable reactive loads, and even nonlinear loads, with electricity provided by generators and distribution and transmission equipment that are not perfectly reliable. See Fig. 8.5 for the wide area synchronous grid of

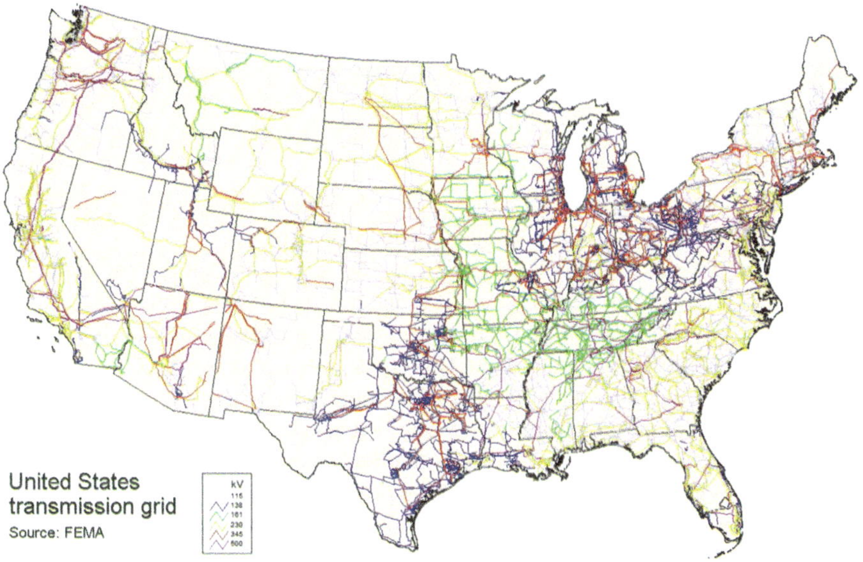

Fig. 8.6 The continental US power transmission grid. (Courtesy of the North American Electric Reliability Corporation (NERC))

Europe, where most members are of the European Transmission System Operation (ETSO) association, and Fig. 8.6 is the presentation of the continental US power transmission grid consisting of about 300,000 km (186,411 mi) of lines operated by approximately 500 companies. The North American Electric Reliability Corporation (NERC) oversees all of them.

An entire grid runs at the same frequency. Where interconnection to a neighboring grid, operating at a different frequency, is required, a frequency converter is required. High-voltage direct current links can connect two grids that operate at different frequencies or that are not maintaining synchronism.

In a synchronous grid, all the generators must run at the same frequency and must stay very nearly in phase with each other and the grid. For rotating generators, a local governor regulates the driving torque, maintaining constant speed as loading changes. Droop speed control ensures that multiple parallel generators share load changes in proportion to their rating. Generation and consumption must be balanced across the entire grid, because energy is consumed as it is produced. Energy is stored in the immediate short term by the rotational kinetic energy of the generators.

Small deviations from the nominal system frequency are very important in regulating individual generators and assessing the equilibrium of the grid as a whole. When the grid is heavily loaded, the frequency slows down, and governors adjust their generators so that more power is output (droop speed control). When the grid is lightly loaded, the grid frequency runs above the nominal frequency, and this is taken as an indication by automatic generation control (AGC) systems across the network that generators should reduce their output.

In addition, there is often central control, which can change the parameters of the AGC systems over timescales of a minute or longer to further adjust the regional network flows and the operating frequency of the grid. For timekeeping purposes, over the course of a day, the nominal frequency will be allowed to vary so as to balance out momentary deviations and to prevent line-operated clocks from gaining or losing significant time.

Transmission networks are complex with redundant pathways. For example, see the map of the US (right) high-voltage transmission network.

The structure or "topology" of a grid can vary depending on the constraints of budget, requirements for system reliability, and the load and generation characteristics. The physical layout is often forced by what land is available and its geology. Distribution networks are divided into two types, radial or network [2].

The simplest topology for a distribution or transmission grid is a radial structure. This is a tree shape where power from a large supply radiates out into progressively lower voltage lines until the destination homes and businesses are reached. However, single failures can take out entire branches of the tree.

Most transmission grids offer the reliability that more complex mesh networks provide. The expense of mesh topologies restricts their application to transmission and medium voltage distribution grids. Redundancy allows line failures to occur, and power is simply rerouted while workmen repair the damaged and deactivated line.

Other topologies used are looped systems found in Europe and tied ring networks.

In cities and towns of North America, the grid tends to follow the classic radially fed design. A substation receives its power from the transmission network; the power is stepped down with a transformer and sent to a bus from which feeders fan out in all directions across the countryside. These feeders carry three-phase power and tend to follow the major streets near the substation. As the distance from the substation grows, the fan-out continues as smaller laterals spread out to cover areas missed by the feeders. This treelike structure grows outward from the substation, but for reliability reasons, it usually contains at least one unused backup connection to a nearby substation. This connection can be enabled in case of an emergency, so that a portion of a substation's service territory can be alternatively fed by another substation.

A wide area synchronous grid or "interconnection" is a group of distribution areas all operating with alternating current (AC) frequencies synchronized (so that peaks occur at the same time). This allows transmission of AC power throughout the area, connecting a large number of electricity generators and consumers and potentially enabling more efficient electricity markets and redundant generation. Interconnection maps are shown of Europe (see Fig. 8.5, below left) and North America (see Fig. 8.6, right).

A large failure in one part of the grid – unless quickly compensated for – can cause the current to re-route itself to flow from the remaining generators to consumers over transmission lines of insufficient capacity, causing further failures.

One downside to a widely connected grid is thus the possibility of cascading failure and widespread power outage. A central authority is usually designated to facilitate communication and develop protocols to maintain a stable grid. For example, the North American Electric Reliability Corporation gained binding powers in the United States in 2006 and has advisory powers in the applicable parts of Canada and Mexico. The US government has also designated National Interest Electric Transmission Corridors, where it believes transmission bottlenecks have developed.

Some areas, for example, rural communities in Alaska, do not operate on a large grid, relying instead on local diesel generators [3].

High-voltage direct current (HVDC) lines or variable-frequency transformers can be used to connect two alternating current interconnection networks which are not necessarily synchronized with each other. This provides the benefit of interconnection without the need to synchronize an even wider area. For example, compare the wide area synchronous grid map of Europe (See Fig. 8.7) with the map of HVDC lines.

Electric utilities across regions are many times interconnected for improved economy and reliability. Interconnections allow for economies of scale, allowing energy to be purchased from large, efficient sources. Utilities can draw power from generator reserves from a different region in order to ensure continuing, reliable power and diversify their loads. Interconnection also allows regions to have access to cheap bulk energy by receiving power from different sources. For example, one region may be producing cheap hydropower during high-water seasons, but in low-water seasons, another area may be producing cheaper power through wind, allowing both regions to access cheaper energy sources from one another during different times of the year. Neighboring utilities also help others to maintain the overall system frequency and also help manage tie transfers between utility regions [4].

There are two types of grids that we can take under consideration, and these are:

Super Grids

Various planned and proposed systems to dramatically increase transmission capacity are known as super or mega grids. The promised benefits include enabling the renewable energy industry to sell electricity to distant markets, the ability to increase usage of intermittent energy sources by balancing them across vast geological regions, and the removal of congestion that prevents electricity markets from flourishing. Local opposition to sitting new lines and the significant cost of these projects are major obstacles to super grids. One study for a European super grid estimates that as much as 750 GW of extra transmission capacity would be required – a capacity that would be accommodated in increments of 5 GW HVDC lines. A recent proposal by TransCanada priced a 1600 km, 3 GW HVDC line at $3 billion USD and would require a corridor wide. In India, a recent 6 GW, 1850-km proposal was priced at $790 million and would require a wide right of way. With 750 GW of new HVDC transmission capacity required for a European super grid, the land and money needed for new transmission lines would be considerable.

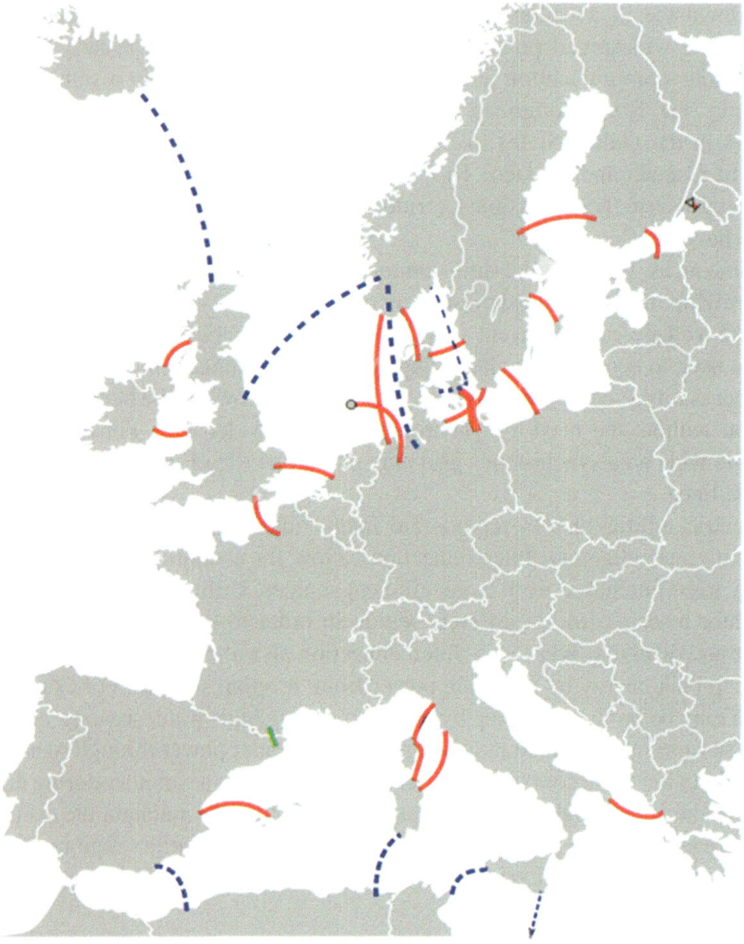

Fig. 8.7 High-voltage direct current interconnection in Western Europe. (Red colors are existing line, green are under construction, and blue are proposed). (Courtesy of Wikipedia)

Smart Grids

The smart grid would be an enhancement of the twentieth-century electrical grid, using two-way communications and distributed so-called "intelligent" devices. Two-way of electricity and information could improve the delivery network. Research is mainly focused on three systems of a smart grid – the infrastructure system, the management system, and the protection system [5].

The infrastructure system is the energy, information, and communication infrastructure underlying of the smart grid that supports:

- Advanced electricity generation, delivery, and consumption.
- Advanced information metering, monitoring, and management.
- Advanced communication technologies.

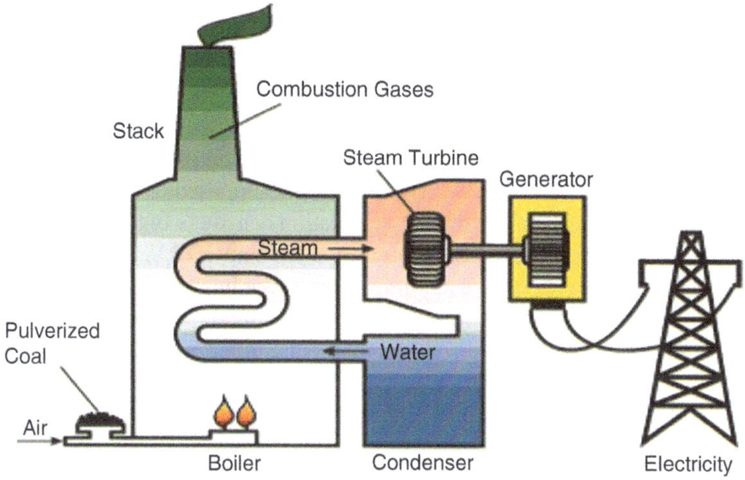

Fig. 8.8 Steam-powered power plant [7]

A smart grid would allow the power industry to observe and control parts of the system at higher resolution in time and space [6]. One of the purposes of the smart grid is real-time information exchange to make operation as efficient as possible. It would allow management of the grid on all time scales from high-frequency switching devices on a microsecond scale to wind and solar output variations on a minute scale and to the future effects of the carbon emissions generated by power production on a decade scale.

8.3 Power Generation

Electrical power usually starts at power plants. Although it may be coal, gas, or even nuclear power, almost every conventional plant produces electrical energy through steam-powered turbines. The fossil fuels are burned in order to make water boil and turn into steam which then enters the turbine and pushes against blades to turn the generator shaft to create electric current (see Fig. 8.8).

Right after the turbine, the steam is usually cooled down and turned into liquid form again in order to increase efficiency.

8.4 Transmission and Distribution

Power plants aren't located right next to your house; they tend to have sites where noise and emissions aren't disturbing issues for the population and near rivers for cooling purposes. Thus, in order to transport the electricity from the plants to the demanding locations, an electrical grid is needed. To minimize dissipation over long distance and to guarantee safety and functionality, different transmission grid types exist using different voltages. The closer the power gets to the consumer, the more the voltage decreases in the following order:

- *High-voltage grid*: Transformer stations transform the power to lower voltages, usually between 60 and 220 kV. These lines are supposed to carry the electricity into different regional areas with high population density or bigger industrial areas.
- *Distribution grid*: Voltages rank between 6 kV and 60 kV. Its main task is to provide major institutions like facilities, schools, or hospitals and the transformer stations responsible for low-voltage grids delivering to private households.
- *Low-voltage grid*: Often referred to as the "last mile," voltages between 110 V and 400 V are common. The low-voltage grid is the last station of the transmission and provides private households with power to use for everyday electronic devices.

8.5 Load Management

As mentioned before, it is necessary to produce the almost exact amount of electricity that is demanded by customers. In order to maintain grid stability, a frequency of 50 or 60 hertz (Hz) (depending on the country's standard) must be generated. Higher deviations (±2.5 Hz) will result in causing damage to the generators.

The curve progression in Fig. 8.9 varies over the day; but, behavior is usually steady and pretty well known for each day and is used as a roadmap for production along general lines. For example, the need for electricity during the night is low, whereas at noon, when everybody starts cooking, it is at its peak.

Analyzing Fig. 8.9 indicates the load curve diagram shows that much of the electricity demand is in fact for continuous 24/7 supply (base load), while some is for a lesser amount of predictable supply for about three quarters of the day and less still for variable peak demand up to half of the time; some of the overnight demand is for domestic hot water systems on cheap tariff. With overnight charging of electric vehicles, it is easy to see how the base-load proportion would grow, increasing the scope for nuclear and other plants which produce it.

The Fig. 8.9 also shows the concept of a load management which is separated into three different types:

- *Base load*: This is the amount of electricity that is demanded and produced at any time. Nuclear, hydroelectric power, or brown coal plants are known and common to use as base-load plants due to the long startup time and/or the low operating and fuel costs.
- *Intermediate load*: Power plants that are easier and faster to regulate are used for the task of middle load. These plants are capable of working within minutes to an hour and have moderate operating costs. Black coal or wind plants are typical of middle loads.

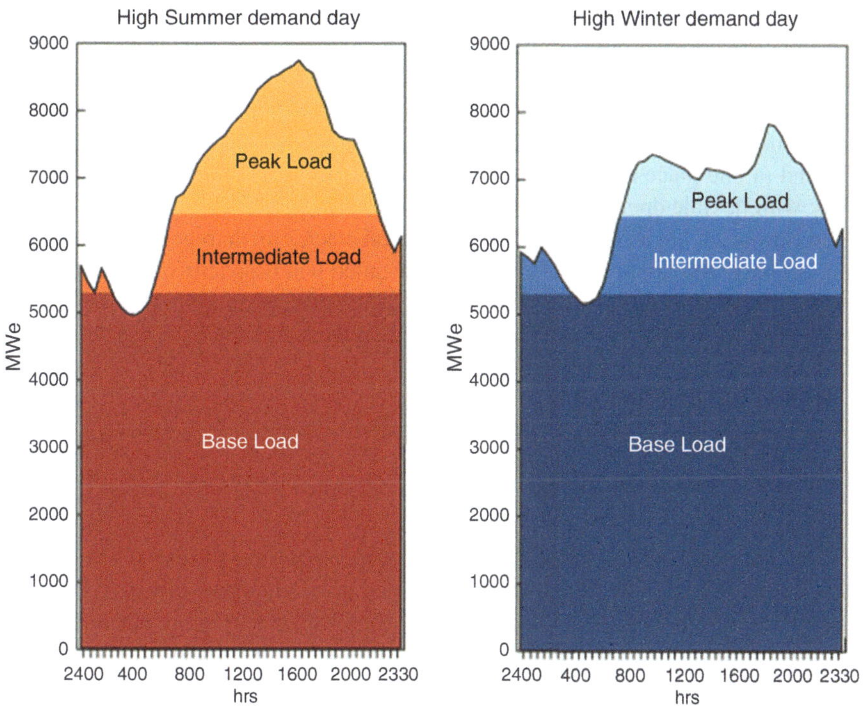

Fig. 8.9 Load curves for typical electric grid [8]. (Courtesy of Vencorp)

- *Peak load*: Peak load is the power demand outside of the daily "roadmap." Different events like unexpected hot and sunny days can lead to an extended use of air conditioners and therefore a higher electricity demand. Peak load plants have a fast response time, which means they're operational within seconds to a few minutes. A typical example would be gas turbine power plants or pumped-storage hydroelectricity.

Hydroelectric power plants are technically qualified for peak load but are used for base load instead because not using the already flowing water would be a waste.

Middle load plants can be and are also used for this task; when not operated under full load, they bear reserves. In some countries like Germany, it is statutory that a certain amount of power plants must have these reserves for supply security reasons. Operating power plants on lower degrees however is to be avoided if possible as the efficiency of the turbines decreases.

Most electricity demand is for continuous, reliable supply that has traditionally been provided by base-load electricity generation. Some is for shorter-term (e.g., peak load) requirements on a broadly predictable basis. Hence if renewable sources are linked to a grid, the question of backup capacity arises; for stand-alone system, energy storage is the main issue. Apart from pumped-storage hydro systems (see later section), no such means exist at present on any large scale [8].

However, a distinct advantage of solar and to some extent other renewable systems is that they are distributed and may be near the points of demand, thereby reducing power transmission losses if traditional generating plants are distant. Of course, this same feature sometimes counts against wind in that the best sites for harnessing it are sometimes remote from population, and the main backup for lack of wind in one place is wind blowing hard in another, hence requiring a wide network with flexible operation [8].

In Europe, at the end of 2014, the five largest electricity markets (UK, Germany, France, Italy, and Spain) had 97.5 GWe of installed wind capacity and 77.6 GWe of installed solar capacity. In the first half of 2015, this produced 107.6 TWh from wind and 60.1 TWh from solar, giving (based on these figures) 25% capacity factor for wind and 17.7% for solar (20.8% in July).

8.6 Types of Storage Technology

Energy storage is the capture of energy produced at one time for use at a later time. A device that stores energy is sometimes called an accumulator or battery. Energy comes in multiple forms including radiation, chemical, gravitational potential, electrical potential, electricity, elevated temperature, latent heat, and kinetic. Energy storage involves converting energy from forms that are difficult to store to more conveniently or economically storable forms. Bulk energy storage is currently dominated by hydroelectric dams, both conventional and pumped. See Fig. 8.10, for the depiction of the Llyn Stwlan dam of the Ffestiniog pumped-storage scheme in Wales. The lower power station has four water turbines which can generate a total of 360 MW of electricity for several hours, an example of artificial energy storage and conversion.

In 2015 hydro supplied about 3988 TWh from 1245 GWe (37% capacity factor), underlining its generally peak use.

Hydroelectric power, using the potential energy of rivers, is by far the best-established means of electricity generation from renewable sources. It may also be large scale – nine of the ten largest power plants in the world are hydro, using dams on rivers. China's Three Gorges Dam leads with 22.5 GWe, then Itaipu in Brazil with 14 GWe, and Xiluodu in China, 13.9 GWe. In contrast to wind and solar generation, hydro plants have considerable mechanical inertia and are synchronous, helping with grid stability.

Hydropower using large storage reservoirs on rivers is not a major option for the future in the developed countries because most major sites in these countries having potential for harnessing gravity in this way are either being exploited already or are unavailable for other reasons such as environmental considerations. Growth to 2030 is expected mostly in China and Latin America. China has commissioned the $26 billion Three Gorges Dam, which produces 22.5 GWe and has a major role in flood control, but it has displaced over 1.2 million people. Brazil is planning to have 25 GWe of new hydro capacity by 2025, involving considerable environmental impact.

Fig. 8.10 Llyn Stwlan dam of the Ffestiniog pumped-storage scheme in Wales

The chief advantage of hydro systems is their capacity to handle seasonal (as well as daily) high peak loads. In practice the utilization of stored water is sometimes complicated by demands for irrigation which may occur out of phase with peak electrical demands.

As part of pumped-storage hydroelectricity (PSH), pumped hydro plants, so far, are considered to be the only possible way to store energy in a huge amount while maintaining a high efficiency and being economical as well and has about 98% share of the total global storage predominant in today's grid. The first plants of this type were built in Switzerland and Italy in the 1890s, making the concept over a 100 years old.

Conceptually, when you lift an object of a certain mass, you overcome gravity. In order to do so, you must supply a force over a height. The force required to lift is defined by the physical law (*m* for mass and *a* for acceleration), but in this case, *a* is replaced by the gravitational acceleration *g* (9.81 meters per square second [m/s^2]). The work, meaning the energy supplied and therefore stored in the object, is defined by $W = F * d$ (in this example the term *d* for distance can be replaced by height). This results in $W = m * g * h$, meaning the energy stored equals the mass multiplied by the gravity and the height (Fig. 8.11).

A PSH plant puts this math into practice. Basically, the system contains two water reservoirs at different elevations. In times of low electricity demand and high production, water is pumped from the lower reservoir into the higher, storing the electricity in the water in the form of potential energy. When needed, for example, on peak demand, the water can be released, flowing down the pipes again and back through the turbine which then generates the electricity. The general formula for the power output is $P = Q * h * \eta * g * \rho$, including the factors of volume flow rate passing the turbines (Q), the hydraulic efficiency of the turbine, and the density of the water ρ.

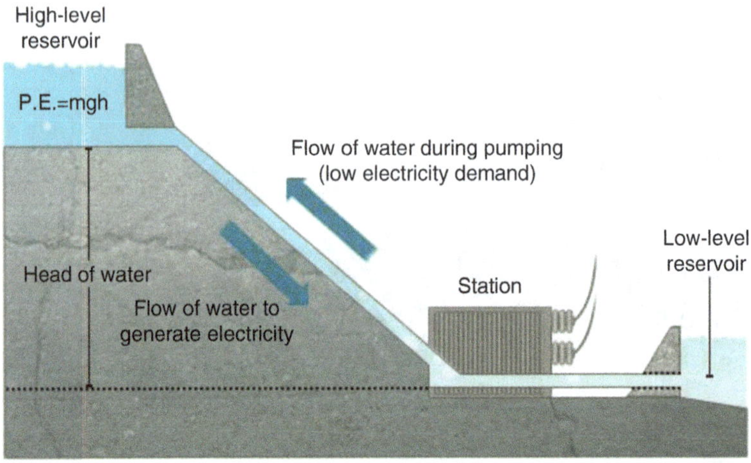

Fig. 8.11 Schematic of a pumped-storage plant

Depending on the height difference, Pelton wheels and Kaplan or Francis turbines are used to maximize efficiency, each reaching roughly about 90%. These turbines are reversible and, therefore, capable of handling both the pumping and generating process. Capacities for PSHs are depending on the location and scale of the reservoirs as well as the altitude difference and can reach from a few MWh to several GWh [7].

Pros and cons are listed here with sign of plus (pros) and minus (cons) in front of them as:

Pros

+ Mature technology, capable of storing huge amounts of energy
+ High overall efficiency (around 70–80%)
+ Fast response time
+ Inexpensive way to store energy

Cons

– Few potential sites
– Huge environmental impacts
– Requires a significant huge water source (Fig. 8.12)

Some technologies provide short-term energy storage, while others can endure for much longer.

A windup clock stores potential energy (in this case mechanical, in the spring tension), a rechargeable battery stores readily convertible chemical energy to operate a mobile phone, and a hydroelectric dam stores energy in a reservoir as gravitational potential energy. Fossil fuels such as coal and gasoline store ancient energy derived from sunlight by organisms that later died, became buried, and over time were then converted into these fuels. Food (which is made by the same process as fossil fuels) is a form of energy stored in chemical form.

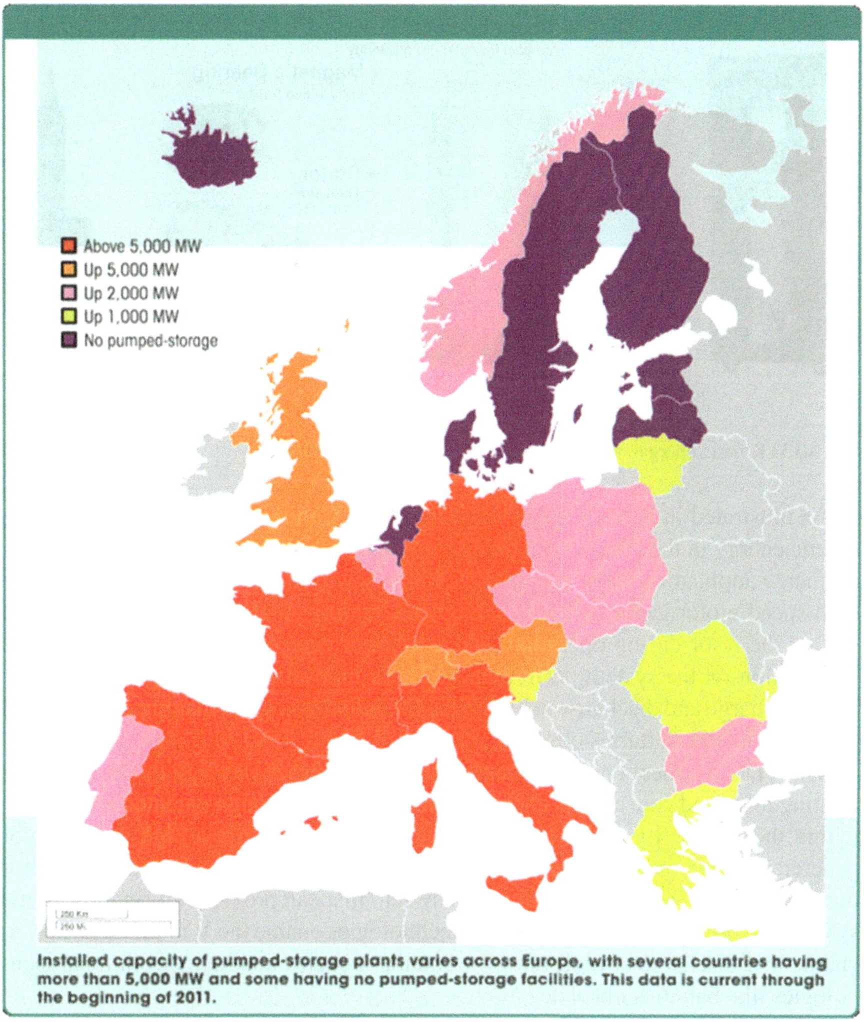

Fig. 8.12 Map of pumped-storage capacities in Europe

8.6.1 Kinetic Energy Storage or Flywheels Concept

The kinetic energy storage (KES) or flywheel functionality system is quite simple, and you may have even played with it when you were a kid. Remember the toy cars that kept going after spinning their wheels? Those were powered by a flywheel. Thus, basically a flywheel is a disk with a certain amount of mass that spins, holding kinetic energy.

Modern high-tech flywheels are built with the disk attached to a rotor in upright position to prevent gravity influence. They are charged by a simple electric motor that simultaneously acts as a generator in the process of discharging.

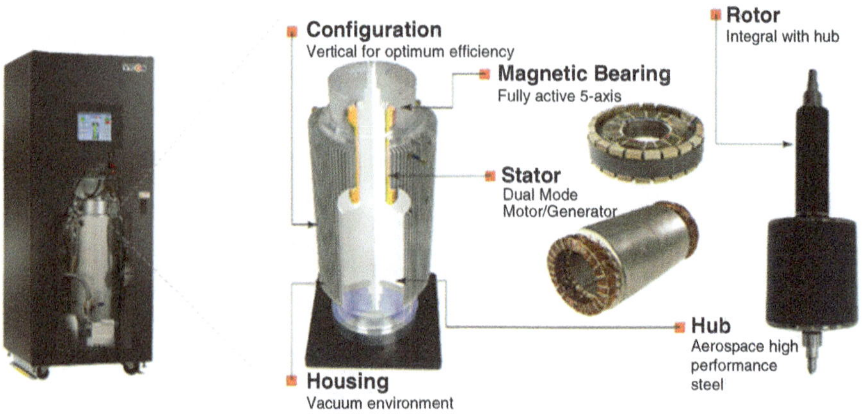

Fig. 8.13 Kinetic energy storage systems. (Courtesy of VYCON)

As illustrated in Fig. 8.13, the VYCON Direct Connect (VDC) ® system stores kinetic energy in the form of a rotating mass and is designed for high-power, short-discharge applications. The patented technology within the VDC system includes a high-speed motor generator, active magnetic bearings that are used to levitate and sustain the rotor during operation, and a superior control system that can provide information on the system performance. These innovative technologies enable the VDC to charge and discharge at high rates for countless cycles making conventional technologies like batteries obsolete.

The VDC flywheel energy storage systems hold kinetic energy in the form of a rotating mass and convert this energy to electric power through patented technology within the flywheel system. Our unique technology includes a high-speed motor generator, active magnetic bearings that are used to levitate and sustain the rotor during operation, and a superior control system that can provide information on the system performance. These innovative technologies enable the VYCON flywheel to charge and discharge at high rates for countless cycles making conventional technologies like batteries obsolete.

This technology is ideal for a variety of applications as a battery-free uninterruptible power supply (UPS) solution, including data centers, healthcare facilities, broadcast stations, and other mission-critical operations that require up to 30–40 s of backup power (see Fig. 8.14).

When dealing with efficiency however, it gets more complicated; as stated by the rules of physics, they will eventually have to deal with friction during operation. Therefore, the challenge to increase that efficiency is to minimize friction. This is mainly accomplished by two measures: the first one is to let the disk spin in a vacuum, so there will be no air friction; and the second one is to bear the spinning rotor on permanent and electromagnetic bearings so it basically floats. The spinning speed for a modern single flywheel reaches up to 16.000 rpm and offers a capacity up to 25 kilowatt hours (kWh), which can be absorbed and injected almost instantly.

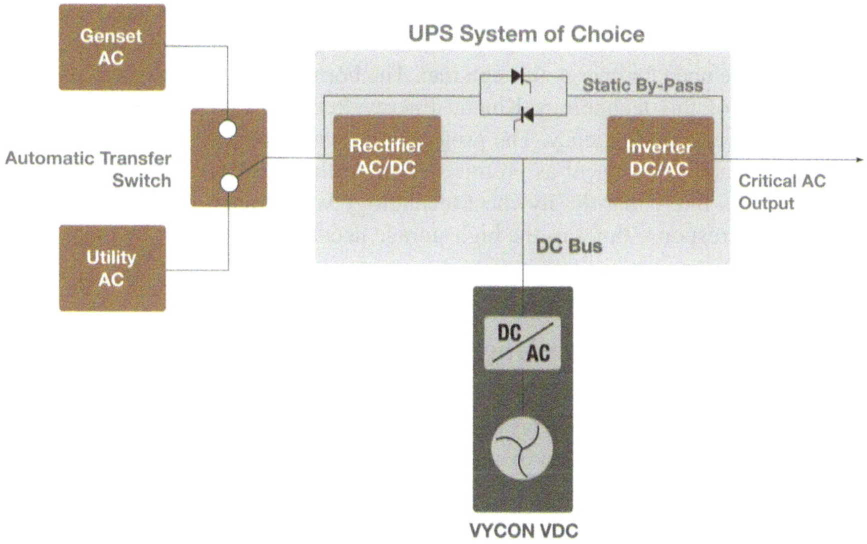

Fig. 8.14 Layout of uninterruptible power supply configuration. (Courtesy of VYCON)

Pros and cons are listed here with sign of plus (pros) and minus (cons) in front of them as:

Pros

+ Low maintenance and long lifespan: up to 20 years
+ Almost no carbon emissions
+ Fast response time
+ No toxic components

Cons

– High acquisitions cost
– Low storage capacity
– High self-discharge (3–20% per hour)

Future prospects of kinetic energy storage systems are stated below here as:

Stephentown, New York, is successfully operating the largest and latest flywheel energy storage system since July 2011. The facility is capable of storing up to 5 megawatt hours (MWh) with its 200 flywheels for several hours and required a budget of roughly $60 million2. This storage system has several advantages compared to others, most notable the low maintenance costs, the fast access to the stored energy, and the fact that you don't need any toxic resources as well as almost no carbon emissions. On the downside stands the low capacity compared to systems like the pumped hydro storage and the high acquisition costs, though compensated by the low maintenance and duration of up to 20 years [9, 10].

On a side note, in lower terms, flywheels could be used in the transport sector to make vehicles more efficient by using their kinetic energy to charge them and therefore lower the need for energy through fuel. The Formula 1 is currently successfully making use of this technology Kinetic Energy Recovery System (KERS) but to improve power, not efficiency. The project in Stephentown will show if flywheels are as good and economical as promised. Due to the high costs and low capacity however, it is likely that the flywheel technology will remain a niche market that requires fast response times as the high storage needs cannot be met by them.

8.6.2 *Superconducting Magnetic Energy Storage*

Superconducting magnetic energy storage (SMES) is a novel technology that stores electricity from the grid within the magnetic field of a coil comprised of superconducting wire with near-zero loss of energy.

SMES is a grid-enabling device that stores and discharges large quantities of power almost instantaneously. The system is capable of releasing high levels of power within a fraction of a cycle to replace a sudden loss or dip in line power. Strategic injection of brief bursts of power can play a crucial role in maintaining grid reliability especially with today's increasingly congested power lines and the high penetration of renewable energy sources, such as wind and solar. See Fig. 8.15 for illustration of top-level depiction of SMES.

The concept of SMES is described in this section as well. The system consists of three major components: the coil, the power conditioning system (PCS), and a cooling system. The idea is to store energy in the form of an electromagnetic field surrounding the coil, which is made of a superconductor. At very low temperatures, some materials lose every electric resistance and thus become superconducting. The superconducting magnetic storage system (SMES) makes use of this phenomenon and – in theory – stores energy without almost any energy loss (practically 90–95% efficiency). See Fig. 8.16 for more details of the components of this system.

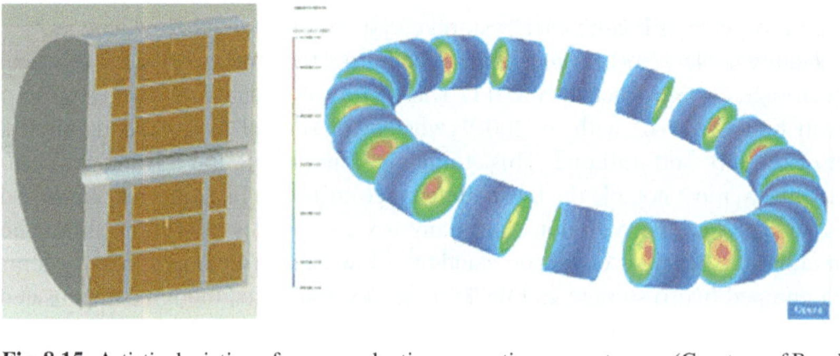

Fig. 8.15 Artistic depiction of superconducting magnetic energy storage. (Courtesy of Brookhaven National Laboratory)

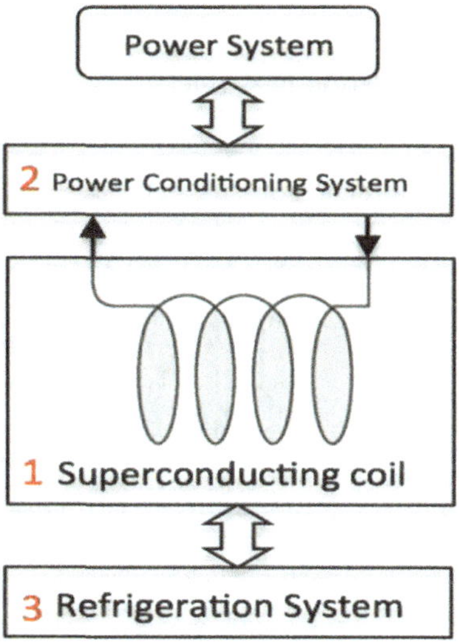

Fig. 8.16 Components of a SMES system

However, since relevant superconducting materials are only known to work below −253 degree centigrade (°C) (20° Kelvin [K]) (e.g., niobium-titanium −264 °C [9 K], niobium-tin −255 °C [18 K]), a system to cool the components down to those temperatures is required. This can be accomplished by liquefying helium; but, it is very expensive and the process lowers the efficiency, especially in standby mode. New high-temperature superconductors have been in development since 1986 reaching the state of superconductivity already at −163 °C (110 K), allowing them to be cooled by liquid nitrogen, and thereby lowering the costs by a factor of 10–20. Known production methods for these materials however make them very brittle and difficult and expensive to process.

The PCS is the interface between the SMES coil and the power system. Its task is to convert alternating current (AC) into direct current (DC) and vice versa since the coil is only capable of storing and releasing the energy in the form of DC.

Figure 8.17 is an illustration of conceptual design of a superconducting coil, which is a toroidal magnet with an outside diameter for the cryostat of ~12 m [11]. A half-size prototype coil was constructed by Toshiba and had been recently tested at the time of the WTEC trip to Japan. The test coil used a forced-flow Nb-Ti cable-in-conduit conductor and demonstrated 20 kA at 2.8 T, which is the rated current for the basic design. The initial testing was conducted at the Japan Atomic Energy Research Institute (JAERI), with further tests planned at Lawrence Livermore National Labs (LLNL) in the United States. Additional information on this SMES program is presented in the ISTEC, Chubu, and Toshiba site reports.

In the United Sates, the most significant program on energy storage in the world is being carried out by Babcock and Wilcox (B&W). This ~$50 million program is cost-shared by industry (70%) and the federal government (30%) through DARPA. B&W will construct and install a 500 kWh SMES primarily to provide spinning reserve to the Anchorage Municipal Light and Power (AML&P) utility. The Anchorage utility is part of the "Alaskan Railbelt System," one of the

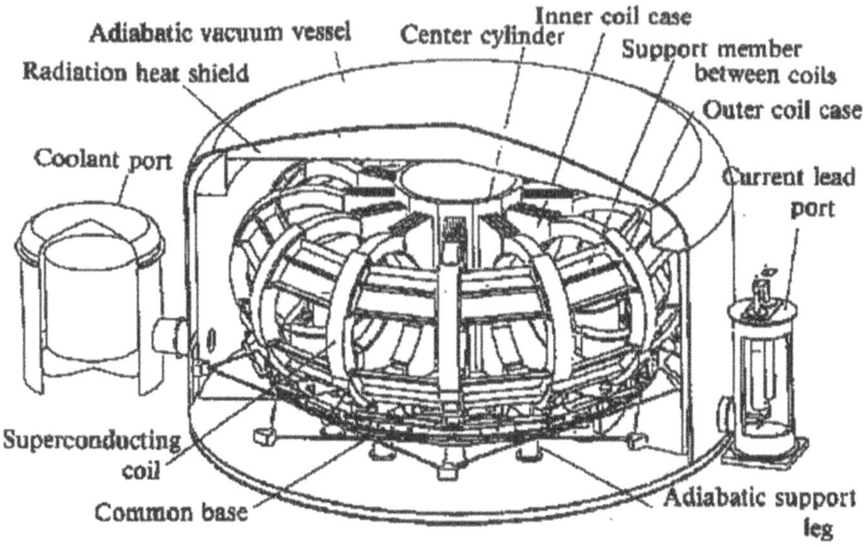

Fig. 8.17 Conceptual design of the ISTEC superconducting coil for the 100 kWh small-scale SMES

most isolated utility networks in the United States. The Anchorage area served by AML&P uses almost half of the railbelt system's peak load, which reaches approximately 600 MW during winter. AML&P is required as part of the railbelt interconnection agreement to designate ~30% of its generating capacity for spinning reserve; part of this reserve is provided by a hydroelectric facility at Bradley Lake. Physical restrictions at Bradley Lake result in dispatch time for this hydro capacity of approximately 1 min or more. This is too long a time to prevent additional load shedding during an event such as a generator outage, which would lead to frequency instability on the system, resulting in programmed load shedding. The planned SMES system will instantly dispatch ~30 MW over a period of ~1 min, which will provide sufficient time to ramp up the hydro capacity and put it on line to prevent further load shedding. As designed, the SMES will store 1800 MJ in a low-aspect solenoid almost 7 m in diameter using an aluminum-stabilized Nb-Ti conductor operating in a "cryostable cooled" mode [12].

Several US companies are producing small SMES systems, called micro-SMES, primarily to provide power quality improvements to selected customers rather than as grid or network solutions. These ~1 MW units with a few MJ stored energy are commercially produced by Superconductivity, Inc. (SI), of Madison, Wisconsin, and by IGC of Latham, New York. SI and IGC have supplied micro-SMES systems to the United States Air Force (USAF) as part of a program to provide uninterruptable power capability and power conditioning, primarily for voltage stabilization, to selected USAF control centers. At present this "power quality" market is also served by battery storage or flywheel systems, especially in Japan and Germany. Outside of the United States, there is no comparable activity for micro-SMES commercialization.

The most active effort on SMES in Germany is carried out at the Forschungszentrum Karlsruhe (FZK) laboratory. FZK is constructing a 250 kJ SMES with a toroidal field design to address a power quality problem due to the frequent startup of large motors at a saw mill. Analysis indicates that a SMES system would be ideal for reducing this flicker problem, due to its fast response capability. FZK is also investigating at the Deutsches Elektronen-Synchrotron (DESY) laboratory for the use of SMES as a pulse power source for providing ~10 GW pulses with a 1.0 ms duration at a 10 Hz rate in order to power RF klystrons [13].

At Siemens, the interest in SMES over the last few years has been mostly restricted to design and evaluation studies, with no current plans for development or construction. Siemens, RWE (the largest German utility), and PreussenElektra have completed an evaluation and conceptual design of a 2 MWh/50 MW SMES for use in providing frequency stabilization to the electric system. SMES continues to be of interest, but recent economic studies by Siemens indicate it may be too expensive compared to other storage technologies [14].

It is thought that using HTS current leads, which are now a commercial reality and can be purchased from a number of manufacturers including Hoechst and ASC, would reduce the heat leak in a SMES system. Use of HTS conductors to fabricate a high field, high density SMES is also projected as an interesting future application. An HTS SMES operating at 10–20 T would require a "composite conductor" capable of carrying thousands of amperes with low AC losses to minimize heating for multiple charge and discharge cycles.

The Technical University of Munich has also been conducting research on SMES and is constructing 1.4 MJ toroidal field system using LTS conductors [15]. At ABB in Switzerland, energy storage is considered an important area; it has been strongly followed in the past, and an LTS SMES has been constructed for experimental evaluation. At the time of the WTEC visit, a major SMES system using LTS conductor under development for the Swiss railroad had been terminated due to realization of an alternative, less costly, solution. ABB had no plans for SC storage but did plan to continue to evaluate the technology.

In summary, the Japanese Super-GM program to develop SC generators represents the major activity worldwide directed at the commercialization of superconductivity in an electric power application. This conclusion is based on the duration of the program, the total money invested by industry and the Japanese government, and the number of institutions and people involved. The future for this program, however, is highly dependent on the "complete" success of the planned testing through 1998 on the three rotor configurations for the generator. The follow-on program to construct a 200 MW-class pilot machine is also highly dependent on a significant improvement in the market forecast for this generator. Without an increase in demand, it is quite likely that even with unqualified success for the generator tests, Japanese industry may not be willing to commit the necessary cost share to go forward with the program. Interest in using HTS conductors in this future program will depend on progress in achieving higher transport properties at higher fields.

The current US DOE SPI program on the development of an HTS synchronous motor is unique. The B&W "spinning reserve" program with Anchorage Municipal Light and Power is the world's largest SMES program and provides an acceptable performance and cost-effective solution to a utility problem.

The benefits of superconducting magnetic energy storage are listed below as:

- Improves power quality for critical loads and provides carryover energy during momentary voltage sags and power outages.
- Improves load leveling between renewable energy sources (wind, solar) and the transmission and distribution network.
- Environmentally beneficial as compared to batteries; superconductivity does not rely on a chemical reaction and no toxins are produced in the process.
- Enhances transmission line capacity and performance – SMES features a high dynamic range, an almost infinite cycling capability, and an energy recovery rate close to 100%.
- Ultrahigh field operation enables long-term storage SMES systems in a compact device with cost advantages in material and system costs.

Pros and cons are listed here with sign of plus (pros) and minus (cons) in front of them as:

Pros

+ Capable of partial and deep discharge
+ Fast response time
+ No environmental hazard

Cons

– High energy losses (~12% per day)
– Very expensive in production and maintenance
– Reduced efficiency due to the required cooling process

Future prospects of superconducting magnetic energy storage are stated below here as:

Future prospects are difficult to determine because they depend on further development in superconducting materials. The discovery of a suitable material with these properties on room temperature would change nearly anything (hence, the consideration for being the Holy Grail of physics) and would make energy storage and transmission easy, safe, and cheap. However, it is uncertain if such a material even exists.

Right now, SMES systems are pretty much like flywheels, considered a niche market, requiring fast response times. Because of the difficult and expensive procedure to process high-temperature superconductors, it is expected that low-temperature materials will come to action in short and medium terms. Right now, the development focus lies on micro-SMES systems with capacities up to 10 kWh, applied mainly for power quality and uninterrupted power sources (UPS) and therefore of no relevant significance for renewable energies right now. Further technological improvements and achievements in processing high-temperature superconductors could change the course and make SMES systems more economical and relevant for energy storage in the future though.

8.6.3 Batteries

Energy storage systems are comprised of three main modules:

- The direct current (DC) battery where energy is stored
- The alternate current (AC) power conversion where the energy is converted from AC to DC
- The control system that manages the operation of any energy storage system

As a battery manufacturer, the researches within the company should have a deep technical understanding of cells, thermal management, and other key battery-related elements that help us to configure the right DC battery technology to meet our customers' needs.

A battery is a device that produces electrical energy from chemical reactions. There are different kinds of batteries with different chemicals. The idea behind them is that the two different chemicals within a battery cell have different loads and are connected with a negative (cathode) and the other with a positive electrode (anode). When connected to an appliance, the negative electrode supplies a current of electrons that flow through the appliance and are accepted by the positive electrode. For the use of storing energy produced by renewable energy sources, only rechargeable batteries are relevant and will be considered. Figure 8.18 is a cross-sectional view of the inside of a lead-acid battery.

Figure 8.18a shows the battery cross section consists of six 2-volt cells connected in series, while Fig. 8.18b is an illustration of each component that is composed of several negative (cathode) and positive (anode) electrodes made of pure spongy lead and lead oxide, respectively. The electrodes, connected in parallel, are immersed in a dilute solution of sulfuric acid.

8.6.3.1 Lead-Acid Batteries

A lead-acid battery is an electrochemical battery that uses lead and lead oxide for electrodes and sulfuric acid for the electrolyte. Lead-acid batteries are the most commonly used in photovoltaic (PV) and other alternative energy systems because their initial cost is lower and because they are readily available nearly everywhere in the world. There are many different sizes and designs of lead-acid batteries, but the most important designation is whether they are deep-cycle batteries or shallow-cycle batteries. See Fig. 8.19 for a diagram of lead-acid battery with its internal components.

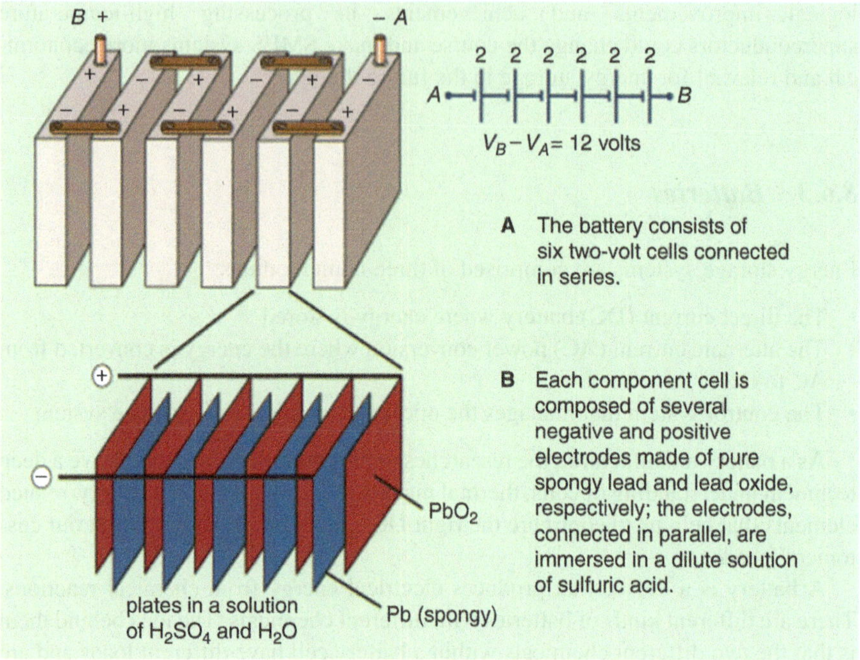

Fig. 8.18 The inside of a lead-acid battery

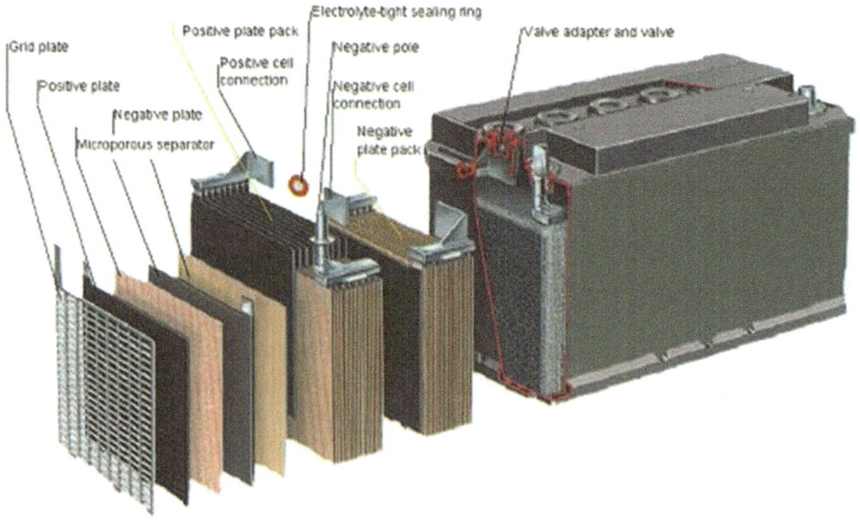

Fig. 8.19 Lead-acid battery

Lead-acid battery is the best solar deal available now – up to $4000 in maximum savings in today's market – and they can be found in three different types of design.

1. *Shallow-cycle batteries*: Like the type used as starting batteries in automobiles, these are designed to supply a large amount of current for a short time and stand mild overcharge without losing electrolyte. Unfortunately, they cannot tolerate being deeply discharged. If they are repeatedly discharged more than 20%, their life will be very short. These batteries are not a good choice for a photovoltaic (PV) system.
2. *Deep-cycle batteries*: These types of batteries are designed to be repeatedly discharged by as much as 80% of their capacity, so they are a good choice for power systems. Even though they are designed to withstand deep cycling, these batteries will have a longer life if the cycles are shallower. All lead-acid batteries will fail prematurely if they are not recharged completely after each cycle. Letting a lead-acid battery stay in a discharged condition for many days at a time will cause sulfating of the positive plate and a permanent loss of capacity.
3. *Sealed deep-cycle lead-acid batteries*: These batteries are maintenance-free. They never need watering or an equalization charge. They cannot freeze or spill, so they can be mounted in any position. Sealed batteries require very accurate regulation to prevent overcharge and over discharge. Either of these conditions will drastically shorten their lives. Sealed batteries are well-suited for remote, unattended power systems.

Conceptually, a lead-acid battery usually has several in-series connected cells, each delivering 2 volts (V) and each consisting several spongy pure lead cathodes, positive loaded lead oxide anodes, and a 20–40% solution of sulfuric acid that acts as an electrolyte. When discharged, both the anode and the cathode undergo a chemical reaction with the electrolyte that progressively changes them into lead sulfate that releases electrical energy in the process. This reaction can be almost completely reversed by supplying the electrodes with electricity, which is the reason a lead-acid battery can be recharged.

The life cycle and the ability to tolerate deep discharges depend on the type. Starting lighting ignition (SLI) batteries used in cars are not designed to be discharged to more than 50% as they have thinner lead plates. Doing so on a regular base will damage them and shorten their life cycle dramatically, whereas deep-cycle batteries with thicker plates can handle this much better but are as a result heavier and bulkier.

Pros and cons are listed here with sign of plus (pros) and minus (cons) in front of them as:

Pros

+ Easy and therefore cheap to produce
+ Mature technology, more than 150 years of experience and development
+ Very high surge-to-weight-ratio, capable of delivering a high jolt of electricity at once, which is why they are so suitable as car starters
+ Easily recyclable

Cons

- Very heavy and bulky
- Rather short lived
- Environmental concerns; although pretty safe, lead is a very toxic element, and exposure can cause severe damage to people and animals
- Corrosion caused by the chemical reactions

Future prospects of superconducting magnetic energy storage are stated below here as:

Lead-acid batteries have pretty much reached the end of the rope in terms of development. It is clear that no significant improvements can be made in capacity, density, or weight. Therefore, resources on future development should concentrate on other battery technologies with higher potentials.

Nonetheless because of the cost-effectiveness, lead-acid batteries are an important part of today's technology systems that can't be denied. Until other battery technologies emerge, they will remain first choice for many applications; however, grid storage is unlikely to be one of them simply because these batteries are not capable of storing huge amount of energy compared to other systems like a pumped-storage hydroelectricity (PSH) plant while staying cost-effective as the energy density is just too low. It is possible to integrate battery banks for few smaller decentralized systems like photovoltaic (PV) systems on rooftops; but, it cannot be used as a definite solution, just for the simple reason that the amount of resources are not available for the required capacity scale. Also, these batteries have a limited life cycle of a few years and therefore have to be replaced by new ones.

Future prospects and outlooks for other battery technologies and development indicate that lead-acid-based batteries will probably play the role of an inexpensive transitory technology.

8.6.3.2 Lithium-Ion Batteries

Lithium-ion batteries are incredibly popular these days. You can find them in laptops, personal digital assistants (PDAs), cell phones, and iPods. They are so common because, pound for pound, they're some of the most energetic rechargeable batteries available. The technology of the lithium battery has been slowly improving to create much more stable products. Learn about plug-in hybrid electric vehicle (PHEV) and lithium battery technology (see Fig. 8.20).

Lithium-ion batteries have also been in the news lately. That is because these batteries have the ability to burst into flames occasionally. It is not very common – just two or three battery packs per million have a problem – but when it happens, it's extreme. In some situations, the failure rate can rise, and when that happens, you end up with a worldwide battery recall that can cost manufacturers millions of dollars.

Fig. 8.20 A portable lithium-ion battery

Conceptually, the lithium is the lightest metal with the highest potential due to its very reactive behavior, which, in theory, makes it very fitting as a compound for batteries. Just as the lead-acid and most other batteries, the lithium-ion battery by definition uses chemical reactions to release electricity. Although all are called lithium-ion batteries, there's a variety of types with slightly different chemical compounds. The construction looks somehow similar to a capacitor, using three different layers curled up in order to minimize space. The first layer acts as the anode and is made of a lithium compound; the second one is the cathode and is usually made of graphite. Between anode and cathode is the third layer – the separator that, as suggested by the name, separates them while allowing lithium ions to pass through. The separator can be made of various compounds allowing different characteristics and with that different benefits and flaws. In addition, the three layers are submerged in an organic solvent – the electrolyte – allowing the ions to move between the anode and the cathode. See Fig. 8.21 for a layout of lithium-ion battery for rechargeable battery charge and rechargeable battery discharge mechanism.

In the charging process, the lithium ions pass through the microporous separator into spaces between the graphite (though not compounded), receiving an electron from the external power source.

During the discharge process, the lithium atoms located between the graphite release its electrons again that migrate over the external circuit to the anode providing a current. The lithium ions move back to the anode as well, parallel to their released electrons.

Because lithium is a very reactive compound and can burst into flames, safety measures have to be included, such as onboard control chips to manage the temperature and prevent a complete discharge.

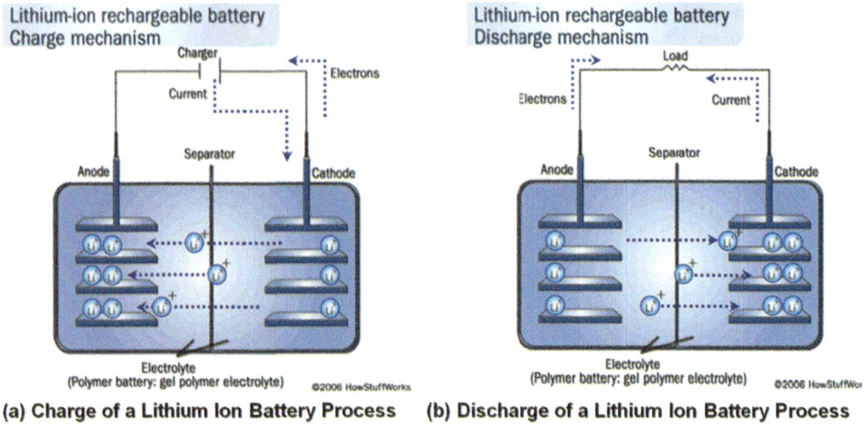

Fig. 8.21 Lithium-ion battery charge and discharge processes. (**a**) Charge of a lithium-ion battery process. (**b**) Discharge of a lithium-ion battery process

Pros and cons are listed here with sign of plus (pros) and minus (cons) in front of them as:

Pros

+ Highest energy density in commercial available batteries with huge potential
+ Provides higher voltage per cell (3.7 V compared to 2.0 V for lead-acid)
+ Low energy loss: only about 5% per month
+ Lithium and graphite as resources are available in large amount

Cons

– Very expensive
– Complete discharge destroys the cells
– Deteriorates even if unused (life cycle of about 5 years)
– Lithium is flammable in contact with atmospheric moisture

Future prospects of superconducting magnetic energy storage are stated below here as:

Lithium-ion batteries would be suitable for storing large amounts of energy if it were not for the costs. The rather expensive processing and the safety measures make them too expensive for commercial use besides small electronic devices like smartphones and laptops. Even for small decentralized systems, competitors like lead-acid batteries are more cost-effective right now, although that will change as they become cheaper.

However, lithium-based batteries have an incredibly huge potential. IBM is currently working on a project called Battery 500. This project's goal is to develop a battery using lithium and the air of the atmosphere as components (both the two lightest elements suitable for this purpose), capable of storing enough energy to power an electric car for 500 miles (~804 km) [16].

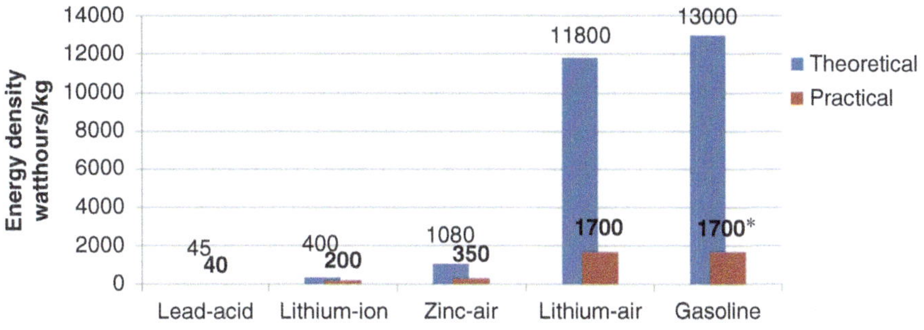

Fig. 8.22 Comparison of energy densities

Commercial use is targeted somewhere between 2020 and 2030 as there are still a lot of obstacles to overcome. Figure 8.22 is a chart of comparison of energy densities among different batteries that commercially are in existence for combustion engine, and companies like IBM are involved with this matter.

Besides the ambitious IBM project, several other companies worldwide are working and experimenting on new suitable compounds for lithium-based batteries, and it is very possible that this technology will reach a point where it becomes cost-effective for storing grid energy.

Despite the ongoing rumor, lithium is not short in supply and, even though in small amounts [0.1–0.2 parts per million (PPM)], it is available in saltwater and can be extracted through technical methods, making the supply almost infinite.

8.6.4 Other and Future Batteries in Development

While smartphones, smart homes, and even smart wearables are growing ever more advanced, they are still limited by power. The battery has not advanced in decades. But we are on the verge of a power revolution.

Big technology and car companies tare are all too aware of the limitations of lithium-ion batteries. While chips and operating systems are becoming more efficient to save power, we are still only looking at a day or two of use on a smartphone before having to recharge. Thankfully, universities are getting involved.

While it may be some time before we get a week's life out of our phones, development is progressing well. We've collected all the best battery discoveries that could be with us soon, from over the air charging to superfast 30 s recharging. Hopefully you'll be seeing this tech in your gadgets soon.

We rarely cover claims of "battery breakthroughs" because it's extremely difficult to distinguish what has the real potential and what is simply a public relation (PR) fluff. He hints at new battery breakthrough on the horizon of this technology, and we are walking into new dawn of smart batteries for smart applications in our day-to-day life as necessity.

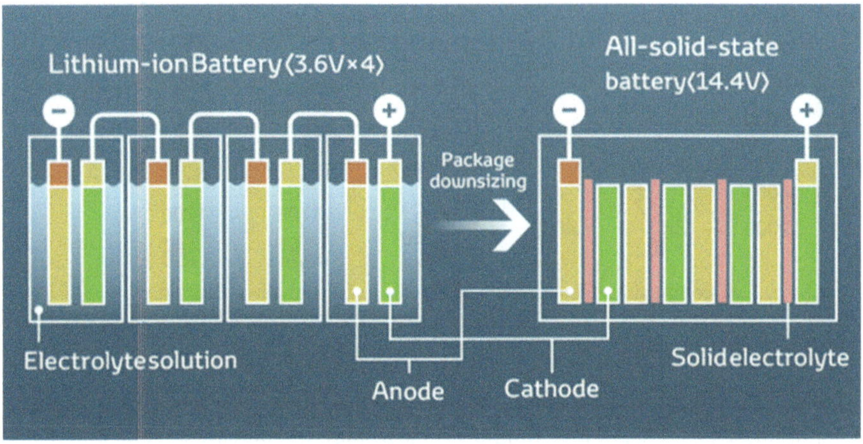

Fig. 8.23 Lithium-ion battery vs. solid-state battery. (Courtesy of Toyota Corporation)

No one has been harder on those claims than Elon Musk, but now Tesla's CEO himself tentatively hints a promising upcoming battery breakthrough on the horizon. As we know, Tesla is a company pioneering the plug-in hybrid electric vehicle (PHEV) and lithium battery technology.

As the world biggest consumers of lithium-ion batteries, Tesla is in a unique position to have access to the latest technology that battery scientists are trying to bring to production.

Tesla previously disclosed that they often receive battery samples and track hundreds of battery research programs to which they assign points based on potential. Previously, CTO JB Straubel, who leads battery development at Tesla, said that very few programs have substantial potential.

One of the most recent battery breakthrough claims that received a lot of attention was a new solid-state battery technology developed by John Goodenough, who is credited as the co-inventor of the lithium-ion (Li-ion) battery cell. See Fig. 8.23, which is the illustration of lithium-ion battery versus solid-state battery layout.

While battery breakthroughs are alluring, they are not actually needed to achieve the battery cost and capacity for all cars to go electric. The incremental improvements to Li-ion batteries that we have seen over the years are believed to be enough to soon be competitive with internal combustion engines before the cost of operation or incentives.

Nonetheless, breakthroughs or other battery cell technologies could be necessary to electrify other modes of transport, like air transport, for example.

Lithium-ion batteries represent a landmark technology that has made the current generation of electric vehicles possible. However, the day of their demise, while it still lies years in the future, is within view. Lithium-ion chemistries have a certain maximum energy density, dictated by those pesky laws of physics, and today's batteries are not so far from that theoretical maximum. If drivers keep demanding longer ranges and faster charging times, then a better technology will have to be found.

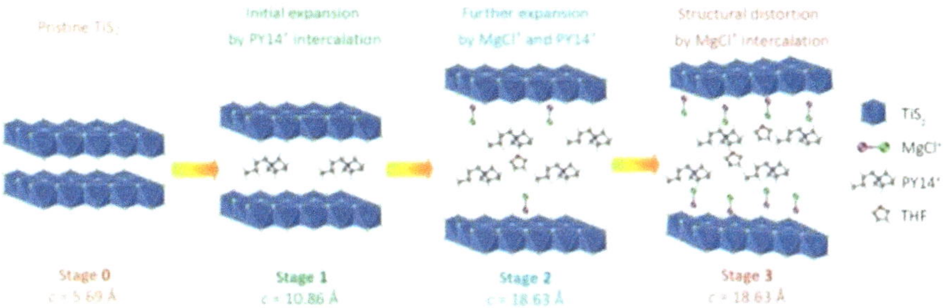

Fig. 8.24 Artistic schematic structure of the magnesium batteries. (Courtesy of the University of Houston)

With progress in development of new generation of Li-ion, safety issues are also growing and have to be taken seriously under consideration. The spectacular explosions and fireballs that some documentary makers revel in are not the norm (when was the last time your phone or computer caught fire?), but Li-ion batteries do have to be handled carefully, and necessary safety features add complexity and cost to battery packs. A new chemistry that is safer could also prove to be cheaper.

Researchers around the world are working on "beyond lithium" projects, and the past year has seen several significant breakthroughs. Of course, advances in the lab take years to make their way to the marketplace, but if and when one of these promising technologies can be commercialized, we could see game-changing improvements in the performance and cost of electric vehicles (EVs).

However, there has been some breakthrough in other types of batteries such is the one in magnesium batteries that was announced by University of Houston researchers on August 24, 2017, where they have claimed that magnesium batteries are safe, and unlike their counterpart lithium-ion batteries, they are not flammable or subject to exploding – but until now their ability to store energy has been limited. Researchers have reported a new design for the battery cathode, drastically increasing the storage capacity and upending conventional wisdom that the magnesium-chloride bond must be broken before inserting magnesium into the host.

Figure 8.24 is the artistic schematic of the structural evolution of titanium disulfide at different stages of intercalation. Interlayers are expanded or distorted as different amounts of pillaring molecules, complex cautions, and solvents are intercalated into the van der Waals gap of a host material at each stage.

Magnesium batteries offer promise for safely powering modern life – unlike traditional lithium-ion batteries, they are not flammable or subject to exploding – but their ability to store energy has been limited.

The work on this type of battery was first conceived by Yao and postdoctoral fellow Hyun Deog Yoo in 2014; the project spanned several years and involved scientists from three universities and three national laboratories, working both experimentally and theoretically.

"Magnesium ion is known to be hard to insert into a host," said Yoo, first author on the paper. "First of all, it is very difficult to break magnesium-chloride bonds.

More than that, magnesium ions produced in that way move extremely slowly in the host. That altogether lowers the battery's efficiency."

The new battery stores energy by inserting magnesium monochloride into a host, such as titanium disulfide. By retaining the magnesium-chloride bond, Yao said, the cathode demonstrated much faster diffusion than traditional magnesium versions.

The researchers report the new battery has storage capacity of 400 mAh/g, compared with 100 mAh/g, for earlier magnesium batteries. Commercial lithium-ion batteries have a cathode capacity of about 200 mAh/g, said Yao, who is also a principal investigator with the Texas Center for Superconductivity at UH.

Voltage of the new battery remains low at about 1 volt. That compares to 3–4 volts for lithium batteries.

The high voltage, coupled with their high-energy density, has made lithium-ion batteries the standard. But lithium is expensive and can develop breaches in its internal structure, a condition known as dendrite growths, which can cause the batteries to catch fire. As an earth-abundant resource, magnesium is cheaper and does not form dendrites. Until now, however, it has been held back by the need for a better cathode – the electrode from which the current flows – and more efficient electrolytes, the medium through which the ionic charge flows between cathode and anode.

The key, Yoo said, is to expand the titanium disulfide to allow magnesium chloride to be inserted – a four-step process called intercalation – rather than breaking the magnesium-chloride bonds and inserting the magnesium alone. Retaining the magnesium-chloride bond doubled the charge the cathode could store.

The magnesium monochloride molecules are too large to be inserted into the titanium disulfide using conventional methods. Building upon their earlier work, the researchers created an open nanostructure by expanding the gaps in the titanium disulfide by 300%, using organic "pillars."

The opening still was small, which increased from 0.57 nanometers to 1.8 nanometers, but Yao said that allowed for the magnesium chloride to be inserted.

"Combined theoretical modeling, spectroscopic analysis, and electrochemical study reveal fast diffusion kinetics of magnesium monochloride cations without scission of magnesium chloride bond," the researchers wrote. The large capacity accompanies excellent rate and cycling performances even at room temperature, opening up possibilities for a variety of effective intercalation hosts for multivalent-ion batteries.

One technology that has been getting a tremendous amount of attention from researchers is the solid-state battery, which uses a solid electrolyte instead of the liquid electrolyte used today. Solid-state batteries could theoretically have double the energy density of current batteries, and last several times longer. They also use a non-flammable electrolyte – usually glass, polymer, or a combination – so they would eliminate the safety issues that plague Li-ion cells.

Lithium-air batteries likewise could offer far greater energy density – maybe as much as 10 times more – but they suffer from poor life cycle. In 2015, Cambridge scientists wowed the battery world with an announcement that they had demonstrated a highly efficient and long-lasting lithium-oxygen battery. Alas, researchers from several universities and national labs have since been unable to duplicate the

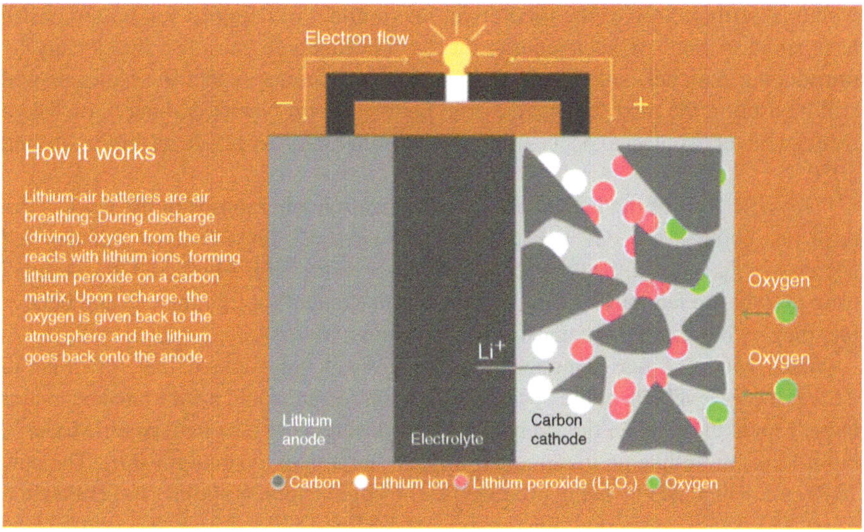

Fig. 8.25 How lithium-air battery works. (Courtesy of Money Incorporation)

original results. Figure 8.25 is an artistic layout of lithium-air battery and the process of how it works.

Other promising battery chemistries use other elements in place of lithium. Sodium batteries powered Jules Verne's futuristic submarine in *20,000 Leagues Under the Sea*. More recently, in 2015, researchers created a prototype sodium-ion battery in the industry-standard 18,650 cylindrical format.

According to a recent article in the Nikkei Asian Review, battery research has seen a big shift in recent years. At one time, nearly half of the presentations at the Battery Symposium in Japan were about fuel cells and Li-ion battery cathode materials. But since 2012, these topics have been supplanted by presentations about solid-state, lithium-air, and non-lithium batteries.

Toyota has been focusing on solid-state and Li-air batteries. At the latest Battery Symposium, battery researcher Shinji Nakanishi discussed a scenario for transitioning from Li-ion batteries to solid-state and then Li-air batteries. "We want our electric cars to go 500 km" on a single charge, he said. "And for this, we want rechargeable batteries that can generate 800–1,000 watt-hours per liter." That would be two to three times the energy density of today's best Li-ion batteries.

Panasonic, Tesla's battery supplier, is also taking a hard look at solid-state technology. "We think the existing technology can still extend the energy density of Li-ion batteries by 20% to 30%," President Kazuhiro Tsuga told Nikkei. "But there is a trade-off between energy density and safety. So if you look for even more density, you have to think about additional safety technology as well. Solid-state batteries are one answer."

Engineers have been pushing the limits of Li-ion technology for decades. Today's best Li-ion cells can reach an energy density of about 300 watts per kilogram, which is getting close to the theoretical maximum. "Existing Li-ion batteries still have room to improve their energy density because you can raise the density by introducing a nickel-based cathode material, so you can expect the batteries will still be used in the next few years," said battery expert Naoaki Yabuuchi of Tokyo Denki University. He expects lithium-ion technology to reach its limits around 2020.

"Tesla has one of the largest cell characterization laboratories in the world – we have just about every cell you can imagine on test," Tesla Product Planner Ted Merendino has stated, back in 2013. However, both Elon Musk and JB Straubel have said that so far, they've seen no viable replacement for lithium-ion, and believe me, they've been asked the question many times.

When Model 3 was announced, some EV-watchers opined that, in order to deliver the new vehicle at the desired price point, Tesla would need to make a major battery breakthrough. In the event, Tesla has developed a new battery for Model 3, but it looks more like an incremental improvement than a paradigm shift. The new 2170 cell, which is now being produced at the Gigafactory, is slightly larger than the trusty 18,650 and can store more energy. According to Elon Musk, it's "the highest energy density cell in the world, and also the cheapest." Advances in the way the cells are assembled into modules and packs are also expected to yield a significant reduction in battery costs.

Figure 8.26 is Tesla's battery pack in the floor pan of the Model S.

Therefore, it appears that lithium will continue its reign for a few more years at least. However, the post-lithium Holy Grail is still out there, and as likely as not, the knights of Tesla's round table will be the ones to bring it home. Battery superstar Jeff Dahn and his colleagues aren't working for Tesla just to make speeches at conferences. It is entirely possible that, at some super-secret facility in California or Nevada, test mules are being powered by solid-state or lithium-air batteries even as we speak.

However we still need to see how this technology is progressing in its development and these type of batteries, so for us consumer to be able to see even a single example of a cell battery working at the laboratory level which is better than the one we have, or the ones that we expect to come out with, said Elon Mask in 2014.

The batteries that are under development and worth to mention here are:

Redox Flow Battery

These batteries technically are similar to conventional batteries, except that the electrolytes (there are different forms, using one or two different fluids) can be exchanged, meaning that if the battery is discharged, the fluids are replaced with loaded ones. This concept could, in theory, become very handy for electric cars as you could charge your car simply by refueling just as you do now. However, the energy density is about 35 watt hours per kilogram (Wh/kg) in the same region as lead-acid batteries right now and therefore considerately low, although the Fraunhofer Institute in Germany claims to have managed to increase density up to the level of lithium-ion batteries (200 Wh/kg). Other advantages are the long lifespan of roughly 40 years and the fact that capacity can be increased by simply increasing the tanks and adding more electrolytes.

Fig. 8.26 Tesla's battery pack in the floor pan of the Model S. (Courtesy of Tesla Corporation)

For the purpose of grid storage, there are commercial available plants; but, the value is limited similar to flywheels, SMES, or other battery storage types due to the yet low-energy density. Pilot projects are in operation, most recently in California for an agricultural processing facility with a capacity of 3.6 MWh [17].

Sodium Battery

The liquid sodium sulfur battery (See Fig. 8.27) is yet another type of battery in development but already operational in some countries like Japan. About 250 megawatts (MW) of sodium battery power have been installed there [18]. Sodium batteries have the advantage of a relatively high density with up to 240 Wh/kg, a long life-span of 10–15 years, and high efficiency (75–90%); but, they need to be operated at high temperatures (350 °C/623 K) to get the sodium liquid, which not only makes it more difficult and expensive to operate but also more dangerous as the liquid sodium reacts easily with the water in the atmosphere. Since the Nippon Tokusyu Togyo Kabushiki-gaisha Co. Ltd. (NGK), the Tokyo Electric Power Togyo Kabushiki-gaisha Co. Ltd. (NGK), and the Tokyo Electric Power Co. Ltd. (TEPCO) began shipping out sodium batteries in 2002, three incidents resulting in fires have occurred, setting the development back.

Zinc-Air Battery

Just like the lithium-air battery, the zinc-air battery uses air as a second component. Zinc-air has been a focus in development for a while because of its safety aspects

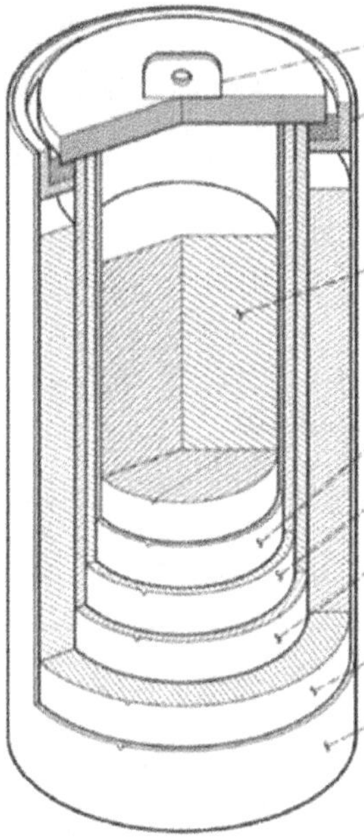

Fig. 8.27 Sodium battery layout

and potential in density but was dropped due to the low efficiency and short life cycles. Two independent companies claim to have solved these problems and plan to begin small-scale field tests this year but have yet to present reliable data [19, 20].

Zinc-air, just like lithium-air, holds a potential for grid storage due to its density and the fact that zinc is a commonly found metal, but as long as no data is presented and no field tests have been made, it remains as an idea with theoretical potential.

Flexible Battery

Flexible batteries are powered by IV cellular fluids. IV and cellular fluids power flexible batteries are recent attention of researchers and scientist in the field of new batteries generation development.

Researchers have engineered bendable batteries that can run on body-inspired liquids such as normal IV saline solution and cell culture medium. The authors designed alternatives to lithium-ion batteries by focusing on the mechanical-stress demands of wearable electronics such as smartwatches and the safety requirements of implantable electronics.

Figure 8.28 is an artistic rendering of fiber-shaped implantable batteries using biocompatible electrolytes.

Researchers in China have engineered bendable batteries that can run on body-inspired liquids such as normal IV saline solution and cell culture medium. In their work, published August 10 in the journal *Chem*, the authors designed alternatives to lithium-ion batteries by focusing on the mechanical-stress demands of wearable electronics such as smartwatches and the safety requirements of implantable electronics.

Fig. 8.28 An artistic rendering of fiber-shaped implantable batteries [21]

"Current batteries like the lithium-ion ones used in medical implants generally come in rigid shapes," says co-senior author Yonggang Wang, a chemistry professor at Fudan University and the Collaborative Innovation Center of Chemistry for Energy Materials. "Additionally, most of the reported flexible batteries are based on flammable organic or corrosive electrolytes, which suffer from safety hazards and poor biocompatibility for wearable devices, let alone implantable ones."

Safety measures for wearable and implantable batteries have generally involved structural reinforcement to prevent hazardous chemicals from leaking out. Instead, the researchers, led by Wang and macromolecular science professor Huisheng Peng, swapped out those toxic and flammable liquids for cheap and environmentally conscious sodium-ion solutions. Among those solutions were two biocompatible ones suitable for implantable devices, given that they pose no harm to the surface or interior of the body. Although electrolyte leakage is still undesired, its danger is minimized by the use of either the normal saline solution pumped into the body in most IV treatments or a cell culture medium that contains amino acids, sugars, and vitamins in addition to sodium ions and thus mimics the fluid that surrounds human cells.

Freed from leakage concerns, which can require so much protective material that batteries become bulky and unbendable, the researchers designed two types of flexible batteries – a 2D "belt"-shaped battery for which they adhered thin electrode films to a net of steel strands and a 1D fiber-shaped battery for which they embedded nanoparticles of electrode material around a carbon nanotube backbone.

Besides testing biocompatible fluids, the authors also tested ordinary sodium sulfate, a safe and fairly inert solution, as a liquid electrolyte suitable for use in external wearable devices.

With sodium sulfate solution as the electrolyte, both battery types outperformed most of the reported wearable lithium-ion batteries in terms of charge-holding capacity (an indicator of how long a battery can function without recharging) and power output for their size. That performance held up when the authors folded and bent the batteries to mimic the impact of wrapping a sensor, watch, or similar device around one's arm. Charge-holding capacity was only marginally reduced for the saline and cell culture-based batteries, most likely because they had slightly lower sodium-ion content than the sodium sulfate solution.

An undesired side reaction involving their fiber-shaped battery is even pointing the researchers toward possible biomedical applications. The same carbon nanotubes that make up the skeleton of the 1D battery can also accelerate the conversion of dissolved oxygen into hydroxide ions, a process that harms battery effectiveness if left uncontrolled but as a stand-alone process boasts therapeutic potential for treating cancer and bacterial infections.

"We can implant these fiber-shaped electrodes into the human body to consume essential oxygen, especially for areas that are difficult for injectable drugs to reach," says Wang. "Deoxygenation might even wipe out cancerous cells or pathogenic bacteria since they are very sensitive to changes in living environment pH. Of course, this is hypothetical right now, but we hope to investigate further with biologists and medical scientists."

8.7 A Battery Inspired Strategy for Carbon Fixation

Scientists working toward the elusive lithium-air battery discovered an unexpected approach to capturing and storing carbon dioxide away from the atmosphere. Using a design intended for a lithium-CO_2 battery, researchers have developed a way to isolate solid carbon dust from gaseous carbon dioxide, with the potential to also separate out oxygen gas through the same method.

Scientists working toward the elusive lithium-air battery discovered an unexpected approach to capturing and storing carbon dioxide away from the atmosphere. Using a design intended for a lithium-CO_2 battery, researchers in Japan and China have developed a way to isolate solid carbon dust from gaseous carbon dioxide, with the potential to also separate out oxygen gas through the same method. Their work appears August 9 in *Joule*, a new interdisciplinary energy journal from Cell Press.

Converting carbon dioxide emissions into other carbon-containing compounds is desirable due to carbon dioxide's contribution to the greenhouse effect and global warming. Examples range from natural processes, such as plants turning CO_2 into oxygen and sugars, to man-made ones, such as injecting carbon dioxide into rock formations to be trapped as carbonate minerals.

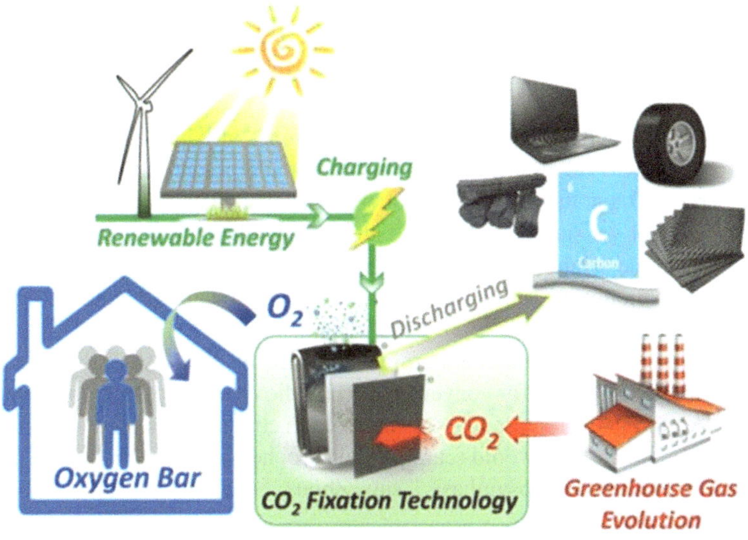

Fig. 8.29 Industrial utilization of Li-CO_2 [21]

"The problem with most physical and chemical pathways for CO_2 fixation is that their products are gases and liquids that need to be further liquefied or compressed, and that inevitably leads to additional energy consumption and even more CO_2 emissions," says senior author Haoshen Zhou of Japan's National Institute of Advanced Industrial Science and Technology and China's Nanjing University. "Instead, we are demonstrating an electrochemical strategy for CO_2 fixation that yields solid carbon products, as well as a lithium-CO_2 battery that can provide the energy necessary for that process."

The researchers encountered the carbon fixation strategy when they tried to recharge a lithium-CO_2 battery prototype. Instead of fully regenerating lithium ions and CO_2 from the lithium carbonate and carbon produced during battery discharge, as would have taken place with a reversible Li-CO_2 battery, the lithium carbonate decomposed, yielding additional carbon, as well as oxygen gas that was not isolated due to rapid reaction with the battery electrolyte. Typically, this kind of buildup causes physical degradation and reduced functional lifespan for a battery, but instead, the deposition of solid carbon boasts a separate advantage, pointing to a promising approach to fix carbon in a stable and easy-to-dispose-of form.

Figure 8.29 is depiction of flowchart of energy storage and carbon fixation using Li-CO_2 technology.

"What is impressive about this work is the possibility to convert one-third of the CO2 species to carbon with high theoretical energy efficiency above 70%," says Joule scientific editor Rahul Malik. "Battery architecture is an unforeseen but intriguing way to look at carbon fixation."

Since generating carbon solids both realizes carbon fixation decreases battery performance, the researchers were not able to simultaneously satisfy both goals within a single device. However, by incorporating a tiny amount of ruthenium metal into their design as a catalyst, they were able to avoid extensive carbon deposition and induce better reversibility, converting their carbon-fixing apparatus into a functioning Li-CO_2 battery.

A remaining challenge for both carbon fixation and battery performance is to move from pure CO_2 to ambient air, a jump that would potentially allow for treating atmospheric CO_2 in the first case and would advance toward the theoretically powerful but not yet stable lithium-air battery technology in the second case. The fixation technique might also be adapted to scrub other harmful or polluting gases such as carbon monoxide, sulfur dioxide, nitric oxide, and nitrogen dioxide from the atmosphere, Zhou says.

Looking ahead, the researchers are also excited by their system's potential to perhaps lead to a pathway for converting carbon dioxide into pure carbon and oxygen gas. "Attaining the release of oxygen gas upon charging, coupled with the accumulation of solid carbon, would realize an electrochemical carbon dioxide fixation strategy analogous to photosynthesis," says Zhou.

8.8 Saliva-Powered Battery

New battery is activated by your spit; saliva-powered battery could be helpful in extreme conditions.

Researchers have developed the next step in microbial fuel cells (MFCs): a battery activated by spit that can be used in extreme conditions where normal batteries do not function.

For the last 5 years, Binghamton University Electrical and Computer Science Assistant Professor Seokheun Choi has focused on developing micro-power sources for the use in resource-limited regions to power point-of-care (POC) diagnostic biosensors; he has created several paper-based bacteria-powered batteries.

"On-demand micro-power generation is required especially for point-of-care diagnostic applications in developing countries," said Choi. "Typically, those applications require only several tens of microwatt-level power for several minutes, but commercial batteries or other energy harvesting technologies are too expensive and over-qualified. Also, they pose environmental pollution issues."

Choi, along with research assistant Maedeh Mohammadifar, created a high-performance, paper-based, bacteria-powered battery by building microbial fuel cells with inactive, freeze-dried exoelectrogenic cells which generates power within minutes of adding saliva. The proposed battery generated reliable power with one drop of saliva, supplying onboard power that could be used by the next generation of disposable, paper-based POC diagnostic platforms.

"The proposed battery has competitive advantages over other conventional power solutions because the biological fluid for on-demand battery activation is readily available even in the most resource-constrained settings, and the freeze-drying technology enables long-term storage of cells without degradation or denaturation," wrote the researchers. Choi is focused on improving the batter's power density so that more applications can be powered.

"Now, our power density is about a few microwatts per centimeter square. Although 16 microbial fuel cells connected in a series on a single sheet of paper generated desired values of electrical current and voltage to power a light-emitting diode (LED), further power improvement is required for other electronic applications demanding hundreds of milliwatts of energy," said Choi.

8.9 Summary

Energy storage, a potential solution for integrating intermittent renewables and improving grid stability, again saw rapid growth this past year. A "transition year" for US energy storage, 2016 saw a more diverse market emerge "both in terms of the types of systems (market segments) deployed and the business models." These trends are expected to continue in the United States with combined residential, commercial, and industrial energy storage deployments predicted to surpass 2 GW by 2021.

Despite this rapid growth, there are those who claim that the adoption of energy storage has been slowed by a "web" of regulations at all levels. Some argue that regulations, which can vary from state to state, are in need of modernization in order to facilitate easier integration of energy storage into our nation's infrastructure. While not necessarily advocating national uniformity for these regulations, energy storage developers hope to avoid a framework as disjointed as that of solar, where they must "navigate 50 different markets with 50 different sets of regulations."

Recently, the Federal Energy Regulatory Commission (FERC) released a policy statement addressing cost recovery for energy storage, stating these facilities are permitted to earn both cost- and market-based revenue streams in light of the multiple services they provide. This decision serves as an example of the type of modernization that energy storage proponents hope for and serves to ease barriers to grid integration. While there is not necessarily opposition to expanding the nation's energy storage capacity, there will likely be challenges and objections to specific rules in the future. FERC's recent statement, for example, was met by some with concerns over equity, possible over-recovery at the expense of the rate payer, and negative impacts on competition within the industry.

References

1. http://www.engineeringnews.co.za/article/electricity-consumption-to-increase-to-over-30-116-b-kwh-globally-in-2030-2009-04-17
2. Abdelhay A. Sallam and Om P. Malik (May 2011). Electric Distribution Systems. IEEE Computer Society Press. p. 21. ISBN 9780470276822
3. Energy profile of Alaska, United States, Editor: Cutler J. Cleveland, Last Updated: July 30, 2008 - Encyclopedia of Earth

4. Glover J. D., Sarma M. S., Overbye T. J. (2010) Power System and Analysis 5th Edition. Cengage Learning. Pg 10
5. Smart Grid - The New and Improved Power Grid: A Survey; IEEE Communications Surveys and Tutorials 2011; X. Fang, S. Misra, G. Xue, and D. Yang; doi:https://doi.org/10.1109/SURV.2011.101911.00087
6. Alexandra Von Meier (2013). Electrical Engineer 137A: Electric Power Systems. Lecture 2, Introduction to Electric Power Systems, Slide 33
7. Bath County Pumped Storage Station, Virginia, USA is currently the largest PSH worldwide with a storage capacity of 30GWh and a 3GW power output
8. http://www.oncor.com/community/knowledgecollege/energy_library/generating/generating01.aspx
9. http://www.world-nuclear.org/info/inf10.html
10. http://investors.beaconpower.com/releasedetail.cfm?ReleaseID=593208
11. Kamiyama, K. 1994. ISTEC Journal 7: 39
12. Huang, X., S.F. Kral, G.A. Lehman, Y.M. Lvovsky, and M. Xu. 1995. *IEEE Trans. on Applied Superconductivity* 5: 428
13. Jungst, K.-P., 1995. *Supraleitung in der Energietechnik II*, VDI-Verlag: 133
14. Prescher, K., W. Nick, H.-E. Vollmar, M. Kleinmaier, and U. Radtke. 1995. *Supraleitung in der Energietechnik II*, VDI-Verlag: 115
15. Lorenzen, H.W., V. Brammer, J. Meinke, F. Rosenbauer, J. Schaller, and R.M. Schottler. 1995. *Supraleitung in der Energietechnik II*, VDI-Verlag: 149
16. http://www.ibm.com/smarterplanet/us/en/smart_grid/article/battery500.html
17. http://www.sustainableplant.com/2012/05/vanadium-flow-battery-to-provide-grid-level-storage-for-gills-onions/
18. http://www.greentechmedia.com/articles/read/is-sodium-the-future-formula-for-energy-storage/
19. http://phys.org/news/2012-01-eos-zinc-battery-recipe-energy.html
20. http://www.treehugger.com/clean-technology/very-promising-zinc-air-battery-could-hold-300-more-energy-than-lithium-ion.html
21. Guo et al. Multi-functional Flexible Aqueous Sodium-Ion Batteries with High Safety. Chem, 2017 DOI: https://doi.org/10.1016/j.chempr.2017.05.004

Index

A
Advanced Fuel Cycle Initiative (AFCI), 224
Advanced gas-cooled reactor (AGR), 19
Alternating current (AC), 254, 267
Anode, 201
Argentina, 29
Atomic Energy Commission (AEC), 70
Automatic generation control (AGC), 253

B
Boiling water reactors (BWR), 9, 16, 17, 122
Brayton topping cycle, 157
Brazil, 29
Brooklyn Microgrid (BMG), 249

C
Canadian CANDU reactor, 12
Cathode, 201
Chernobyl, 10
Closed-cycle cooling system, 104
Coal power plants (CPPs), 170
Combined cycle (CC), 87, 110, 113, 159, 163, 171
Combined cycle Brayton-Rankine cycles, 161
Combined cycle gas turbine (CCGT), 118, 139
Cooling towers, 8
Critical point, 169

D
Department of Energy (DOE), 6, 134, 211
Deutsches Elektronen-Synchrotron (DESY), 269
Direct current (DC), 267
Dry region, 169

E
Electric hydrolysis, 201
Electric vehicles (EVs), 279
Emergency planning zone (EPZ), 80
Energy and Industrial Technology Development Organization (NEDO), 198
Energy Information Administration (EIA), 191, 245
Enriched fuel, 9
Environmental Protection Agency (EPA), 120
Euratom, 29
Europe (EU), 27
European Nuclear Energy Forum (ENEF), 60
European pressurized-water reactor (EPR), 57
European Sustainable Nuclear Industrial Initiative, 59
European Transmission System Operation (ETSO), 253
Evaluation Methodology Group, 53, 54

F
Fast-breeder reactors (FBR), 7, 20
Fast-neutron reactors (FNR), 20
Federal Energy Regulatory Commission (FERC), 230, 233, 289
Fermi, E., 3
First-of-a-kind (FOAK), 30
Fission, 2, 6, 67
Florida Reliability Coordinating Council (FRCC), 236

B. Zohuri, *Small Modular Reactors as Renewable Energy Sources*,
https://doi.org/10.1007/978-3-319-92594-3

Fluoride salt-cooled high-temperature reactors (FHR), 38, 77
Forschungszentrum Karlsruhe (FZK), 269
France, 29
French Institute for Radiological Protection and Nuclear Safety (IRSN), 211
French Nuclear Safety Authority (ASN), 211, 222
Fuel cell vehicle (FCV), 198
Fuel Cycle Crosscut Group (FCCG), 54

G
Gas-cooled reactor (GCR), 17, 19
Gas fast reactor (GFR), 44, 190
Gas turbine combined cycle (GTCC), 110
Gas turbine modular helium reactor (GT-MHR), 19
General Design Criteria (GDC), 77
Generation II, 27
Generation III, 27
Generation IV, 6, 28, 29, 32, 53
Generation IV International Forum (GIF), 29, 34, 73, 190, 218
Generation Nuclear Plant (NGNP), 32
Generic Design Assessment (GDA), 208
Global Actinide Cycle International Demonstration (GACID), 223
Global Business Network (GBN), 120
Graphite-moderated reactor (GMR), 10
Greenhouse gas (GHG), 197

H
Hanford, 5
Heat exchanger (HX), 131
Heat recovery steam generator (HRSG), 85, 106, 108, 109, 140, 156, 158
Heavy water (D_2O), 7
Heavy-water reactors (HWR), 9–11
High-temperature gas-cooled reactor (HTGR), 19, 122, 134
High-temperature reactors (HTR), 32
High-voltage direct current (HVDC), 255
Hydrogen energy and fuel cells, 60

I
Idaho National Laboratory (INL), 134
Independent System Operator (ISO), 238
Inertial confinement fusion (ICF), 88
Innovative Nuclear Reactors and Fuel Cycles (INPRO), 29
Instrumentation and control (I&C), 223
Intercooled combined cycle (IC-CC), 114
Intermediate heat exchanger (IHX), 130
International Energy Agency (IEA), 120

J
Japan, 29
Japan Atomic Energy Research Institute (JAERI), 267

K
Kinetic Energy Recovery System (KERS), 266
Kinetic energy storage (KES), 263

L
Lawrence Livermore National Labs (LLNL), 267
Lead-bismuth eutectic (LBE), 45
Lead-cooled fast reactor (LFR), 45, 190
Licensing Technical Support (LTS), 100
Light water (H_2O), 7
Light-water reactors (LWR), 10, 77, 95, 98, 111, 166
Liquid-metal fast-breeder reactor (LMFBR), 23, 98, 111, 122, 157
Liquified natural gas (LNG), 234
Lithium-ion (Li-ion), 278
Load-Serving Entity (LSE), 238
Long-run marginal cost (LRMC), 232
Los Alamos, 5

M
Manhattan project, 5
Microbial fuel cells (MFCs), 288
Million tons of uranium (MtU), 31
Minimum Offer Price Rule (MOPR), 238
Molten salt reactor (MSR), 37, 190
Multinational Design Evaluation Program (MDEP), 222

N
Natural gas, 2
New Mexico, 5
Next generation nuclear plant (NGNP), 6, 127, 132, 134, 139
North American Electric Reliability Corporation (NERC), 236, 253
Nuclear air Brayton combined cycle (NACC), 110
Nuclear criticality, 5

Nuclear fuel cycle, 21
Nuclear power plants (NPPs), 7–9, 14, 27, 28, 32, 56, 167, 168, 170, 190, 218
Nuclear Regulatory Commission (NRC), 33, 99, 134, 156, 211
Nuclear steam supply systems (NSSS), 122

O
Oak Ridge, 5
Office for Nuclear Regulation (ONR), 211
Open-air cooling system, 104
Organization for Economic Co-operation and Development (OECD), 179
Organization of Petroleum Exporting Countries (OPEC), 175
Organizations/Independent System Operators (RTOs/ISOs), 237

P
Parts per million (PPM), 277
Pennsylvania, Jersey, and Maryland (PJM), 234
People's Republic of China, 29
Personal digital assistants (PDAs), 274
Photovoltaic (PV), 274
Pinch point, 141, 149, 158, 159
Plant-life management technologies and plant license extension practices (PLIM/PLEX), 30
Plug-in hybrid electric vehicle (PHEV), 274, 278
Point-of-care (POC), 288
Power conditioning system (PCS), 266
Power conversion system (PCS), 127–129
Pressure swing adsorption (PSA), 198
Pressurized heavy water reactor (PHWR), 14
Pressurized water reactors (PWR), 9, 14, 16, 95, 121, 167
Probability of risk analysis (PRA), 96
Project Arrangements (PA), 222
Public relation (PR), 277
Pumped-storage hydroelectricity (PSH), 261, 274

R
Reactor kinetics, 80
Reactor Safety Study (RSS), 80
Recuperated Brayton cycles, 161
Research and development (R&D), 198
Return on investment (ROI), 165
Roadmap Integration Team (RIT), 53
Roosevelt, F., President, 5
Russian Federation, 29

S
Safety Advisory Group Report (NSAG) 10, 218
Saturation dome, 168
SERC Reliability Corporation (SERC), 236
Short-run marginal cost (SRMC), 232
Small modular reactors (SMRs), 208, 218, 230
Sodium-cooled fast reactor (SFR), 39, 57, 190
South Africa, 29
Southern Company (SOCO), 236
Starting lighting ignition (SLI), 273
Steam reforming, 201
Superconducting magnetic energy storage (SMES), 266
Supercritical water-cooled reactor (SCWR), 40, 190
Switzerland, 29
System Steering Committee (SSC), 29
Szilard, L., 3

T
Technical working group (TWG), 53
Tennessee, 5
Tennessee Valley Authority (TVA), 236
Thermal energy, 9
Thermal reactor, 8, 10, 22, 23
Tokyo Electric Power Co. Ltd. (TEPCO), 283
Topping cycles, 161
Total cost of ownership (TCO), 165
Transmission Organizations (RTOs), 238
Tristructural-isotropic (TRISO) fuel, 221

U
Uninterruptible power supply (UPS), 264, 270
United Kingdom, 29
United States, 1, 5, 9, 29
United States Air Force (USAF), 268
University of Chicago, 3, 5, 10

V
Very-high temperature reactors (VHTR), 32, 33, 35, 36, 96, 189, 190
VYCON Direct Connect (VDC), 264

W
Washington, 5
Water cooled graphite moderated, 122
Western European Nuclear Regulators' Association (WENRA), 218

The manufacturer's authorised representative in the EU is Springer Nature Customer Service Centre GmbH, Europaplatz 3, 69115 Heidelberg, Germany. If you have any concerns regarding our products, please contact ProductSafety@springernature.com

Printed and bound by CPI Group (UK) Ltd, Croydon, CR0 4YY
30/09/2026
02225658-0010